Finite Geometries

Developments in Mathematics

VOLUME 3

Series Editor:

Krishnaswami Alladi, *University of Florida, U.S.A.*

Aims and Scope

Developments in Mathematics is a book series publishing

(i) Proceedings of Conferences dealing with the latest research advances,

(ii) Research Monographs, and

(iii) Contributed Volumes focussing on certain areas of special interest.

Editors of conference proceedings are urged to include a few survey papers for wider appeal. Research monographs which could be used as texts or references for graduate level courses would also be suitable for the series. Contributed volumes are those where various authors either write papers or chapters in an organized volume devoted to a topic of special/current interest or importance. A contributed volume could deal with a classical topic which is once again in the limelight owing to new developments.

Finite Geometries

Proceedings of the Fourth Isle of Thorns Conference

Edited by

A. Blokhuis
University of Eindhoven

J.W.P. Hirschfeld
University of Sussex

D. Jungnickel
University of Augsburg

and

J.A. Thas
University of Ghent

KLUWER ACADEMIC PUBLISHERS
DORDRECHT / BOSTON / LONDON

Library of Congress Cataloging-in-Publication Data

Isle of Thorns Conference (4th : 2000 : University of Sussex)
Finite geometries : proceedings of the Fourth Isle of Thorns Conference / edited by A. Blokhuis ... [et al.].
p. cm. -- (Developments in mathematics ; v. 3)
Includes index.
ISBN 0-7923-6994-7
1. Finite geometries--Congresses. I. Blokhuis, A. (Aart) II. Title. III. Series.

QA167.2 .I74 2000
516'.13--dc21

2001029800

ISBN 0-7923-6994-7

Published by Kluwer Academic Publishers,
P.O. Box 17, 3300 AA Dordrecht, The Netherlands.

Sold and distributed in North, Central and South America
by Kluwer Academic Publishers,
101 Philip Drive, Norwell, MA 02061, U.S.A.

In all other countries, sold and distributed
by Kluwer Academic Publishers,
P.O. Box 322, 3300 AH Dordrecht, The Netherlands.

Printed on acid-free paper

Printed in the Netherlands.

Contents

Preface

When?

These are the proceedings of *Finite Geometries*, the Fourth Isle of Thorns Conference, which took place from Sunday 16 to Friday 21 July, 2000. It was organised by the editors of this volume.

The Third Conference in 1990 was published as *Advances in Finite Geometries and Designs* by Oxford University Press and the Second Conference in 1980 was published as *Finite Geometries and Designs* by Cambridge University Press.

The main speakers were A.R. Calderbank, P.J. Cameron, C.E. Praeger, B. Schmidt, H. Van Maldeghem.

There were 64 participants and 42 contributions, all listed at the end of the volume.

Conference web site http://www.maths.susx.ac.uk/Staff/JWPH/

Why?

This collection of 21 articles describes the latest research and current state of the art in the following inter-linked areas:

- combinatorial structures in finite projective and affine spaces, also known as Galois geometries, in which combinatorial objects such as blocking sets, spreads and partial spreads, ovoids, arcs and caps, as well as curves and hypersurfaces, are all of interest;

- geometric and algebraic coding theory;

- finite groups and incidence geometries, as in polar spaces, generalized polygons and diagram geometries;

- algebraic and geometric design theory, in particular designs which have interesting symmetric properties and difference sets, which play an important role, because of their close connections to both Galois geometry and coding theory.

In recent year, some outstanding results have been proved on the nonexistence of maximal arcs, on the existence of difference sets, on the size of blocking sets, on the classification of generalized polygons and on the classification of cyclic two-weight codes.

In the last decade, new journals which include the area are Advances in Geometry, Designs Codes and Cryptography, Finite Fields and their Applications, Journal of Algebraic Combinatorics, Journal of Combinatorial Designs.

Other recent conferences which include the field have been the Fifth Conference on Finite Fields and Applications (Augsburg 1999), the First and Second Pythagorean Conferences on Geometries, Combinatorial Designs and Related Structures (1996, 1999), the Third Finite Geometry and Combinatorics Conference (Deinze, 1997), Combinatorics '98 (Palermo), and Combinatorics 2000 (Gaeta).

What?

A distinct feature of the material at the conference and in Finite Geometry generally is its wide range and overlapping connections between the different areas. The articles in this collection are a mixture of survey articles and original research papers.

Diagram geometries

Blok and Pasini investigate Lie incidence geometries defined over a field and the relation between such a geometry and that defined over a subfield.

Cardinali and Pasini study geometries of points, lines and planes whose point and plane residues are respectively a projective space of dimension at least three and its dual, and classify them completely.

Cara completely classifies the residually connected, flag-transitive geometries of the alternating group A_8 subject to two further, natural conditions.

Configurations in finite projective spaces

Buekenhout presents a generalization of plane cubic curves with the aim of axiomatising them.

Cossidente and King show that the only ovoids in $PG(3,q)$ fixed by a fairly small group are the known ones, namely the elliptic quadric and the Suzuki–Tits ovoid.

Giulietti, Hirschfeld and Korchmáros investigate the Desarguesian plane of order 13 with a view to showing that the second largest complete arc has size less than $q-1$ for $q > 13$; they find an optimal plane curve of genus 10 with 54 rational points and another sextic with 12 rational points such that every line of the plane meets it in at most four points.

Hirschfeld and Storme extend their 1995 survey on arcs, caps, blocking sets to the present day, including almost MDS codes and minihypers; this includes both arcs of higher degree and blocking sets. The best results on the Main Conjecture for MDS codes follow from those for arcs of degree two.

Luyckx and J.A. Thas link a variety of different structures, namely, the non-singular quadric in $PG(6,q)$, flocks in $PG(4,q)$, internal and external points of a conic, rational normal cubic scrolls in order to classify a class of 1-systems.

Wehlau characterizes many classes of caps in $PG(n,2)$.

Generalized quadrangles

Ghinelli and Löwe show the uniqueness of the generalized quadrangle derivable from a pencil of quadrics in $PG(3,q)$ with empty rational base.

Payne studies a recently-discovered q-clan and considers in particular the novelty for $q > 27$ of the associated translation planes, flocks and generalized quadrangles.

K. Thas gives new characterizations of translation generalized quadrangles.

Generalized polygons

Govaert and Van Maldeghem construct a generalized quadrangle in a generalized hexagon using a particular spread of the hexagon, and show that the quadrangle characterizes the spread.

Gramlich and Van Maldeghem make a detailed study of the existence of epimorphisms of generalized polygons.

Mühlherr and Van Maldeghem consider embeddings of generalized polygons and use them for characterization theorems.

Design theory

Blokhuis, Jungnickel and Schmidt investigate designs that are almost projective planes. They are symmetric divisible designs such that for any point P there is exactly one other point P' with the property that there are precisely two lines joining P and P', but any other point is joined to P by precisely one line; the dual property also holds.

Praeger surveys the line-transitivity of designs to find when they may also be linear spaces.

Difference sets

Schmidt gives a survey of the known exponent bounds for difference sets and relative difference sets.

Coding theory

Landjev studies the links between linear codes and configurations in projective spaces.

Permutation groups

Cameron gives a variety of results on fixed points and cycles in permutation groups arising in disparate areas.

Algorithms

Enge gives a framework for a discrete logarithm algorithm with subexponential running time.

Aart Blokhuis
James W.P. Hirschfeld
Dieter Jungnickel
Joseph A. Thas

POINT-LINE GEOMETRIES WITH A GENERATING SET THAT DEPENDS ON THE UNDERLYING FIELD

R.J. Blok

Department of Mathematics
Michigan State University
E. Lansing, MI 48824
United States of America
rblok@math.msu.edu

A. Pasini

Dipartimento di Matematica
Università degli Studi di Siena
53100 Siena
Italy
pasini@unisi.it

Abstract

Suppose Γ is a Lie incidence geometry defined over some field $\mathbb{F}$ having a Lie incidence geometry Γ_0 of the same type but defined over a subfield $\mathbb{F}_0 \leq \mathbb{F}$ as a subgeometry. We investigate the following question: how many points (if any at all) do we have to add to the point-set of Γ_0 in order to obtain a generating set for Γ? We note that if Γ is generated by the points of an apartment, then no additional points are needed. We then consider the long-root geometry of the group $\mathrm{SL}_{n+1}(\mathbb{F})$ and the line-grassmannians of the polar geometries associated to the groups $\mathrm{O}_{2n+1}(\mathbb{F})$, $\mathrm{Sp}_{2n}(\mathbb{F})$ and $\mathrm{O}^+_{2n}(\mathbb{F})$. It turns out that in these cases the maximum number of points one needs to add to Γ_0 in order to generate Γ equals the maximal number of roots one needs to adjoin to $\mathbb{F}_0$ in order to generate $\mathbb{F}$. We prove that in the case of the long-root geometry of the group $\mathrm{SL}_{n+1}(\mathbb{F})$ the point-set of Γ_0 does not generate Γ. As a by-product we determine the generating rank of the line grassmannian of the polar geometry associated to $\mathrm{Sp}_{2n}(\mathbb{F})$ $(n \geq 3)$, if $\mathbb{F}$ is a prime field of odd characteristic.

A. Blokhuis et al. (eds.), Finite Geometries, 1–25.

1. Introduction

1.1. The problem studied in this paper

In a recent paper [10] Cooperstein determines the generating rank for the long-root geometries of $SL_{n+1}(\mathbb{F})$, $O^+_{2n}(\mathbb{F})$, $O_{2n+1}(\mathbb{F})$ and $O^-_{2n+2}(\mathbb{F})$ for a prime field $\mathbb{F}$ of any characteristic. The result of a simple computation on the long-root geometry Γ of the group $SL_3(\mathbb{F})$, but now with $\mathbb{F}$ a proper extension over its prime field $\mathbb{F}_0$, was confirmed by an observation of Smith and Völklein (see Proposition 1.2 of [14]): the set considered by Cooperstein in [10] only generates the long-root subgeometry Γ_0 of $SL_3(\mathbb{F}_0)$ but it does not generate Γ. However, by adding one well-chosen point to this set, one obtains a generating set for Γ.

In the present paper we investigate the following problem: suppose Γ is a Lie incidence geometry Γ defined over some field $\mathbb{F}$ having a Lie incidence geometry Γ_0 of the same type but defined over a subfield $\mathbb{F}_0 \leq \mathbb{F}$ as a subgeometry. How many points (if any at all) do we have to add to the point-set of Γ_0 in order to obtain a generating set for Γ? We are mainly concerned with the following Lie incidence geometries: the long-root geometry associated to the group $SL_{n+1}(\mathbb{F})$ and the line-grassmannians of the polar geometries associated to the groups $O_{2n+1}(\mathbb{F})$, $Sp_{2n}(\mathbb{F})$ and $O^+_{2n}(\mathbb{F})$ (these geometries are described in the following subsection).

1.2. Contents and main results

In Section 2, after having recalled a few basic notions on incidence geometries, we provide the general framework for the problem studied in this paper: for a Lie incidence geometry Γ defined over some field $\mathbb{F}$, we describe how one obtains a Lie incidence geometry Γ_0 of the same type as Γ, but defined over a subfield $\mathbb{F}_0 \leq \mathbb{F}$, as a subgeometry of Γ in the case that Γ is obtained from a building Δ associated to a non-twisted Chevalley group $G(\mathbb{F})$; Γ_0 is the Lie incidence geometry obtained from the sub-building Δ_0 of Δ associated to the Chevalley group $G(\mathbb{F}_0)$ viewed as a subgroup of $G(\mathbb{F})$.

In Section 3 we give a simple application of the results of Section 2, showing that if Γ is a Lie incidence geometry that is generated by the points of an apartment, then Γ_0 generates Γ.

In Section 4, Γ is the long-root geometry associated to the group $SL_{n+1}(\mathbb{F})$ with $n \geq 2$. We regard the building Δ as the geometry of subspaces of $PG(n, \mathbb{F})$. For Δ_0 we can take the geometry of subspaces of $PG(n, \mathbb{F}_0)$ viewed as a subgeometry of $PG(n, \mathbb{F})$ and Γ_0 becomes the subgeometry of Γ whose points belong to Δ_0. The geometries Γ and Γ_0 can be described as follows: the points of Γ and Γ_0 are the point-hyperplane flags of $\mathrm{PG}(n,\mathbb{F})$ and $\mathrm{PG}(n,\mathbb{F}_0)$. The lines of Γ and Γ_0 are the flags (X,Y) with $X \subseteq Y$ of codimension $n-2$ in Y; thus $X = Y$, when $n = 2$. A point (p,H) lies on a line (X,Y) whenever $p \subseteq X$ and $Y \subseteq H$.

According to the above description, we call Γ the *point-hyperplane geometry* of $\mathrm{PG}(n,\mathbb{F})$.

Theorem 1.1 *Given a field* $\mathbb{F}$*, let* $\mathbb{F}_0$ *be a subfield of* $\mathbb{F}$*. For* $F = \mathbb{F}$ *and* $F = \mathbb{F}_0$*, let* Γ *and* Γ_0 *be the long-root geometry of* $\mathrm{SL}_{n+1}(F)$*, viewed as the point-hyperplane geometry of* $\mathrm{PG}(n,F)$*. Then the following hold.*

(i) *The span* $\langle\Gamma_0\rangle_\Gamma$ *of* Γ_0 *in* Γ *is the set of point-hyperplane flags* (P,H) *of* $\mathrm{PG}(n,\mathbb{F})$ *with* $P \subseteq E \subseteq H$ *for some element* E *of* $\mathrm{PG}(n,\mathbb{F}_0)$.

(ii) *Suppose* S *is a set of points of* Γ *with* $\langle S\rangle_\Gamma \supseteq \Gamma_0$*. Then* $\mathbb{F}$ *contains an extension* $\mathbb{F}_1$ *of* $\mathbb{F}_0$ *with the following property: the elements of* $\mathrm{PG}(n,\mathbb{F}_1)$ *are the elements* E *of* $\mathrm{PG}(n,\mathbb{F})$ *such that all* $(P,H) \in \Gamma$ *with* $P \subseteq E \subseteq H$ *are in* $\langle S\rangle_\Gamma$.

(iii) *In* (ii)*, if* S *is a generating set of* Γ_0*, then* $\mathbb{F}_1 = \mathbb{F}_0$.

Proof Part (i) is Theorem 4.5. Part (ii) is Theorem 4.7. Part (iii) is Theorem 4.6. □

Corollary 1.2 *With the notation of Theorem* 1.1*, if* $\mathbb{F}_0$ *is a proper subfield of* $\mathbb{F}$*, then* Γ_0 *does not generate* Γ.

Proof If, on the contrary, Γ_0 does generate Γ, then by part (ii) of Theorem 1.1, we have $\mathrm{PG}(n,\mathbb{F}_0) = \mathrm{PG}(n,\mathbb{F})$, which is clearly absurd. □

Problem Corollary 1.2 does not imply that the generating rank of Γ (see Subsection 2.1) is larger than the generating rank of Γ_0. Indeed, possibly there exists a generating set for Γ, not contained in Γ_0 and having as many points as a minimal generating set of Γ_0.

In Section 5 we consider the line-grassmannian Γ of the polar geometry associated to the group $\mathrm{O}_{2n+1}(\mathbb{F})$, $\mathrm{O}^+_{2n}(\mathbb{F})$ or $\mathrm{Sp}_{2n}(\mathbb{F})$ with $n \geq 3$. We view the building Δ as the polar geometry of subspaces of a vector space V of dimension $2n+1$ or $2n$ that are totally singular with respect to a non-degenerate quadratic form of Witt index n, or the polar geometry of subspaces of a vector space V of dimension $2n$ with respect to a non-degenerate symplectic form. The points (lines) of Γ are the lines (point-plane flags) of Δ and the incidence relation is inherited from the incidence relation between elements and flags of Δ.

In each of these cases, for a suitable choice of the quadratic or symplectic form involved, for Δ_0 we can take the geometry of elements of Δ that are defined over $\mathbb{F}_0$ and Γ_0 becomes the subgeometry of Γ whose points belong to Δ_0.

Note The line-grassmannian of the polar geometry associated with the group $O_{2n+1}(\mathbb{F})$ or $O_{2n}^{+}(\mathbb{F})$ is precisely the long-root geometry of that group. Also for the group $SL_{n+1}(\mathbb{F})$, we study its long-root geometry here. The only long-root geometry studied by Cooperstein in [10] that is not considered here is the long-root geometry of the group $O_{2n+2}^{-}(\mathbb{F})$, which is the line-grassmannian of the polar geometry associated to this group. The reason is that we do not want to go into the complications that arise in this case when constructing a subgeometry of the same type defined over a subfield, due to the fact that $O_{2n+2}^{-}(\mathbb{F})$ is a twisted Chevalley group.

Theorem 1.3 *Given a field $\mathbb{F}$, let $\mathbb{F}_0$ be a subfield of $\mathbb{F}$. For $F = \mathbb{F}$ and $F = \mathbb{F}_0$, let Γ and Γ_0 be the line-grassmannian of the polar geometry $\Delta(F)$ associated to the group $O_{2n+1}(F)$, $Sp_{2n}(F)$ or $O_{2n}^{+}(F)$ with $n \geq 3$. Let S be a set of points of Γ with $\langle S\rangle_\Gamma \supseteq \Gamma_0$. Then $\mathbb{F}$ contains an extension $\mathbb{F}_1$ of $\mathbb{F}_0$ such that the elements of $\Delta(\mathbb{F}_1)$ are precisely those elements of $\Delta(\mathbb{F})$ that are generated by 1-elements E with the following property:*

(∗) *all lines L of $\Delta(\mathbb{F})$ with $L \subseteq E^{\perp}$ and $L \cap E \neq \{0\}$ belong to $\langle S\rangle_\Gamma$.*

In fact, all elements E of $\Delta(\mathbb{F}_1)$ satisfy (∗) and conversely, the i-elements E of $\Delta(\mathbb{F})$ with $i < n$ having property (∗) all belong to $\Delta(\mathbb{F}_1)$.

The proof is given in Section 5.

Problems (1) Drop the condition $i < n$ in the last claim of Theorem 1.3.

(2) Computations in some cases with $n = 3$ suggest that if S is a generating set for Γ_0, then $\mathbb{F}_1 = \mathbb{F}_0$ and $\langle S\rangle_\Gamma$ is precisely the collection of points L of Γ for which there exists an i-element E of $\Delta(\mathbb{F}_0)$ such that $L \subseteq E^{\perp}$ (in $\Delta(\mathbb{F})$) and $L \cap E \neq \emptyset$. However, the arguments we have used in the examined cases do not seem to carry over to arbitrary n.

Theorem 1.4 *Let the field $\mathbb{F}$ be generated by adjoining k elements to some subfield $\mathbb{F}_0$. For $F = \mathbb{F}$ and $F = \mathbb{F}_0$, let Γ and Γ_0 be the long-root geometry of the group $SL_{n+1}(F)$ or the line-grassmannian of the polar geometry associated to the group $O_{2n+1}(F)$, $Sp_{2n}(F)$ or $O_{2n}^{+}(F)$. Then Γ can be generated by adding at most k points to Γ_0, viewed as a subgeometry.*

Proof This follows from Corollaries 4.8 and 5.4. □

The previous results entail several consequences for generating ranks, which are defined in Subsection 2.1. We first state the following proposition, which is similar to the results of Cooperstein [10], although the geometry considered here is not a long-root geometry.

Proposition 1.5 *For any prime field* $\mathbb{F}$ *with characteristic* $\mathrm{Char}(\mathbb{F})$ *different from 2, the line-grassmannian of the natural polar geometry for the group* $\mathrm{Sp}_{2n}(\mathbb{F})$ *has generating rank* $2n^2 - n - 1$.

The proof is given at the end of Section 5.

Corollary 1.6 *Given a prime power* q *and an integer* $n \geq 3$, *let* Γ *be the long-root geometry of the group* $G = \mathrm{SL}_{n+1}(q)$ *or the line-grassmannian of the polar geometry associated to the group* $G = \mathrm{O}_{2n+1}(q)$, $\mathrm{Sp}_{2n}(q)$ *or* $\mathrm{O}^+_{2n}(q)$, *with* q *odd when* $G = \mathrm{Sp}_{2n}(q)$ *or* $\mathrm{O}^+_{2n}(q)$. *Then* Γ *has generating rank* r *with* $r_0 \leq r \leq r_0 + 1$ *and* r_0 *as follows:*

G	$\mathrm{SL}_{n+1}(q)$	$\mathrm{O}_{2n+1}(q)$	$\mathrm{Sp}_{2n}(q)$	$\mathrm{O}^+_{2n}(q)$
r_0	$n^2 - 1$	$2n^2 + n$	$2n^2 - n - 1$	$2n^2 - n$

Proof Given a prime number p let Γ_0 be the long-root geometry associated to $G_0 = \mathrm{SL}_{n+1}(p)$ or the line-grassmannian of the polar geometry for $G_0 = \mathrm{O}_{2n+1}(p)$ or $\mathrm{O}^+_{2n}(p)$. By Theorems 4.1, 5.1 and 6.1 of Cooperstein [10], Γ_0 has a generating set of size $m = n^2 - 1$, $2n^2 + n$ or $2n^2 - n$, respectively. Furthermore, Γ_0 admits an embedding in $\mathrm{PG}(m-1, p)$, except possibly when $G_0 = \mathrm{O}^+_{2n}(2)$ with $n > 3$. Therefore, in these cases, Γ_0 has generating rank $r_0 = m$. The finite field of order q is at most a simple extension over its prime field. The result follows from Theorem 1.4.

By the same argument, except for exploiting Proposition 1.5 instead of the results of Cooperstein [10], we obtain the conclusion in the case of $G = \mathrm{Sp}_{2n}(q)$. □

In a forthcoming paper we will determine the precise generating rank for some of these geometries.

Note The case of $G = \mathrm{Sp}_{2n}(q)$ with q even is not considered in the above corollary, in view of the classical isomorphism between the groups $\mathrm{Sp}_{2n}(2^e)$ and $\mathrm{O}_{2n+1}(2^e)$ and the corresponding isomorphism between their polar geometries.

The case of $G = \mathrm{O}^+_{2n}(q)$ with q even is also missing in Corollary 1.6, but for a different reason. With the notation used in the proof of

Corollary 1.6, let $G_0 = O_{2n}^+(2)$. By Theorem 4.1 of Cooperstein [10], Γ_0 admits a generating set of size $m = 2n^2 - n$. Therefore, the generating rank of Γ_0 is at most m. However, no embedding of Γ_0 in $PG(m-1,2)$ is known when $n > 4$, whereas an embedding of Γ_0 in $PG(m-2,2)$, which we shall call the *natural embedding* of Γ_0, is induced by the natural embedding ε of the line-grassmannian of $PG(2n-1,2)$ in $PG(m-1,2)$. (Note that Γ_0 is a subgeometry of the polar geometry for $Sp_{2n}(2)$ and the ε-image of the latter spans a hyperplane of $PG(m-1,2)$; compare [4, Section 3.3].) Accordingly, when $n > 4$ we can only claim the generating rank r_0 of Γ_0 is at least $m-1$ and at most m.

When $n = 3$, Γ_0 is isomorphic to the long-root geometry of $SL_4(2)$, the generating rank of which is $r_0 = 15$ $(= m)$. When $n = 4$, we have three natural embeddings of Γ_0, permuted by the triality of the D_4-building, but none of them is universal, by the second claim of Corollary 5.6 of [1]; see also [2]. Consequently, the universal embedding of Γ_0, which exists by Ronan [12, Corollary 2], is 27-dimensional and so Γ_0 has generating rank $r_0 = 28$ $(= m)$.

Therefore, by Theorem 1.4, when q is even, the line-grassmannian of the polar geometry for $O_{2n}^+(q)$ has generating rank r with $r_0 \le r \le r_0+1$ where r_0 is as follows:

n	r_0
3	$15\ (= 2n^2 - n)$
4	$28\ (= 2n^2 - n)$
$n > 4$	$2n^2 - n - 1 \le r_0 \le 2n^2 - n$

2. Preliminaries

2.1. Terminology

Following Tits [16], by a *geometry* Δ of *rank* n we will mean a quadruple $(\mathcal{O}, \star, \tau, I)$ consisting of a set $\mathcal{O}$ of elements, a symmetric relation $\star$ on $\mathcal{O}$ called the *incidence relation*, a set I of size n, called the *set of types*, and a surjective mapping $\tau\colon \mathcal{O} \longrightarrow I$ (called the *type-map*) with the property that, if x, y are distinct incident objects, then x and y have different type (namely, $\tau(x) \neq \tau(y)$).

A *flag* of a geometry Δ is a set of pairwise incident elements of Δ. The *type* of a flag F is its image $\tau(F)$ by τ. For $J \subseteq I$, the flags of type J are also called *J-flags*. The flags of type I are called *chambers*. The *panels* are the flags of type $I \setminus \{i\}$, for $i \in I$. Two flags F_1, F_2 are said to be *incident* when $F_1 \cup F_2$ is a flag.

Taking the flags as cells, we may also regard a geometry as a cell complex where every vertex is given a colour and type in such a way that no two vertices of the same colour or type belong to the same cell (compare Tits [15]).

A *subgeometry* of $\Delta = (\mathcal{O}, \star, \tau, I)$ is a geometry $\Delta_0 = (\mathcal{O}_0, \star_0, \tau_0, I_0)$, where $\mathcal{O}_0 \subseteq \mathcal{O}$, and for all $X, Y \in \mathcal{O}_0$ we have $\tau_0(X) = \tau(X)$, and $X \star_0 Y$ if and only if $X \star Y$.

A *point-line geometry* is a pair $\Gamma = (\mathcal{P}, \mathcal{L})$ where $\mathcal{P}$ is a set whose elements are called 'points' and $\mathcal{L}$ is a collection of subsets of $\mathcal{P}$ called 'lines' with the property that any two points belong to at most one line.

A *subgeometry* Γ_0 of Γ is a point-line geometry $(\mathcal{P}_0, \mathcal{L}_0)$ such that $\mathcal{P}_0 \subseteq \mathcal{P}$ and, for each $l_0 \in \mathcal{L}_0$ there is an $l \in \mathcal{L}$ such that $l_0 \subseteq l$.

A *subspace* of Γ is a subset $X \subseteq \mathcal{P}$ such that any line containing at least two points of X entirely belongs to X. Clearly, every subspace X can also be regarded as a subgeometry $(X, \mathcal{L}_X)$ of Γ, where $\mathcal{L}_X$ is the set of lines of Γ contained in X.

The *span* of a set $S \subseteq \mathcal{P}$ is the smallest subspace containing S; it is the intersection of all subspaces containing S and is denoted by $\langle S \rangle_\Gamma$. We say that S *spans* (or *generates*) $\langle S \rangle_\Gamma$ or that S is a *generating set* for $\langle S \rangle_\Gamma$.

The *generating rank* of Γ is the minimal size of a generating set of Γ.

Note Clearly, a point-line geometry is a geometry of rank 2. We may take the words 'point' and 'line' as its types, or the integers 0 and 1, or whatever pair of symbols we like.

It should also be noted that the above definition of geometry of rank n is weaker than other definitions existing in the literature. For instance, Buekenhout [6] also requires all maximal flags to be chambers. In Buekenhout [5] and Pasini [11] residual connectedness and firmness are also assumed. However, as we only need a terminological framework in this paper, where to put certain well-known structures (buildings, polar geometries, and so on), our lax definition is sufficient for our purposes.

2.2. The geometries considered in this paper

The point-line geometries considered in this paper are shadow geometries of certain buildings.

We will view a building as a geometry belonging to a Coxeter diagram, as in Tits [15]; see also Tits [16] or Pasini [Chapter 13][11]. For a description of a building as a chamber system see Tits [16] or Ronan [13]. The buildings we study in this paper are of irreducible spherical type, have rank $n \geq 3$ and arise from a (non-twisted) Chevalley group defined over a field $\mathbb{F}$.

We recall that the (adjoint) non-twisted Chevalley groups of rank at least 3 are the following: $\mathrm{PSL}_{n+1}(\mathbb{F})$, $\mathrm{P}\Omega_{2n+1}(\mathbb{F})$, $\mathrm{PSp}_{2n}(\mathbb{F})$, $\mathrm{P}\Omega_{2n}^{+}(\mathbb{F})$, $E_6(\mathbb{F})$, $E_7(\mathbb{F})$, $E_8(\mathbb{F})$ and $F_4(\mathbb{F})$, with Dynkin diagrams A_n, B_n, C_n, D_n, E_6, E_7, E_8 and F_4, respectively. The corresponding buildings are said to be *defined over* $\mathbb{F}$ and to have type A_n, B_n, C_n, etc. Note that the type of a building, as defined above, is not simply the name of its Coxeter diagram, as the Dynkin diagrams B_n and C_n, which are different, correspond to the same Coxeter diagram, often called C_n in the literature.

In this paper, we are mainly interested in the diagrams A_n, B_n, C_n and D_n, but E_6 and E_7 will also be considered in Theorem 3.1. We take the positive integers $1, 2, 3, \ldots$ as types for these diagrams, labelling the nodes as follows:

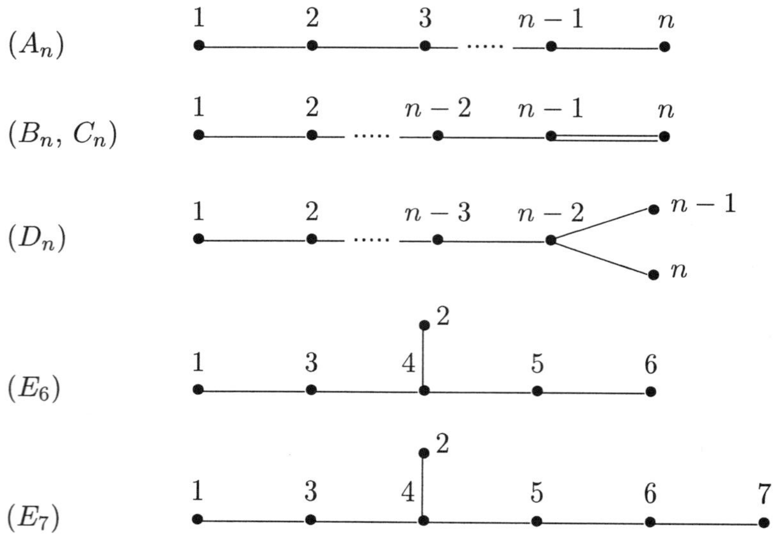

Henceforth, when referring to labels of the nodes of one of the above diagram, we shall mean the labels that diagram has been given here.

2.2.1 Shadow geometries. Given a building Δ, let M be its diagram and let I be its set of types. Given a non-empty $J \subseteq I$, we denote $\mathrm{Sh}_J(\Delta)$ the J-shadow geometry of Δ (Tits [15, Chapter 12]; see also Pasini [11, Chapter 5], where these geometries are called J-Grassmann geometries). We regard $\mathrm{Sh}_J(\Delta)$ as a point-line geometry: The J-flags of Δ are taken as points; for every $j \in J$ and every flag F of Δ of type $(J \cup M(j)) \setminus \{j\}$, with $M(j)$ denoting the set of types joined to j in M, the collection of J-flags of Δ incident to F is a line of $\mathrm{Sh}_J(\Delta)$ and all lines of $\mathrm{Sh}_J(\Delta)$ are defined in this way.

In particular, when $J = \{i\}$ the points of $\mathrm{Sh}_J(\Delta)$ are the i-elements of Δ and the lines of $\mathrm{Sh}_J(\Delta)$ correspond to the flags of Δ of type $M(i)$. If furthermore i is an end-node of M and j is the unique node joined to it in M (as when M is as in one of the previous pictures, $i = 1$ and $j = 2$), then the lines of $\mathrm{Sh}_i(\Delta)$ correspond to the j-elements of Δ.

With M and J as above, we say that $\mathrm{Sh}_J(\Delta)$ *is of type* M_J (of type M_i when $J = \{i\}$).

2.2.2 Geometries of type $A_{n,i}$ and $A_{n,\{1,n\}}$. Given a field $\mathbb{F}$, let Δ be the building of type A_n ($n \geq 3$) defined over $\mathbb{F}$. It is well known that Δ is isomorphic to the projective geometry $\mathrm{PG}(n, \mathbb{F})$. For $i = 2, 3, ..., n-1$, the i-shadow geometry $\mathrm{Sh}_i(\Delta)$ is called the *i-grassmannian* of $\mathrm{PG}(n, \mathbb{F})$ (*line-grassmannian* when $i = 2$). Its points are the $(i-1)$-spaces of $\mathrm{PG}(n, \mathbb{F})$ (namely, the i-spaces of $V(n+1, \mathbb{F})$) and its lines are the collections of $(i-1)$-spaces contained in a given i-space and containing a given $(i-2)$-space.

Now let $J = \{1, n\}$. Then $\mathrm{Sh}_J(\Delta)$ is isomorphic to the *long-root geometry* of $\mathrm{SL}_{n+1}(\mathbb{F})$. Its points are the point-hyperplane pairs of $\mathrm{PG}(n, \mathbb{F})$. A line of $\mathrm{Sh}_J(\Delta)$ is the collection of point-hyperplane flags (p, H) where either $H = H_0$ and $p \in L$ for a given hyperplane H_0 and a given line L of H_0, or $H \supset S$ and $p = p_0$ for a given $(n-2)$-space S of $\mathrm{PG}(n, \mathbb{F})$ and a given point p_0 of S. Accordingly, two points (p_1, H_1) and (p_2, H_2) of $\mathrm{Sh}_J(\Delta)$ are collinear if and only if either $p_1 = p_2$ or $H_1 = H_2$.

In particular, let $n = 3$. The 2-grassmannian of $\Delta = \mathrm{PG}(3, \mathbb{F})$ is isomorphic to the polar geometry, say Π, associated to $\mathrm{O}_6^+(\mathbb{F})$, the elements of which are the subspaces of $V = V(6, \mathbb{F})$ that are totally singular for a given non-singular quadratic form of Witt index 3 over V:

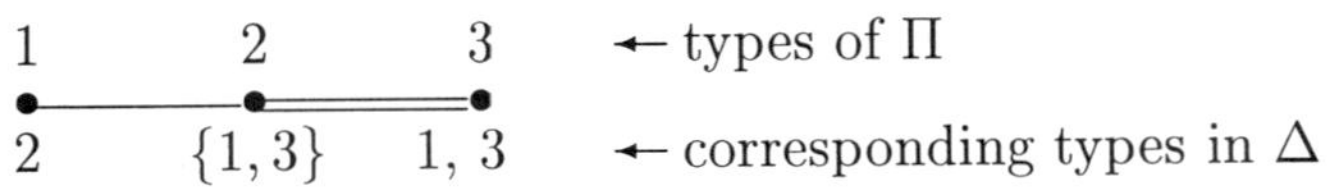

The $\{1, 3\}$-shadow geometry of Δ is isomorphic to the 2-shadow geometry of Π, which we shall call the *line-grassmannian* of Π.

2.2.3 Line-grassmannians of polar spaces. Let Δ be the building of type B_n ($n \geq 3$) defined over a given field $\mathbb{F}$. Namely, Δ is the geometry of subspaces of $V(2n+1, \mathbb{F})$ that are totally singular for a given non-singular quadratic form of Witt index n. The 2-shadow geometry $\mathrm{Sh}_2(\Delta)$ of Δ is a subgeometry of the 2-grassmannian of $\mathrm{PG}(2n, \mathbb{F})$. Its points are the lines (2-elements) of the polar space Δ and its lines correspond to the point-plane flags (namely, $\{1, 3\}$-flags) of Δ. We call $\mathrm{Sh}_2(\Delta)$ the *line-grassmannian* of Δ.

Now, let Δ be the building of type C_n $(n \geq 3)$ defined over $\mathbb{F}$, namely the polar geometry of subspaces of $V(2n, \mathbb{F})$ that are totally isotropic for a given non-degenerate symplectic form of Witt index n. The 2-shadow geometry of Δ (type $C_{n,2}$) is called the *line-grassmannian* of Δ.

Finally, let Δ be the building of type D_n $(n \geq 4)$ defined over $\mathbb{F}$. The 1-shadow geometry $\Pi = \mathrm{Sh}_1(\Delta)$ of Δ, viewed as a geometry of rank n, is the polar geometry associated to $\mathrm{O}^+_{2n}(\mathbb{F})$ (Tits [15, Chapter 7]), the elements of which are the subspaces of $V(2n, \mathbb{F})$ that are totally singular for a given non-singular quadratic form of Witt index n. As types for the elements of Π we take their dimensions as subspaces of $V(2n, \mathbb{F})$, thus obtaining the same labelling as for the Dynkin diagrams B_n and C_n.

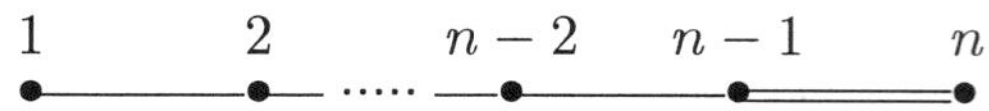

Clearly, $\mathrm{Sh}_2(\Delta) \cong \mathrm{Sh}_2(\Pi)$. We call $\mathrm{Sh}_2(\Delta)$ the *line-grassmannian* of Δ (also, of Π).

Note When Δ is of type B_n or D_n, its line-grassmannian is isomorphic to the long-root geometry for $\mathrm{O}_{2n+1}(\mathbb{F})$ or $\mathrm{O}^+_{2n}(\mathbb{F})$, respectively.

2.3. Subgeometries defined over a subfield

The main purpose of this subsection is the following: given a field $\mathbb{F}$, a subfield $\mathbb{F}_0$ of $\mathbb{F}$, a Dynkin diagram M $(= A_n, B_n, C_n, D_n, E_6, E_7, E_8, F_4$ or $G_2)$ and a building Δ of type M defined over $\mathbb{F}$, associated to a non-twisted Chevalley group G of type M, we want to define a sub-building Δ_0 of Δ defined over $\mathbb{F}_0$ that also has type M. To that goal, we regard Δ as obtained from the natural (B, N)-pair of G (see Tits [15, Chapter 3]; also Carter [8, Chapter 8]).

First some properties of (non-twisted) Chevalley groups will be discussed. Most information on Chevalley groups used below can be found in Carter [8].

Given a Dynkin diagram M defined over the set of types I, we denote by $M(\mathbb{F})$ and $\bar{M}(\mathbb{F}))$ the adjoint and universal Chevalley group of type M defined over $\mathbb{F}$; thus, $A_n(\mathbb{F})$ stands for $PSL_{n+1}(\mathbb{F})$, $\bar{A}_n(\mathbb{F})$ for $SL_{n+1}(\mathbb{F})$, and so on. Let Φ_M be the root system with Dynkin diagram M. It is known (Carter [8, Theorem 12.1.1]) that, if $M \neq A_1$, the group $\bar{M}(\mathbb{F})$ is generated by a set of generators $\bar{x}_r(t)$, one for each $r \in \Phi_M$ and $t \in \mathbb{F}$, subject to the following relations:

$$(1) \quad \bar{x}_r(t_1)\bar{x}_r(t_2) = \bar{x}_r(t_1 + t_2),$$

$$(2) \qquad \bar{x}_s(u)\bar{x}_r(t) \;=\; \prod_{i,j>0} \bar{x}_{ir+js}(C_{ijrs}(-t)^i u^j)$$

for certain integers C_{ijrs} and

$$(3) \qquad \bar{h}_r(t_1)\bar{h}_r(t_2) \;=\; \bar{h}_r(t_1 t_2)$$

for any non-zero elements $t_1, t_2 \in \mathbb{F}$ and with

$$\bar{h}_r(t) := \bar{n}_r(t)\bar{n}_r(-1), \qquad \bar{n}_r(t) := \bar{x}_r(t)\bar{x}_{-r}(-t^{-1})\bar{x}_r(t).$$

Furthermore, there is a surjective homomorphism $\bar{M}(\mathbb{F}) \longrightarrow M(\mathbb{F})$ sending a (B,N)-pair $(\bar{B},\bar{N})$ for $\bar{M}(\mathbb{F})$ to the (B,N)-pair (B,N) for $M(\mathbb{F})$ whose kernel is precisely the centre $\bar{Z}$ of $\bar{M}(\mathbb{F})$. Moreover, $\bar{Z} \leq \bar{B} \cap \bar{N}$. Hence, the buildings associated to $\bar{M}(\mathbb{F})$ and $M(\mathbb{F})$ (and all intermediate quotients) are the same.

Often, any quotient of $\bar{M}(\mathbb{F})$ by a subgroup of $\bar{Z}$ is referred to as a Chevalley group (defined over $\mathbb{F}$).

As the above discussion fails to hold when $M = A_1$, henceforth we assume $M \neq A_1$.

Lemma 2.1 *For every subfield $\mathbb{F}_0$ of $\mathbb{F}$, the Chevalley group $\bar{M}(\mathbb{F})$ contains $\bar{M}(\mathbb{F}_0)$ as a subgroup.*

Proof We put $\bar{G} := \bar{M}(\mathbb{F})$, $\bar{G}_0 := \bar{M}(\mathbb{F}_0)$ and denote all subgroups of $\bar{G}_0$ with a subscript 0. Define a map $\phi\colon \bar{G}_0 \longrightarrow \bar{G}$ by sending $\bar{x}_r(t) \in \bar{G}_0$ to the 'same' element in $\bar{G}$, for each $r \in \Phi_M$ and $t \in \mathbb{F}_0$. We can extend this to a homomorphism since the relations holding in $\bar{G}_0$ also hold in $\bar{G}$. Since $\bar{G}_0$ is an extension by its centre $\bar{Z}_0$ of the simple group $G(\mathbb{F}_0)$, the kernel of ϕ is contained in $\bar{Z}_0$. Put $\bar{H}_0 = \bar{B}_0 \cap \bar{N}_0$. Then $\bar{Z}_0 \leq \bar{H}_0$ (see the discussion above). In the proof of Theorem 12.1.1. of [8] it is proved that $\bar{H}_0$ is equal to the direct product $\prod_{i=1}^n \bar{H}_{0,p_i}(\mathbb{F}_0)$, where $\bar{H}_{0,p_i}(\mathbb{F}_0) = \langle \bar{h}_{0,p_i}(t) \mid t \in \mathbb{F}_0 \rangle$ and $\{p_i\}_{i=1}^n$ is a basis for Φ_M.

From the relations at the beginning of this subsection we can see that the kernel of $\phi\colon \bar{H}_{0,p_i} \longrightarrow \bar{H}_{p_i}$ is trivial, so $\phi\colon \bar{G}_0 \longrightarrow \bar{G}$ is an isomorphism onto its image. □

Lemma 2.2 *Let $\mathbb{F}_0$ be a subfield of $\mathbb{F}$, as in Lemma 2.1. For any non-empty $J \subseteq I$ let $\bar{P}_J(\mathbb{F})$ and $\bar{P}_J(\mathbb{F}_0)$ be the standard parabolic subgroup of type J of $\bar{M}(\mathbb{F})$ and $\bar{M}(\mathbb{F}_0)$ respectively and let $(\bar{B},\bar{N})$ be the (B,N)-pair for $\bar{M}(\mathbb{F})$. Then*

$$\bar{B}\bar{P}_J(\mathbb{F}_0)\bar{B} = \bar{P}_J(\mathbb{F}).$$

Proof We use the notation as in the proof of Lemma 2.1. From the relations given at the beginning of this subsection it is clear that, for each $r \in \Phi_M$, all elements $\bar{n}_r(t)$ $(t \in \mathbb{F})$ belong to the same $\bar{H}$-coset in $\bar{N}$. Thus they represent the same element of the Weyl group W of Φ_M. Since every element of W is represented by some $\bar{n}_r(t)$ and $\bar{B}_0 \leq \bar{B}$ and $\bar{N}_0 \leq \bar{N}$, we can write

$$\bar{B}\bar{P}_{J0}\bar{B} = \bar{B}\bar{B}_0 W_J \bar{B}_0 \bar{B} = \bar{B} W_J \bar{B} = \bar{P}_J.$$

□

Definitions Let Δ and Δ_0 be the buildings associated to the groups $\bar{M}(\mathbb{F})$ and $\bar{M}(\mathbb{F}_0)$, respectively. Using Lemmas 2.1 and 2.2, we can identify the building Δ_0 with a sub-building of Δ. We call Δ_0 a sub-building of Δ *defined over* $\mathbb{F}_0$ or, an $\mathbb{F}_0$-*sub-building* of Δ, for short.

With I the type-set of M, let J be a non-empty subset of I and $\Gamma = \mathrm{Sh}_J(\Delta)$, $\Gamma_0 = \mathrm{Sh}_J(\Delta_0)$ (see 2.2). As Δ_0 can be regarded as a subgeometry of Δ, we may also view the points of Γ_0 as points of Γ and every line of Γ_0 is contained in a unique line of Γ. Thus we can regard Γ_0 as a subgeometry of Γ. We will call Γ_0 an $\mathbb{F}_0$-*subgeometry* of Γ.

Example Let $M = A_n$. Then $\bar{M}(\mathbb{F}) = \mathrm{SL}_{n+1}(\mathbb{F})$ and elements $\bar{x}_r(t)$ are precisely the standard transvection matrices with respect to some chosen basis. By taking only those elements $\bar{x}_r(t)$ with $t \in \mathbb{F}_0$ we get $\mathrm{SL}_{n+1}(\mathbb{F}_0)$. The building Δ_0 is the geometry of subspaces of $\mathrm{PG}(n, \mathbb{F}_0)$. By Lemmas 2.1 and 2.2, we can also view it as the sub-building of Δ containing only the subspaces of $\mathrm{PG}(n, \mathbb{F})$ that are $\mathbb{F}_0$-rational. Note that the latter description depends on the choice of a basis of $V(n+1, \mathbb{F})$.

2.4. Apartments and opposition

Given a Dynkin diagram M and a Chevalley group G of type M, let (B, N) be a (B, N)-pair of G and let Δ be the building associated to G, arising from (B, N). Let W be the Weyl group of (B, N), with generator set $\{r_i\}_{i \in I}$. For $w \in W$, let $l(w)$ denote the length of a reduced expression for w with respect to $\{r_i\}_{i \in I}$; let w_0 be the longest word of W in this sense. For $J, K \subset I$, let $P_J = BW_JB$ and $P_K = BW_KB$ be the standard parabolic subgroups of G of type J and K. We recall that the left cosets of P_J and P_K correspond to the flags of Δ of type $I \setminus J$ and $I \setminus K$. The *distance* between flags gP_J and hP_K $(g, h \in G)$ is the shortest w such that $P_J g^{-1} h P_K = P_J w P_K$. These flags are called *opposite* when $P_J w P_K = P_J w_0 P_K$.

We recall that an *apartment* of Δ is the collection of elements of Δ that, viewed as left cosets of maximal parabolics, have a non-empty intersection with a given left coset of N. The apartments of Δ are isomorphic to the Coxeter complex of type M and each pair of flags is contained in some apartment (Tits [15]).

Lemma 2.3 (Tits [17]) *For any apartment A and any chamber c there exists a chamber d on A opposite to c.*

For a subfield $\mathbb{F}_0$ of the ground field $\mathbb{F}$ of G, let Δ_0 be the $\mathbb{F}_0$-sub-building of Δ. Then the apartments of Δ_0 are also apartments of Δ.

Corollary 2.4 *For any flag F of Δ and any non-empty $J \subseteq I$, there is a J-flag F_0 in Δ_0 opposite to F.*

Proof This follows from Lemma 2.3 by taking for A an apartment of Δ_0 and for c a chamber on F. □

3. Geometries generated by $\mathbb{F}_0$-subgeometries

Theorem 3.1 *For $M = A_n, B_n, C_n, D_n, E_6$ or E_7 and a field $\mathbb{F}$, let Δ be the building with diagram M defined over $\mathbb{F}$. Let $i \in I$ be the label of a node of M and $\mathbb{F}_0$ a subfield of $\mathbb{F}$. Then the i-shadow geometry of Δ is spanned by the points of its $\mathbb{F}_0$-subgeometry, if its type M_i is one of the following*:

(i) *$A_{n,i}$ with $i \in \{1, 2, \ldots, n\}$ and $n \in \mathbb{N}_{\geq 1}$;*

(ii) *$D_{n,1}$, $D_{n,n-1}$, $D_{n,n}$ with $n \in \mathbb{N}_{\geq 3}$;*

(iii) *$E_{6,1}$, $E_{6,6}$, $E_{7,7}$;*

(iv) *$B_{n,n}$ with $n \in \mathbb{N}_{\geq 2}$, provided that* $\mathrm{Char}(\mathbb{F}) \neq 2$;

(v) *$C_{n,1}$, with $n \in \mathbb{N}_{\geq 2}$, provided that* $\mathrm{Char}(\mathbb{F}) \neq 2$.

Proof By Theorem 1 of Blok and Brouwer [3], in the above cases the i-shadow geometry Γ of Δ is spanned by the points of an apartment of Δ. The $\mathbb{F}_0$-subgeometry of Γ is the i-shadow geometry of the sub-building Δ_0 of Δ defined over $\mathbb{F}_0$. The points on an apartment of Δ_0 are also the points on an apartment of Δ. □

Proposition 3.2 *For $M = B_n, C_n$ ($n \in \mathbb{N}_{\geq 2}$) and a field $\mathbb{F}$, let Δ be the building of type M defined over $\mathbb{F}$. Let i be a type of M and $\mathbb{F}_0$ a subfield of $\mathbb{F}$. Then the i-shadow geometry of Δ is spanned by the points of its $\mathbb{F}_0$-subgeometry, if its type M_i is one of the following:*

(i) *$B_{n,1}$;*

(ii) *$C_{n,1}$ if* $\mathrm{Char}(\mathbb{F}) = 2$.

Proof We assume that the $B_{n,1}$ geometry is defined by the form f described in Section 5 and show that it is generated by its $\mathbb{F}_0$-subgeometry

$B_{n,1}(\mathbb{F}_0)$ defined by the same form. Clearly, the $B_{n,1}$ geometry contains the $D_{n,1}$ geometry defined by the form described in Section 5 as a hyperplane induced by this embedding. By Theorem 3.1, this $D_{n,1}$ geometry is generated by the points of its $\mathbb{F}_0$-subgeometry $D_{n,1}(\mathbb{F}_0)$ which is a hyperplane of $B_{n,1}(\mathbb{F}_0)$. Thus $D_{n,1}(\mathbb{F}_0)$, together with any point p of $B_{n,1}(\mathbb{F}_0)$ outside $D_{n,1}(\mathbb{F}_0)$, generates $B_{n,1}$: if q is any other point outside $D_{n,1}(\mathbb{F}_0)$ then either it is collinear to p and we are done or we can find lines l, m on p, q respectively with $l^{\perp} \cap m = \{0\}$, where $\perp$ denotes collinearity. Now, every point of l is collinear to a unique point of m; so we can find a point r on l that is collinear to a point s of m outside $D_{n,1}(\mathbb{F}_0)$, assuming that $|\mathbb{F}_0| \geq 2$. Now we see that r, s and hence q is generated by p together with $D_{n,1}(\mathbb{F}_0)$.

The second assertion follows from the first, by noting that the isomorphism from $B_{n,1}$ to $C_{n,1}$, which is projection from the nucleus, carries $B_{n,1}(\mathbb{F}_0)$ to $C_{n,1}(\mathbb{F}_0)$. □

4. The long-root geometry of $\mathrm{SL}_{n+1}(\mathbb{F})$

In this section, $\mathbb{F}$ is a given field and Δ is the building of type A_n defined over $\mathbb{F}$, namely the projective geometry of linear subspaces of $V = V(n+1, \mathbb{F})$. Furthermore, $\Gamma = (\mathcal{P}, \mathcal{L})$ is the $\{1, n\}$-shadow geometry of Δ. As noted in Subsection 2.2, Γ is the long-root geometry associated to the group $\mathrm{SL}_{n+1}(\mathbb{F})$.

By the words 'point' and 'line' we will mean a point and a line of Γ, unless specified otherwise. The i-elements of Δ will be regarded as i-spaces of V and denoted by capital letters. Given a line l of Γ, we denote $\mathcal{P}(l)$ its set of points. We also denote the set of points of Γ incident to an element E of Δ by $\mathcal{P}(E)$. Hence, $\mathcal{P}(E)$ is the set of $\{1, n\}$-flags of Δ incident to E, usually called the $\{1, n\}$-*shadow* of E. We also extend this notation to $\{0\}$ and V, by setting $\mathcal{P}(\{0\}) = \mathcal{P}(V) = \mathcal{P}$, by convention.

Given a set $S \subseteq \mathcal{P}$ of points of Γ, we say that an element E of Δ is *full* with respect to S (also, *S-full*, for short) if $\mathcal{P}(E) \subseteq \langle S \rangle_{\Gamma}$. We denote the collection of elements that are full with respect to S as $\mathcal{F}(S)$ and the incidence structure induced on it by Δ as $\Delta(\mathcal{F}(S))$.

For a subfield $\mathbb{F}_0$ of $\mathbb{F}$, the sub-building Δ_0 of Δ defined over $\mathbb{F}_0$ is isomorphic to $\mathrm{PG}(n, \mathbb{F}_0)$ and its $\{1, n\}$-shadow geometry Γ_0 (which is the long-root geometry of $\mathrm{SL}_{n+1}(\mathbb{F}_0)$) is the $\mathbb{F}_0$-subgeometry of Γ.

The geometry Γ is preserved under the duality that interchanges i and $n+1-i$ elements for all $i = 1, 2, \ldots, n$. We will often make use of this fact, for instance by stating the dual part of a result but not proving it explicitly.

Lemma 4.1 *Let E_1, E_2 be elements of Δ.*

(i) *For a line l of Γ, if $\emptyset \neq \mathcal{P}(l) \cap \mathcal{P}(E_1) \neq \mathcal{P}(l) \cap \mathcal{P}(E_2) \neq \emptyset$, then* $\mathcal{P}(l) \subseteq \mathcal{P}(E_1 \cap E_2) \cup \mathcal{P}(\langle E_1, E_2 \rangle_V)$.

(ii) *Every point in* $\mathcal{P}(E_1 \cap E_2) \cup \mathcal{P}(\langle E_1, E_2 \rangle_V) \setminus (\mathcal{P}(E_1) \cup \mathcal{P}(E_2))$ *lies on a line meeting both* $\mathcal{P}(E_1)$ *and* $\mathcal{P}(E_2)$.

Proof Set $Z = E_1 \cap E_2$ and $Z' = \langle E_1, E_2 \rangle_V$.

(i) For $i = 1, 2$, let $(A_i, H_i) \in \mathcal{P}(l) \cap \mathcal{P}(E_i)$ be distinct. By duality we may assume that $A_1 = A_2 = A$. Then, $Z \neq \{0\}$ and $A \subseteq Z \subseteq H_1, H_2$, so $\mathcal{P}(l) \subseteq \mathcal{P}(Z)$.

(ii) Let $(A, H) \in \mathcal{P}(Z') \setminus (\mathcal{P}(E_1) \cup \mathcal{P}(E_2))$. Then A lies on some 2-space L of V meeting E_1 and E_2 in 1-spaces A_1 and A_2 respectively, and $H \supseteq E_1, E_2$. Clearly, (L, H) is the required line of Γ. The case $(A, H) \in \mathcal{P}(Z) \setminus (\mathcal{P}(E_1) \cup \mathcal{P}(E_2))$ is settled dually. □

Lemma 4.2 *Let X be a subspace of Γ.*

(i) *Given an n-element H of Δ, suppose that H is spanned by 1-spaces $A_1, A_2, ..., A_n$ of V such that $(A_i, H) \in X$ for every $i = 1, 2, ..., n$. Then H is full with respect to X. Dually, let A be a 1-element of Δ with $A = \cap_{i=1}^{n} H_i$ for n-spaces H_1, H_2,..., H_n of V such that $(A, H_i) \in X$ for $i = 1, 2, ..., n$. Then A is full with respect to X.*

(ii) *If an element E of Δ is spanned by 1-spaces of V that are full with respect to X, then E is full with respect to X. Dually, if an element E of Δ is the intersection of n-elements that are full with respect to X, then E is full with respect to X.*

(iii) *Let L and H be a 2- and an n-element of Δ such that $L \not\subseteq H$ and let $A = L \cap H$. Suppose that L is full with respect to X and $(A, H) \in X$. Then A is full with respect to X.*

Proof (i) This follows from the fact that $\mathcal{P}(H)$ and $\mathcal{P}(A)$ are subspaces (see Lemma 4.1(i)) isomorphic to the projective geometry $\mathrm{PG}(n-1, \mathbb{F})$.

(ii) For every n-element $H \supseteq E$ of Δ and every $j = 1, 2, ..., i$, $(A_j, H) \in X$ because A_j is full. Hence $(A, H) \in X$ for any 1-element $A \subseteq E$, by (i). Therefore, E is X-full.

(iii) Let $H_1, H_2, ..., H_{n-1}$ be n-elements with $L = \cap_{i=1}^{n-1} H_i$. Then $A = H_1 \cap ... \cap H_{n-1} \cap H$. Furthermore, $(A, H) \in X$ by assumption and $(A, H_i) \in X$ for every $i = 1, 2, ..., n-1$ because L is X-full. The conclusion follows from (i). □

Lemma 4.3 *Let X be a subspace of Γ containing Γ_0.*

(i) *Every element of Δ_0 is full with respect to X.*

(ii) *Let A_0, A_1, A_2 be distinct 1-elements of Δ not contained in the same 2-element. Suppose that both A_1 and A_2 are full with respect to X. Then there exists an n-element of Δ that is full with respect to X and contains both A_1 and A_2 but not A_0.*

Proof (i) We first prove the statement for a 1-element A of Δ_0. Clearly, A is the intersection of n-elements $H_1, H_2, ..., H_n$ of Δ_0 and $(A, H_i) \in \Gamma_0$ for all $i = 1, 2, ..., n$. By Lemma 4.2(i), A is X-full. The conclusion now follows from Lemma 4.2(ii).

(ii) Clearly, there exist 1-elements $A_3, A_4, ..., A_n$ of Δ_0 which, together with A_1 and A_2, span a hyperplane H of V not containing A_0. The elements A_1, A_2 are X-full by assumption whereas $A_3, A_4, ..., A_n$ are X-full by (i). Hence H is X-full, by Lemma 4.2(i). □

Lemma 4.4 *Let X be a subspace of Γ containing Γ_0. Then, for every element E of $\Delta \setminus \Delta_0$ full with respect to X, there is a chamber C of Δ containing E and such that all elements of C are full with respect to X and do not belong to Δ_0.*

Proof Let E be as in the hypotheses of the lemma and let i be its dimension in V. Suppose $i > 1$. We shall show that there exists an $(i-1)$-element E' of $\Delta \setminus \Delta_0$ that is also X-full.

Take an $(n+1-i)$-element $E_0 \in \Delta_0$ opposite to E, that is, such that $E \cap E_0 = \{0\}$. By Corollary 2.4, E_0 exists. Clearly $E = \langle H \cap E \rangle_V$, where H ranges over an independent set of n-elements of Δ_0 containing E_0. As $E \notin \Delta_0$, one of these n-elements, say H', satisfies $H' \cap E \notin \Delta_0$. By Lemma 4.3(i), H' is full. Hence $E' = H' \cap E$ is the required $(i-1)$-element, by Lemma 4.1(ii).

The case $i < n$ is handled dually. The conclusion now follows by repeating the above arguments $n-1$ times. □

Theorem 4.5 $\langle \Gamma_0 \rangle_\Gamma = \bigcup_{E \in \Delta_0} \mathcal{P}(E)$.

Proof As for every two elements E_1, E_2 of Δ_0 also $E_1 \cap E_2$ and $\langle E_1, E_2 \rangle_V$ are elements of Δ_0, it follows from the first part of Lemma 4.1 that the right hand side of the above formula is a subspace of Γ. This proves '$\subseteq$'.

By Lemma 4.3(i), every 1-element of Δ_0 is full with respect to Γ_0. An i-element ($i > 1$) of Δ_0 is spanned in V by an $(i-1)$-element and a 1-element of Δ_0. Thus, by inductively applying the second part of Lemma 4.1, we obtain that $\mathcal{P}(E) \subseteq \langle \Gamma_0 \rangle_\Gamma$. □

Theorem 4.6 $\Delta(\mathcal{F}(\Gamma_0)) = \Delta_0$.

Proof The inclusion '$\supseteq$' in Theorem 4.5 implies the inclusion '$\supseteq$' here.

Now suppose there exists an i-element H that is full with respect to Γ_0, but is not in Δ_0. We assume that $i = n$, using Lemma 4.4. As H is full, for every 1-element $A \subseteq H$, the point (A, H) of Γ belongs to $\langle\Gamma_0\rangle_\Gamma$. For any such 1-element A there exists $E \in \Delta_0$ with $(A, H) \in \mathcal{P}(E)$ or, equivalently, $A \subseteq E \subseteq H$, by Theorem 4.5. Hence $H \in \Delta_0$, a contradiction. □

Theorem 4.7 *Suppose X is a subspace of Γ containing Γ_0. Then $\mathbb{F}$ contains an extension $\mathbb{F}_1$ of $\mathbb{F}_0$ such that $\Delta(\mathcal{F}(X)) = \Delta(\mathbb{F}_1)$.*

Proof For $i = 1, 2, ..., n$, let $\mathcal{F}(X)_i$ denote the collection of X-full i-elements of Δ. We will show that $\mathcal{F}(X)_1$ and $\mathcal{F}(X)_2$ are the collections of points and lines of a non-degenerate projective space. Note first that, by Lemma 4.2(ii), if $A, B \in \mathcal{F}(X)_1$ and L is the 2-space of V spanned by A and B, then $L \in \mathcal{F}(X)_2$.

We now prove that the point-line geometry $\Pi = (\mathcal{F}(X)_1, \mathcal{F}(X)_2)$, with the natural incidence relation (inherited from Δ) satisfies the axiom of Pasch. Given $A \in \mathcal{F}(X)_1$, let L_1, L_2 be distinct 2-elements on A and let B_1, C_1 and B_2, C_2 be distinct 1-elements in L_1 and L_2, different from A and belonging to $\mathcal{F}(X)_1$. Let L be the 2-element on B_1 and B_2, and let L' be the 2-element on C_1 and C_2. Then $L_1, L_2, L, L' \in \mathcal{F}(X)_2$, as each of them is spanned by pairs of full 1-elements. Let $P = L \cap L'$. Clearly, $P = H \cap L$ for any n-element H on L' not containing L. By Lemma 4.3(ii), such an H can be chosen in $\mathcal{F}(X)_n$. Hence $P \in \mathcal{F}(X)_1$, by Lemma 4.2(iii). That is, L and L' meet in Π.

Thus, Π is a (possibly degenerate) projective space and, using Lemma 4.2(ii), we find that $\Delta(\mathcal{F}(X))$ contains the geometry of subspaces of Π. Since $\Gamma_0 \subseteq X$, the dimension of Π is n. We now show that Π is not degenerate. Given $L \in \mathcal{F}(X)_2$, take an $(n-1)$-element $E_0 \in \Delta_0$ opposite to L. That is, $L \cap E_0 = \{0\}$. Then the n-elements of Δ_0 containing E_0 intersect L in distinct X-full 1-elements, by Lemmas 4.3(i) and 4.2(iii). However, there are at least three such n-elements. Hence L has at least three points in Π.

Now note that Π is a non-degenerate projective space contained in the Desarguesian projective space Δ and containing the Desarguesian projective space Δ_0. It is easy to see that Π is Desarguesian as well. Since a Desarguesian projective geometry determines the algebraic structure of the field over which it is defined, Π is defined over some field $\mathbb{F}_1$ with $\mathbb{F}_0 \leq \mathbb{F}_1 \leq \mathbb{F}$. Now the second part of Lemma 4.2 implies '$\supseteq$'.

Next we prove the inclusion '$\subseteq$'. From Lemma 4.4 applied with $\mathbb{F}_0 = \mathbb{F}_1$, it follows that if there is an X-full i-element in $\Delta \setminus \Delta(\mathbb{F}_1)$, then there is a 1-element also exists with the same properties. However, this contradicts the fact that all X-full 1-elements belong to $\Delta(\mathbb{F}_1)$. □

Theorem 4.7 shows that $X \supseteq \Gamma(\mathbb{F}_1)$. However in general we do not have equality here.

Corollary 4.8 *Suppose that $\mathbb{F}$ can be generated by adjoining k elements to $\mathbb{F}_0$. Then Γ can be generated by adding at most k points to its subgeometry Γ_0.*

Proof Let $k = 1$. We shall show that we can choose a point x of Γ such that there is a 1-element A of Δ not defined over $\mathbb{F}_0$, but full with respect to $\Gamma_0 \cup \{x\}$.

For clarity, let $\mathbb{F}_0$ be the prime field of $\mathbb{F}$. By Theorem 4.6, $\langle\Gamma_0\rangle_\Gamma$ is a proper subspace of Γ. Let $x = (A, H)$ where $A = \langle(\alpha, 1, \dots, 1)\rangle_V$ and H is any n-element from $\Delta \setminus \Delta_0$ containing A. Since the 2-element $L = \langle(1, 0, \dots, 0), (0, 1, \dots, 1)\rangle_V \in \Delta_0$ is full with respect to Γ_0 and A is the intersection of H and L, so A is full with respect to $X = \Gamma_0 \cup \{x\}$, by Lemma 4.2(iii).

By Theorem 4.7, $\Delta(\mathcal{F}(X)) = \Delta(\mathbb{F}_1)$ for some intermediate field $\mathbb{F}_0 \leq \mathbb{F}_1 \leq \mathbb{F}$, but since $A \notin \Delta_0$, apparently we have $\mathbb{F}_1 = \mathbb{F}$. The corollary now follows by induction on k. □

5. The line-grassmannians of the buildings of type B_n, C_n and D_n

Henceforth, $\mathbb{F}$ is a given field, M is the Dynkin diagram of rank $n \geq 3$ and type B_n, C_n or D_n and $V = V(m, \mathbb{F})$, with $m = 2n + 1, 2n$ or $2n$ respectively. When $M = B_n$ or C_n, we denote by Δ the building of type M defined over $\mathbb{F}$. When $M = D_n$ and $n \geq 3$, we change the notation used at the end of Subsection 2.2, denoting by Δ the polar geometry associated to a non-singular quadratic form of Witt index n over V. Thus, in any case, the i-elements of Δ are i-spaces of V.

Note As noted in Subsection 2.2, when $M = D_3$ the line-grassmannian of Δ is isomorphic to the long-root geometry for $SL_4(\mathbb{F})$, considered in the previous section. Thus, the case of $M = D_3$ might be omitted here, but we prefer to keep it, for the sake of uniformity of exposition.

We recall that, after choosing a suitable basis $\{e_i\}_{i=1}^m$ of V and rescaling (in the B_n case), the (quadratic or symplectic) form f on V that gives rise to Δ can be expressed as follows:

$$(B_n) \quad f(x_1, x_2, \dots, x_{2n+1}) = \textstyle\sum_{i=1}^n x_i x_{i+n} + x_{2n+1}^2;$$
$$(C_n) \quad f((x_1, \dots, x_{2n}), (y_1, \dots, y_{2n})) = \textstyle\sum_{i=1}^n (x_i y_{i+n} - y_i x_{i+n});$$
$$(D_n) \quad f(x_1, x_2, \dots, x_{2n}) = \textstyle\sum_{i=1}^n x_i x_{i+n}.$$

The set $\{e_i\}_{i=1}^{2n}$ is called a *hyperbolic set* of vectors. In all cases, the apartments of Δ arise from the hyperbolic sets of vectors: given a hyperbolic set of vectors $\{e_i\}_{i=1}^{2n}$, the elements of the corresponding apartment are the elements of Δ spanned by subsets of that set.

From now on, we assume to have chosen a basis $B = \{e_i\}_{i=1}^{m}$ of V as above. We call B a *hyperbolic basis* because it contains a hyperbolic set.

With B as above, the form f is defined over the prime subfield of $\mathbb{F}$. For any subfield $\mathbb{F}_0 \leq \mathbb{F}$, for Δ_0 we can simply take the building whose elements are the elements of Δ that are defined over $\mathbb{F}_0$ with respect to the basis B. Clearly, Δ_0 is a building of the same type as Δ, but defined over the subfield $\mathbb{F}_0$ of $\mathbb{F}$.

In each case, we denote by Γ the line-grassmannian of Δ; see Subsection 2.2. Accordingly, Γ_0 is the line-grassmannian of Δ_0. Clearly Γ_0 is a subgeometry (but not a subspace) of Γ.

Henceforth, the words 'point' and 'line' will refer to points and lines of Γ, unless otherwise specified. The set of points of Γ will be denoted $\mathcal{P}$. We shall now denote 1-elements and 2-elements of Δ by small letters, keeping capital letters for the remaining elements of Δ or for a generic element of Δ.

Let $\perp$ be the orthogonality relation defined by f on V. For any element E of Δ, we set

$$\mathcal{P}^*(E) := \{l \in \mathcal{P} \mid l \subseteq E^{\perp} \text{ and } l \cap E \neq \{0\}\}.$$

Note that $\mathcal{P}^*(E) \supseteq \mathcal{P}(E)$ and that equality holds only when E is a 1-element. Note also that $\mathcal{P}^*(E)$ is a subspace of Γ.

Given $S \subseteq \mathcal{P}$, we say that an element E of Δ is *full* with respect to S (also, *S-full*, for short) if $\mathcal{P}^*(E) \subseteq \langle S\rangle_{\Gamma}$. We denote the collection of elements that are full with respect to S as $\mathcal{F}(S)$ and the incidence structure induced on it by Δ as $\Delta(\mathcal{F}(S))$.

Lemma 5.1 *Let $S \subseteq \mathcal{P}$ and suppose that $E_1, E_2 \in \Delta$ are full with respect to S.*

(i) *If E_1 is a 1-element and E_2 is an i-element with $1 < i < n$, then $E_1^{\perp} \cap E_2$ ($\neq \{0\}$) is full with respect to S.*

(ii) *If $E_1 \perp E_2$, then $\langle E_1, E_2\rangle_V$ is full with respect to S.*

Proof (i) If $E_1^{\perp} \cap E_2 = E_2$ there is nothing to prove. Suppose this is not the case and $E_3 = E_1^{\perp} \cap E_2$. Then E_3 is an $(i-1)$-element of Δ. Let $l \in \mathcal{P}^*(E_3)$. If $l \subseteq E_3$, then $l \subseteq E_2$ and we are done.

Suppose $l \not\subseteq E_3$ and let $p = l \cap E_3$. The family $E_3^{\perp}/p$ of elements of Δ containing p and contained in $E_3^{\perp}$ is a (possibly degenerate) polar space

Π of rank $r = n - 1$. The radical $R(\Pi)$ of Π is the family of 2-elements of Δ contained in E_3 and containing p. Thus, $R(\Pi) = \emptyset$ if and only if $p = E_3$ and, if $R(\Pi) \neq \emptyset$, the rank of $R(\Pi)$ in Π is $r_0 = i - 2$. As $r - r_0 = n - 1 - (i - 2) = n - i + 1 \geq 2$, if E and E' are elements of Δ containing E_3 (hence contained in $E_3^{\perp}$) with dimensions i and 2 respectively in V and $E' \not\subseteq E^{\perp}$, then Π is spanned by its points contained $E^{\perp} \cup E'$. The elements $E = E_2$ and $E' = \langle p, E_1 \rangle_V$ satisfy the previous hypotheses. Therefore, the 2-elements of Δ containing p and contained in $E_2^{\perp} \cup \langle E_1, p \rangle_V$, viewed as points of Π, span Π. However, since E_1 and E_2 are full, all those 2-elements belong to $\langle S \rangle_\Gamma$. Furthermore, two points l_1 and l_2 of Π are collinear in Π if and only if $\langle l_1, l_2 \rangle_V$ is a 3-element of Δ. It is now clear that two points of Π are collinear in Π if and only if they are collinear as points of Γ. As Π is spanned by points belonging to $\langle S \rangle_\Gamma$, all points of Π belong to $\langle S \rangle_\Gamma$. In particular, $l \in \langle S \rangle_\Gamma$. This shows that E_3 is S-full.

(ii) Suppose $E_1 \perp E_2$. Let $E_3 = \langle E_1, E_2 \rangle_V$ and let $l \in \mathcal{P}^*(E_3) \setminus (\mathcal{P}^*(E_1) \cup \mathcal{P}^*(E_2))$. Suppose p is a 1-element in $l \cap E_3$ and $q \neq p$ is a 1-element in $l \cap E_3^{\perp} = E_1^{\perp} \cap E_2^{\perp}$.

Since $l \notin \mathcal{P}^*(E_1) \cup \mathcal{P}^*(E_2)$ we have $p \subseteq E_3 \setminus (E_1 \cup E_2)$. Hence there are 1-elements $p_1 \subseteq E_1$ and $p_2 \subseteq E_2$ such that $p \subseteq \langle p_1, p_2 \rangle_V$. Then $(q, \langle q, p_1, p_2 \rangle_V)$ is a line of Γ containing $l_1 = \langle p_1, q \rangle_V \in \mathcal{P}^*(E_1)$, $l_2 = \langle p_2, q \rangle_V \in \mathcal{P}^*(E_2)$ and l. Furthermore, $l_1, l_2 \in \langle S \rangle_\Gamma$, because E_1 and E_2 are full, by assumption. Hence $l \in \langle S \rangle_\Gamma$; we are done. □

Lemma 5.2 *Let X be a subspace of Γ containing Γ_0. Then all elements of Δ_0 are X-full.*

Proof Let p be a 1-element of Δ_0. Then $p^{\perp}$ is a degenerate polar space of rank n, with radical $R(\Pi) = p$. Accordingly, the systems Π and Π_0 of all elements of Δ and Δ_0 contained in $p^{\perp}$ and properly containing p is a non-degenerate polar space of rank $n-1 \geq 2$. Let $\mathcal{L}$ be a spanning set of points of Π_0. As Δ and Δ_0 have the same Dynkin type, $\mathcal{L}$ also spans Π. However, for every 3-element E of Δ on p, the flag (p, E) is a line of Γ. As X is a subspace of Γ containing Γ_0, X contains every point of Π. That is, p is X-full.

Thus, we have proved that all 1-elements of Δ_0 are X-full. As every element of Δ_0 is spanned (as an element of Δ) by 1-elements belonging to Δ_0, the conclusion follows from the above and Lemma 5.1(ii). □

Lemma 5.3 *Let X be a subspace of Γ containing Γ_0. Then, for every $i = 2, 3, ..., n-1$ and every i-element E of $\Delta \setminus \Delta_0$, full with respect to X, there is a flag $F = (E_1, E_2, ..., E_i = E)$ of Δ such that, for every $j = 1, 2, ..., i$, the element E_j has type j, is full with respect to X and does not belong to Δ_0.*

Proof Let $E = E_i$ be an i-element of $\Delta \setminus \Delta_0$, full with respect to X, with $1 < i < n$. We shall show that there exists an $(i-1)$-element E' of $\Delta \setminus \Delta_0$ that is also X-full. Take an i-element $E_0 \in \Delta_0$ opposite to E, that is, such that $E_0^{\perp} \cap E = \{0\}$. Such an element exists, by Corollary 2.4. Clearly $E = \langle a^{\perp} \cap E \rangle_V$, where a ranges over a set of 1-elements of Δ_0 contained in E_0 and spanning E_0. As $E \notin \Delta_0$, one of these 1-elements, say a', satisfies $a'^{\perp} \cap E \notin \Delta_0$. Since a' is X-full (Lemma 4.2), $E' = a'^{\perp} \cap E$ is the required $(i-1)$-element, by Lemma 5.1(i). □

We are now ready to prove Theorem 1.3.

Proof of Theorem 1.3 For $i = 1, 2, \ldots, n$ let $\mathcal{F}(S)_i$ denote the collection of S-full i-elements of Δ. We will show first that $\mathcal{F}(S)_1$ and $\mathcal{F}(S)_2$ are the collections of points and lines of a non-degenerate polar geometry, using the well-known Buekenhout–Shult axioms [7]. Throughout the proof we refer to 1-elements and 2-elements as 'points' and 'lines' respectively.

Let $l \in \mathcal{F}(S)_2$. Assume that there exists $q \in \mathcal{F}(S)_1$ such that $q^{\perp} \not\supseteq l$. Then $p = l \cap q^{\perp} \in \mathcal{F}(S)_1$ by Lemma 5.1(i). Hence we are dealing with a possibly degenerate polar geometry. We call this polar geometry Π.

We now show that Π is not degenerate. We will use the fact that all elements of $\Delta(\mathbb{F}_0)$ are S-full (Lemma 5.2) and the following principle:

(1) if l and m are opposite lines each containing at least two S-full points, then there is a one-to-one correspondence between the S-full points on l and the S-full points on m given by collinearity.

Proof of (1): First note that since l contains two elements from $\mathcal{F}(S)_1$ it belongs to $\mathcal{F}(S)_2$; see Lemma 5.1. Let p be a point on l and let q be the unique point on m collinear to p. As we saw above, if q is S-full then $p = q^{\perp} \cap l$ is S-full. Thus (1) is proved.

Given a line l of Π there is a line $m \in \Delta(\mathbb{F}_0)$ opposite to l (such an element exists by Corollary 2.4). Since m contains at least three S-full points, it follows from principle (1) that also l contains at least three S-full points. Thus, Π is non-degenerate.

Since the points and lines of Π are points and lines of Δ with the same incidence relation, the i-elements (singular subspaces of rank i) of Π can be identified as those i-elements of Δ that are generated by points from Π.

Recall that we have chosen a basis B such that the form f that gives rise to Δ is defined over the prime field of $\mathbb{F}$ (and hence over any subfield). Assume that $\Delta(\mathbb{F}_0)$ is the $\mathbb{F}_0$-sub-building of Δ defined by f. We will show that $\Pi = \Delta(\mathbb{F}_1)$, where the latter is the $\mathbb{F}_1$-sub-building of Δ defined by f, for some subfield $\mathbb{F}_1$ of $\mathbb{F}$ containing $\mathbb{F}_0$.

We first prove $\Pi \subseteq \Delta(\mathbb{F}_1)$ by showing that every point p of Π belongs to $\Delta(\mathbb{F}_1)$.

Consider $E = \langle e_1, e_2, \dots, e_n \rangle$. This is an n-element of Π because all points of $\Delta(\mathbb{F}_0)$ belong to Π. The points of Π on E form a projective space containing the points of $\Delta(\mathbb{F}_0)$ on E. Hence there exists a subfield $\mathbb{F}_1$ of $\mathbb{F}$ containing $\mathbb{F}_0$ that coordinatizes all points of Π on E. Thus we have the property:

(2) all points and other elements of Π on E belong to $\Delta(\mathbb{F}_1)$.

Now let E' be an n-element of $\Delta(\mathbb{F}_0)$ disjoint from (opposite to) E and let p be a point of Π on E'. Then $p^\perp \cap E$ is an element of Π and hence belongs to $\Delta(\mathbb{F}_1)$. Since E, E' and f are defined over $\mathbb{F}_0$, also p must belong to $\Delta(\mathbb{F}_1)$. Thus (2) holds for E' as well.

Consider the graph on the n-elements of $\Delta(\mathbb{F}_0)$ in which two such elements are joined by an edge whenever they are disjoint as subspaces of V. This graph is connected and so we have (2) for every n-element of $\Delta(\mathbb{F}_0)$.

Now let p be any point of Π. For every n-element E of the polar space Π also $p^\perp \cap E$ belongs to Π. By the preceding, if E belongs to $\Delta(\mathbb{F}_0)$ then $p^\perp \cap E$ belongs to $\Delta(\mathbb{F}_1)$. We prove that $p = \cap_E (p^\perp \cap E)^\perp$, where E runs over all n-elements of $\Delta(\mathbb{F}_0)$. Then it follows that p belongs to $\Delta(\mathbb{F}_1)$, as f is defined over $\mathbb{F}_1$.

Let $(p_i)_{i=1}^m$ and $(q_i)_{i=1}^m$ be non-zero vectors in p and q and suppose that $p^\perp \cap E = q^\perp \cap E$ for every n-element of $\Delta(\mathbb{F}_0)$. We prove that $p = q$. First we note that it follows from the assumption that $p^\perp \cap E = q^\perp \cap E$ for every element of $\Delta(\mathbb{F}_0)$. We use the fact that for each element e of the hyperbolic basis B there is a unique element $\widehat{e}$ that is not orthogonal to it; for $i \in \{1, 2 \dots, 2n+1\}$ let $\hat{\imath}$ be such that $\widehat{e_i} = e_{\hat{\imath}}$.

By considering $p^\perp \cap E$ and $q^\perp \cap E$ for all 2-elements E of the form $\langle e_i, e_j \rangle$, $\langle e_i, e_{n+i} \rangle$, $\langle e_{n+i}, e_{n+j} \rangle$ with $i, j \in \{1, 2 \dots, n\}$, we find in the D_n case that $p_k \neq 0$ if and only if $q_k \neq 0$, and moreover that $p_k/p_l = q_k/q_l$ whenever $p_k \neq 0 \neq p_l$ (and $q_k \neq 0 \neq q_l$) $(k, l \in \{1, 2, \dots, m\})$. Thus we conclude that $p = q$. We reach the same conclusion in the C_n case by considering, in addition, all 2-elements of the form $\langle e_i + e_j, e_{\hat{\imath}} - e_{\hat{\jmath}} \rangle$ and also in the B_n case by considering, in addition, the 2-elements of the form $\langle e_i, e_j - e_{\hat{\jmath}} + e_{2n+1} \rangle$ $(i, j \in \{1, 2, \dots, n\})$.

This concludes the proof that $\Pi \subseteq \Delta(\mathbb{F}_1)$.

Next, we prove $\Pi \supseteq \Delta(\mathbb{F}_1)$ by showing that all points of $\Delta(\mathbb{F}_1)$ belong to Π. Suppose p and q are points of $\Pi \cap \Delta(\mathbb{F}_1)$ lying on a line l of $\Delta(\mathbb{F}_1)$.

If l belongs to $\Delta(\mathbb{F}_0)$, then we already know that the points of Π on l are precisely the points of $\Delta(\mathbb{F}_1)$ on l. If l is any other line, we can find a line m of $\Delta(\mathbb{F}_0)$ opposite to l. Then the one-to-one correspondence between the points of l and m given by collinearity relates the points of $\Delta(\mathbb{F}_1)$ on l with the points of $\Delta(\mathbb{F}_1)$ on m. By principle (1), the same holds for the points of Π on l and m. It follows that the points of Π on l are precisely the points of $\Delta(\mathbb{F}_1)$ on l.

As the points of $\Delta(\mathbb{F}_0)$ are contained in Π and generate the geometry of points and lines of $\Delta(\mathbb{F}_1)$ (see Theorem 3.1 and Proposition 3.2), all points of $\Delta(\mathbb{F}_1)$ belong to Π. This concludes the proof that $\Pi \supseteq \Delta(\mathbb{F}_1)$.

Next, we prove the last part of the theorem. All elements of $\Delta(\mathbb{F}_1)$ are S-full because they are generated by S-full 1-elements; see Lemma 5.1. As for the converse, we must show that every S-full i-element E of Δ with $i < n$ belongs to $\Delta(\mathbb{F}_1)$. The case $i = 1$ was proved above. Now let $1 < i < n$. From Lemma 5.3 applied with Δ replaced by $\Delta(\mathbb{F}_1)$, it follows that if there is an S-full i-element in $\Delta \setminus \Delta(\mathbb{F}_1)$, then there is a 1-element with the same properties. But this contradicts the fact that all S-full 1-elements belong to $\Delta(\mathbb{F}_1)$. □

Corollary 5.4 *Suppose that $\mathbb{F}$ is generated by adjoining k elements to its subfield $\mathbb{F}_0$. Then Γ can be generated by adding at most k points to its subgeometry Γ_0.*

Proof Let $k = 1$ and suppose that $\mathbb{F} = \mathbb{F}_0(\alpha)$ for some element $\alpha \in \mathbb{F}$. Take a point l of Γ_0 containing 1-elements $a, b \in \Delta_0$ and let $p \subset l$ not be in Δ_0 (for example, $\langle \alpha e_1 + e_2\rangle_V$). Let l_0 be a point in Γ_0 opposite to l and let q be a 1-element on l_0 such that $p = q^{\perp} \cap l$. Put $x = \langle p, q\rangle_V$. Since $l \in \Delta_0$ is full with respect to Γ_0, p is full with respect to $X = \Gamma_0 \cup \{x\}$. By Theorem 1.3, the X-full points of Δ are precisely those of $\Delta(\mathbb{F}_1)$ for some extension $\mathbb{F}_1$ of $\mathbb{F}_0$ contained in $\mathbb{F}$. However, since $p \notin \Delta_0$, we have $\mathbb{F}_1 = \mathbb{F}$. The corollary now follows by induction on k. □

Proof of Proposition 5 Suppose Δ is of type C_n and let $B = \{e_i, e_{n+i}\}_{i=1}^{n}$ be the chosen hyperbolic basis. For $J, K \subseteq I = \{1, 2, ..., n\}$, put $E_{J,K} = \langle e_j, e_{n+k} \mid j \in J, k \in K\rangle_V$. The collection of totally isotropic subspaces of this form is an apartment A.

Let e be a 1-element of Δ contained in $E_{I,\emptyset}$ but not in $E_{J,\emptyset}$ for any $J \subset I$. Let S be the collection of 2-elements of A, together with any $n-1$ 2-elements on e that span an n-element meeting $E_{I,\emptyset}$ only in e. Then S is a generating set for Γ.

First we recall that, if $\mathrm{Char}(\mathbb{F}) \neq 2$, then the 1-elements of Δ contained in a given apartment span Δ (see Theorem 3.1; also Cooperstein and Shult [9]). As this also applies to the polar space formed by the elements of Δ on a given 1-element, we get that all 1-elements of A are S-full. In particular, all 2-elements $\langle e_i, e\rangle_V$ are in the span of S, whence

e is S-full. Similarly, every 1-element of Δ contained in $E_{i,i}$ is S-full, for every $i \in I$.

Consider the geometry Θ of pairs (p, H), where p is a 1-element and H is an $(n-1)$-element with $p \subset H \subset E_{I,\emptyset}$. This geometry is isomorphic to the long-root geometry associated to $\mathrm{SL}_n(\mathbb{F})$. It is not difficult to see that it is also isomorphic to the subgeometry of Γ containing those 2-elements of Δ that meet both $E_{I,\emptyset}$ (in p) and $E_{\emptyset,I}$ (in $H^{\perp} \cap E_{\emptyset,I}$).

Note that the 1-elements of A in $E_{I,\emptyset}$ together with e form a spanning set for Θ (compare Cooperstein [10, Theorem 4.1]). Hence, for any 1-element $p \subset E_{I,\emptyset}$, all 2-elements on p meeting $E_{\emptyset,I}$ are in the span of S. By an argument similar to the one used to prove that e is S-full, it follows that all 1-elements in $E_{I,\emptyset}$ are S-full. The remainder of the proof is based on the following principle (see the proof of Theorem 1.3, principle (1)):

(∗) Suppose that a 2-element l contains two S-full 1-elements. If there exists a 2-element m opposite to l such that all 1-elements in m are S-full, then all 1-elements in l are S-full.

As all 1-elements in $E_{I,\emptyset}$ are S-full, it follows from (∗) that, if a 2-element l meets $E_{i,i}$ and $E_{j,j}$ in 1-elements different from $E_{i,\emptyset}$ and $E_{j,\emptyset}$ for distinct $i, j \in I$, then all 1-elements of l are S-full. (The 2-element $\langle e_i, e_j \rangle_V$ can be given the role of m in (∗).)

By considering now the 2-elements $\langle e_i + e_{n+i}, e_j + e_{n+j} \rangle_V$, for any distinct $i, j \in I$, it follows in turn that all 1-elements contained in a 2-element of A are S-full. As the 1-elements of A span Δ and every 2-element of Δ is opposite to some 2-element of A, we find that in fact all 1-elements of Δ are full.

Thus, Γ is spanned by $2n^2 - n - 1$ points if $\mathbb{F}$ is a prime field of odd characteristic. It has a natural embedding of dimension $2n^2 - n - 1$ into a hyperplane of the natural embedding for the line-grassmannian of the ambient projective space $\mathrm{PG}(2n-1, \mathbb{F})$. It follows that its generating rank is $2n^2 - n - 1$. □

Acknowledgements

The first author was supported by a grant of the Istituto Nazionale di Alta Matematica (INDAM).

References

[1] N. Batens and A. Pasini, On certain hyperplanes and projective embeddings of D_4-buildings, *Int. J. of Math. Game Theory Algebra* **8** (1998), 137–166.

[2] N. Batens and A. Pasini, On certain hyperplanes and projective embeddings of D_4-buildings (Errata), *Int. J. of Math. Game Theory Algebra*, to appear.

[3] R. J. Blok and A. E. Brouwer, Spanning point-line geometries in buildings of spherical type, *J. Geom.* **62** (1998), 26–35.

[4] R.J. Blok and A. Pasini, On absolutely universal embeddings, submitted.

[5] F. Buekenhout, The basic diagram of a geometry, *Geometries and Groups* (eds. M. Aigner and D. Jungnickel), Springer, Berlin, 1981.

[6] F. Buekenhout, Foundations of incidence geometry, Chapter 3 of *Handbook of Incidence Geometry* (ed. F. Buekenhout), North-Holland, Amsterdam, 1995.

[7] F. Buekenhout and E.E. Shult, On the foundations of polar geometry, *Geom. Dedicata* **3** (1974), 155–170.

[8] R.W. Carter, *Simple Groups of Lie Type*, Wiley, London, 1972.

[9] B.N. Cooperstein and E.E. Shult, Frames and bases of Lie incidence geometries, *J. Geom.* **60** (1997), 17–46.

[10] B. N. Cooperstein, Generating long root subgroup geometries of classical groups over finite prime fields, *Bull. Belg. Math. Soc. Simon Stevin* **5** (1998), 531–548.

[11] A. Pasini, *Diagram Geometries*, Oxford University Press, Oxford, 1994.

[12] M.A. Ronan, Embeddings and hyperplanes of discrete geometries, *European J. Combin.* **8** (1987), 179–185.

[13] M. A. Ronan, *Lectures on Buildings*, Academic Press, New York, 1989.

[14] S.D. Smith and H. Völklein, A geometric presentation for the adjoint module of $SL_3(k)$, *J. Algebra* **127** (1989), 127–138.

[15] J. Tits, *Buildings of Spherical Type and Finite BN-Pairs*, First edition, Springer, Berlin, 1974.

[16] J. Tits, A local approach to buildings, *The Geometric Vein* (eds. C. Davis et al.), Springer, Berlin, 1981, 519–547.

[17] J. Tits, Twin buildings and groups of Kac–Moody type, *Groups, Combinatorics and Geometry* (eds. M.W. Liebeck and J. Saxl), Cambridge University Press, Cambridge, 1992, 249–286.

ON A CLASS OF SYMMETRIC DIVISIBLE DESIGNS WHICH ARE ALMOST PROJECTIVE PLANES

Aart Blokhuis
Department of Mathematics
Technical University of Eindhoven
Den Dolech 2
5600 MB Eindhoven
The Netherlands
aart@win.tue.nl

Dieter Jungnickel
Institut für Mathematik
Universität Augsburg
86135 Augsburg
Germany
jungnickel@math.uni-augsburg.de

Bernhard Schmidt
Institut für Mathematik
Universität Augsburg
86135 Augsburg
Germany
schmidt@math.uni-augsburg.de

Abstract We consider a class of symmetric divisible designs $\mathcal{D}$ which are almost projective planes in the following sense: Given any point p, there is a unique point p' such that p and p' are on two lines, whereas any other point is joined to p by exactly one line; and dually. We note that either the block size k or $k-2$ is a perfect square, and exhibit examples for $k=3$ and $k=4$. Then we add the condition that $\mathcal{D}$ should admit an abelian Singer group, so that we may study the associated divisible difference sets. Under this additional assumption, we show that k is a

A. Blokhuis et al. (eds.), Finite Geometries, 27–34.

square (unless $k = 3$) and that the only possible prime divisors of k are 2 and 3.

1. Preliminaries

About 15 years ago, Zoltan Füredi suggested considering a class of symmetric divisible designs $\mathcal{D}$ which are in some sense as close to projective planes as possible. More precisely, $\mathcal{D}$ should be a square 1-design satisfying the following two axioms:

(A1) Given any point p, there is a unique point p' such that p and p' are on two lines, whereas any other point is joined to p by exactly one line.

(A2) Given any line L, there is a unique line L' such that L and L' intersect in two points, whereas any other line intersects L uniquely.

For the purposes of this paper, we shall denote such a structure as an APP(k) (for "almost projective plane") if it has line size k.

Example 1.1 Developing the start block $\{0, 1, 3\}$ modulo 6 gives an APP(3); similarly, an APP(4) arises from the start block $\{0, 1, 6, 10\} \subseteq \mathbb{Z}_{12}$.

In the language of design theory, an APP(k) is just a symmetric divisible design with parameters

$$m = k(k-1)/2,\ n = 2,\ k,\ \lambda_1 = 2 \text{ and } \lambda_1 = 1; \tag{1.1}$$

see [2] or [8]. In view of the fact that the axioms (A1) and (A2) are so close to those for projective planes, we prefer to speak of "lines" instead of "blocks". We shall also call the point x' in axiom (A1) the *mate* of the point x, and similarly for lines.

Of course, we hoped to find some more interesting examples than the ones given above, maybe even for values of k where $k-1$ is not a power of a prime. Unfortunately, we did not succeed with this, so our results will all be about non-existence. We begin with the following necessary condition which is a special case of the Bose-Connor theorem [3].

Proposition 1.2 *An APP(k) can only exist if either k or $k-2$ is a perfect square. More precisely, k has to be a square if $k \equiv 0$ or $1 \bmod 4$; if additionally $m = k(k-1)/2 \equiv 2 \bmod 4$, then $k-2$ has to be the sum of two squares. For $k \equiv 2$ or $3 \bmod 4$, $k-2$ must be a square, and the equation*

$$kx^2 + (-1)^{m(m-1)/2} \cdot 2y^2 = z^2 \tag{1.2}$$

has a non-trivial solution in integers.

Proof For the convenience of the reader, we shall sketch a proof of the fact that either k or $k-2$ is a square, as this will be the part that we will mainly require in what follows. To this end, let A be an incidence matrix for an APP(k), where points and lines are arranged into consecutive pairs of mates. Then, by (A1) and (A2),

$$AA^T = A^TA = \begin{pmatrix} k & 2 & 1 & 1 & \dots & 1 & 1 \\ 2 & k & 1 & 1 & \dots & 1 & 1 \\ 1 & 1 & k & 2 & \dots & 1 & 1 \\ 1 & 1 & 2 & k & \dots & 1 & 1 \\ & & & & \ddots & & \\ 1 & 1 & 1 & \dots & 1 & k & 2 \\ 1 & 1 & 1 & \dots & 1 & 2 & k \end{pmatrix}$$

$$= (k-1)I + J + E,$$

where E denotes the direct sum of $\binom{k}{2}$ copies of the matrix $\begin{pmatrix} 0 & 1 \\ 1 & 0 \end{pmatrix}$.

It is a bit tedious but routine to calculate the determinant of this matrix, either directly or by determining its eigenvalues. The result is

$$(\det A)^2 = \det AA^T = (k-2)^{(k^2-k)/2} k^{(k^2-k+2)/2}, \qquad (1.3)$$

which implies the assertion, as the two exponents appearing in (1.3) always have opposite parity. □

Let us give the following sample application of Proposition 1.2 which will be useful later.

Corollary 1.3 *An* APP(3^b) *with* $b \neq 1$ *can only exist if* b *is even or* $b = 3$.

Proof let b be odd, so that $3^b - 2$ is a perfect square. By a result of Ljunggren [5] (rediscovered by Nagell [6]), the only solution of the Diophantine equation $3^x - 2 = y^2$ indeed occurs for $x = 3$. □

In the next section, we add the condition that $\mathcal{D}$ should admit an abelian Singer group G, so that we may study the associated divisible difference sets with parameters $(k(k-1)/2, 2, k, 2, 1)$. Recall that a k-element subset D of a group G of order $v = mn$ is called a *divisible difference set* with parameters $(m, n, k, \lambda_1, \lambda_2)$, if the list of differences $(d - d' : d, d' \in D, d \neq d')$ covers every element in $G \backslash N$ exactly λ_2-times, and the elements in $N\setminus\{0\}$ exactly λ_1 times, where N is a specified subgroup of G of order n. We provided two small cyclic examples above; see [4] or [8] for background on divisible difference sets. Under this additional assumption, we will prove that the only possible prime divisors of k are 2 and 3. The proof will involve a novel trick which is of independent interest and might turn out to be useful in other situations, too. Finally, we shall also obtain some further restrictions using standard tools like the Mann test [1]; in particular, it turns out that k has to be a square provided that $k \neq 3$.

2. Divisible difference sets for APP's

We now consider an APP(k) $\mathcal{D}$ admitting a *Singer group* G, that is, a group of automorphisms which acts regularly on points and thus also on lines, as $\mathcal{D}$ has full rank over $\mathbb{Q}$ by equation (1.3). Note that we exhibited cyclic examples for the cases $k = 3$ and $k = 4$ is Section 1. Our results will provide strong evidence for the conjecture that only these two cases can occur if we assume G to be abelian; parts of these results would carry over to non-abelian groups, but we will not bother stating this explicitly. As usual, we proceed by studying the group ring equation characterizing the associated divisible difference set D, as summarized in the following lemma; see [4] or [8] for background and details. In particular, we write G multiplicatively from now on and make use of the standard convention of identifying subsets of G with the corresponding formal sums of their elements in $\mathbb{Z}G$. We also require the notation

$$A^{(t)} := \sum_{g\in G} a_g g^t, \quad \text{where } A = \sum_{g\in G} a_g g,$$

and where t is some integer.

Lemma 2.1 *Let $\mathcal{D}$ be an APP(k) admitting a Singer group G. Then $\mathcal{D}$ is the development of a divisible difference set D with parameters (1.1) in G; in particular, G has a unique subgroup N of order 2. Moreover, D may be any subset of G for which the associated group ring element $D = \sum_{d\in D} \in \mathbb{Z}G$ satisfies the equation*

$$DD^{(-1)} = (k-2) + G + N. \tag{2.1}$$

Proof The assertions are merely special cases of known basic results on divisible difference sets; see [8]. The only part which might need some comment is the part concerning N. This is an easy consequence of (A1) which shows that the point classes (which are just the cosets of the special subgroup N appearing in the divisible difference set condition) have size 2. Hence only one group ring element can appear as a difference twice, and thus G has a unique involution, as the number of difference representations $g = d(d')^{-1}$ of an element $g \in G$ always agrees with that of the inverse element g^{-1}. □

The following theorem gives a strong restriction on the possible values of k. It is our major result, and its proof should be of special interest as it contains a novel trick. First, it seems to be only the second example where a group ring of a characteristic p dividing the order of G is applied to study difference sets (the first case being work of Pott [7] concerning extraneous multipliers of planar difference sets, see also [2]). More importantly, the trick of computing intersection numbers of a certain auxiliary subset mod p and then using square counting in characteristic 0 to obtain a contradiction is certainly new and might well have other interesting applications.

Theorem 2.2 *Let D be a divisible difference set with parameters (1.1) in an abelian group G. Then $k = 2^a 3^b$ for some non-negative integers a and b.*

Proof By Lemma 2.1, G has a unique subgroup N of order 2. We apply the canonical epimorphism π from G onto $H = G/N$ to equation (2.1) and obtain the following identity in the group ring $\mathbb{Z}H$:

$$\widetilde{D}\widetilde{D}^{(-1)} = k + 2H, \tag{2.2}$$

where we write $\widetilde{X}$ for the image of $X \in \mathbb{Z}G$ under the canonical extension of π to an epimorphism from $\mathbb{Z}G$ to $\mathbb{Z}H$. We now select a prime divisor p of k and consider equation (2.2) as an identity in the group algebra $\mathbb{Z}_pH$. Using this in conjunction with the fact $X^p = X^{(p)}$ (see [2, Lemma VI.3.7]), we get the validity of the following computation in $\mathbb{Z}_pH$:

$$\begin{aligned}\widetilde{D}^{(p)}\widetilde{D}^{(-1)} &= \widetilde{D}^p\widetilde{D}^{(-1)} = \widetilde{D}^{p-1}(\widetilde{D}\widetilde{D}^{(-1)}) \\ &= \widetilde{D}^{p-1}(k+2H) = k\widetilde{D}^{p-1} + 2k^{p-1}H = 0,\end{aligned}$$

as p divides k. This shows that all coefficients appearing in the group ring element

$$A = \widetilde{D}^{(p)}\widetilde{D}^{(-1)} = \sum_{h \in H} a_h h \in \mathbb{Z}H$$

are divisible by p. We will now use square counting for the a_h. Trivially,

$$\sum_{h \in H} a_h = k^2. \tag{2.3}$$

Next note that $\sum_{h\in H} a_h^2$ equals the coefficient of 1 in the group ring element $AA^{(-1)}$. But

$$\begin{aligned} AA^{(-1)} &= \widetilde{D}^{(p)}\widetilde{D}^{(-1)}\widetilde{D}\widetilde{D}^{(-p)} = \widetilde{D}^{(p)}\widetilde{D}^{(-p)}(k+2H) \\ &= \left[\widetilde{D}\widetilde{D}^{(-1)}\right]^{(p)}(k+2H) = k\left[\widetilde{D}\widetilde{D}^{(-1)}\right]^{(p)} + 2k^2H \\ &= k(k+2H^{(p)}) + 2k^2H. \end{aligned}$$

Thus the coefficient of 1 in $AA^{(-1)}$ is at most $k(k+2k)+2k^2$, depending on the number of elements of order p in H which can be at most $k-1$, as the Sylow p-subgroup of H has order at most k. This gives us the estimate

$$\sum_{h\in H} a_h^2 \le 5k^2. \tag{2.4}$$

Using (2.3) and (2.4) and the fact that all coefficients a_h are divisible by p, we get the following inequality:

$$0 \le \sum_{h\in H} a_h(a_h - p) \le 5k^2 - pk^2. \tag{2.5}$$

Clearly the preceding inequality gives a contradiction for $p \ge 7$. Now assume $p=5$ and that (2.5) holds so that we have equality throughout. In particular, $a_h \in \{0,5\}$ for all $h \in H$. Moreover, equality in (2.5) also requires equality in (2.4). As pointed out above, this implies that k is a power of 5 and that the Sylow 5-subgroup of H is elementary abelian. But then $\widetilde{D}^{(5)} \subseteq H^{(5)}$, which is a group of order $(k-1)/2$. Hence at least one coefficient of $\widetilde{D}^{(5)}$ must be ≥ 3, as D was a k-subset. But $\widetilde{D}^{(-1)}$ has one coefficient 2 (namely that of the element $dN = d'N$ of $H = G/N$, where d and d' are the two elements of D appearing in the "difference" representation of the unique involution in G from D), and so $A = \widetilde{D}^{(5)}\widetilde{D}^{(-1)}$ contains at least one coefficient $a_h \ge 6$. This contradiction rules out the case $p=5$, too, and finishes the proof. □

3. Some further restrictions

In this final section, we provide some further restrictions on divisible difference sets associated with APP's, mainly using a standard tool, namely the Mann test; see [1] and [8].

Theorem 3.1 *Let D be a divisible difference set with parameters (1.1) in an abelian group G, and assume $k \neq 3$. Then k is a square of the form $k = 2^{2a}3^{2b}$ for some non-negative integers a and b.*

Proof By Proposition 1.2, either k or $k-2$ is a square. It suffices to show that the latter case cannot arise; assume otherwise. Now 2 and 3 are the only primes that can divide k, by Theorem 2.2; hence either $k = 2 \cdot 3^b$ or $k = 3^b$, where b is odd.

In the first case, we apply the Mann test by selecting a subgroup U of G for which $H = G/U$ has order $u = 3$; then $N \subseteq U$. As 2 divides k and is self-conjugate modulo u (note $2 \equiv -1 \bmod 3$), the Mann test implies that 2 should divide k to an even power, a contradiction.

In the second case, Corollary 1.3 gives $k = 27$. We will rule out this possibility by an ad hoc argument involving the Mann test and intersection numbers. To this end, we select a subgroup U of G of order 13 and hence of index $u = 54$; thus $N \not\subseteq U$. As 5 divides $k-2$ and is self-conjugate modulo u (note $5^9 \equiv -1 \bmod 54$), the Mann test implies that 5 has to divide $k-2$ to an even power, which is, of course, true. Let us write $H = G/U$, and denote the image of our hypothetical divisible difference set in the group ring $\mathbb{Z}H$ by $\widetilde{D}$. By the proof of the extension of the standard Mann test given by Pott in [8, Theorem 2.4.6], the coefficients of $\widetilde{D} = \sum_{h\in H} a_h h$ are constant modulo 5 on the cosets of the image $\widetilde{N}$ of N. (Note that the conclusion of Pott's result is satisfied; thus we do not get an immediate contradiction but have to analyse the situation more closely.) The image of equation (2.1) in $\mathbb{Z}H$ in our special case is given by

$$\widetilde{D}\widetilde{D}^{(-1)} = 25 + 13H + \widetilde{N}. \tag{3.1}$$

Now $\widetilde{D}$ cannot be absolutely constant on on the cosets of $\widetilde{N}$ since otherwise $\widetilde{D} = \widetilde{N}X$ for some $X \in \mathbb{Z}H$, contradicting equation (3.1). Thus $\widetilde{D}$ has at least one coefficient ≥ 5. Equating the coefficient of 1 in equation (3.1), we obtain

$$\sum_{h\in H} a_h^2 = 39. \tag{3.2}$$

But $\sum a_h = 27$ and the minimum of $\sum a_h^2$ under the conditions $a_h \geq 5$ for at least one h and $\sum a_h = 27$ is $5^2 + 22 = 47$, contradicting (3.2). □

We conclude with the following further restriction.

Proposition 3.2 *Let D be a divisible difference set with parameters (1.1) in an abelian group G, where $k \neq 3$ is a power of 3. Then $k = 3^{4a}$ for some positive integer a.*

Proof We once again use the Mann test. By Proposition 1.2, k is a square; hence it suffices to rule out the possibility $k = 3^{4a+2}$. In this case, $k \equiv 9 \bmod 16$ and hence $k-2 \equiv 7 \bmod 16$. In particular, there has to exist some prime $p \equiv 3 \bmod 4$ dividing the square-free part of $k-2$. As k is a square, we see that 2 is a square modulo p, and thus in fact $p \equiv 7 \bmod 8$, by a well-known result from number theory. Now select a subgroup U of G for which $H = G/U$ has order $u = 8$; then $N \not\subseteq U$, as $v = k(k-1) \equiv 8 \bmod 16$. As p divides $k-2$ and is self-conjugate modulo u (note $p \equiv 7 \equiv -1 \bmod 8$), the Mann test implies that p has to divide $k-2$ to an even power, a contradiction. □

Acknowledgements

The authors thank Robert Calderbank for mentioning Füredi's problem to them and for letting them see his unpublished notes about this topic. They are also indebted to J.H.E. Cohn for pointing out references [5] and [6]. Most of the research for this paper was done while the second author visited the Department of Mathematics and Computer Science of the Technical University of Eindhoven; he gratefully acknowledges the hospitality and financial support provided by TUE.

References

[1] K.T. Arasu, D. Jungnickel and A. Pott, The Mann test for divisible difference sets, *Graphs Comb.* **7** (1991), 209–217.

[2] T. Beth, D. Jungnickel and H. Lenz, *Design Theory*, Second edition, Cambridge University Press, Cambridge, 1999.

[3] R.C. Bose and W.S. Connor, Combinatorial properties of group divisible incomplete block designs, *Ann. Math. Stat.* **23** (1952), 367–383.

[4] D. Jungnickel, On automorphism groups of divisible designs, *Canad. J. Math.* **34** (1982), 257–297.

[5] W. Ljunggren, Über einige Arcustangensgleichungen, die auf interessante unbestimmte Gleichungen führen, *Ark. Mat. Astr. Fys.* **29A** (1943), no.13.

[6] T. Nagell, Verallgemeinerung eines Fermatschen Satzes, *Archiv Math.* **5** (1954), 153–159.

[7] A. Pott, On abelian difference sets with multiplier −1, *Archiv Math.* **53** (1989), 510–512.

[8] A. Pott, *Finite Geometry and Character Theory*, Lecture Notes in Mathematics **1601**, Springer, Berlin, 1995.

GENERALIZED ELLIPTIC CUBIC CURVES, PART 1

Francis Buekenhout
Université Libre de Bruxelles
Département de Mathématiques - C.P. 216
Boulevard du Triomphe
B-1050 Brussels
Belgium
fbueken@ulb.ac.be

Abstract

We define the concept of Generalized Elliptic Cubic Curve (GECC) which is not necessarily embedded in a projective plane and which appears as an Incidence Geometry. We develop foundations and raise several problems. All GECCs with up to 8 points are classified.

1. Introduction

1.1. Points and tangents

Algebraic curves in projective planes over fields, and more generally algebraic varieties embedded or not in a projective space, provide fascinating objects that have been studied extensively over the last 350 years. A need for a purely geometric understanding has arisen at least since 1950, especially with the work of B. Segre. The rise of Incidence Geometry as a unifying principle of Geometry (Buekenhout [11]) asks for more developments in that direction. Today, the second degree objects, namely conics and quadrics, are particularly well understood in this direction even if exciting problems remain open. The geometric concepts corresponding to them are the ovals, ovoids and especially the quadratic sets that may live in any projective space including non-Desarguesian planes; see Buekenhout [9] and Beutelspacher and Rosenbaum [3]. A quadratic set appears as a set of points: tangents are lines that intersect the object in either one or all points; the set of points provides the tangents. When this theory is extended in the direction of polar spaces, it is necessary to give both the points and the tangents, as is clear from

A. Blokhuis et al. (eds.), Finite Geometries, 35–48.

symplectic polar spaces. I am convinced that this is a necessity also for algebraic curves and hypersurfaces from the third degree on if we want to develop a general valid theory, despite the existence of numerous deep results considering cubics as sets of points. In my opinion, we need to give the points, the tangent at each point, and the multiplicity of the contact to show if the point is ordinary or an inflexion point.

1.2. Intrinsic curves

An algebraic projective variety has an intrinsic structure induced by the surrounding space. This intrinsic structure deserves a formalization that can occur at different levels of sophistication. It is a good question to ask for the various embeddings of a (generalized) variety. My study of generalized cubics in 1962 was inspired by such views. For quadrics and more generally polar spaces there is now a deep theory starting with work of Veldkamp going back to 1959 and pursued by various authors since then. Of course, conics look a little meagre in this context; I was developing a study of intrinsic ovals in my thesis, [8]. These are now called B-ovals by various authors. This study of intrinsic ovals was much influenced by my former unpublished paper on cubics of 1962, [7]. This paper was amplified in [10] but the work was not completed.

1.3. Basic references on cubic curves and their combinatorics

For a pleasant introduction to cubics, Bix [4] is recommended. A thorough study of the literature on finite cubics and their combinatorics can be found in Hirschfeld [17]. This book provides in particular a deep classification of cubics in a given plane up to equivalence under projectivity. It is based primarily on the configuration of inflexion points and inflexion tangents. The classification is a synthesis of work due to various authors beginning with Cicchese [12].

1.4. Setting

The subject of this paper is Generalized Elliptic Cubic Curves or GECCs, their embedding in various projective planes, different questions that are raised and the small examples with no more than 8 points. The aim is not to avoid classical cubics but to show that GECCs may help to understand them better and to find new interesting questions about them.

In Part 2, further questions related to GECCs shall be discussed; these are also dealt with in [7]. Here is a short description. On a

GECC, every point determines a commutative loop provided with a particular 'associative' element. The GECC is entirely under control within this structure. Any two loops obtained in this way are related by a 'symmetric isotopy'. Every point of a GECC determines an involution on the set of points. This set of involutions is sharply transitive. From this point of view, a GEC namely a generalized elliptic curve is a set of points provided with a sharply transitive set of involutions. In the case of genuine elliptic curves, the involutions are automorphisms. The group they generate has a regular normal subgroup of index 2 or 1. This acts transitively on the various structures of a GECC underlying the same set of involutions.

There exist GECCs whose set of involutions does not derive from some GECC. The smallest example has 6 points. In [7], all GECCs of at most 6 points were obtained.

All the work of [7] was obtained under the influence of my promoter, P. Libois, and of the great book of Enriques [13]. The standard classical idea here is the structure of abelian group related to a cubic curve. On this matter especially the finite case, see the beautiful work in Hirschfeld and Voloch [18, 19], and in Ghinelli, Melone and Ott [15], Voloch [30], Rück [25] and Schoof [28]. For the history of that remarkable structure see Schappacher [27].

2. Generalized elliptic cubic curves

2.1. First example

In order to illustrate the concept of a GECC, here is a picture; see Figure 1. It describes a cubic of 8 points taken from the plane of order 5. It has a unique inflexion point. Its equation is

$$X^3 - XZ^2 - Y^2Z = 0.$$

2.2. Definitions

A GECC consists of a set of *points* and a set of unordered triples called *lines* such that any pair $\{a, b\}$ of not necessarily distinct points is contained in a unique line $\{a, b, c\}$. If p is a point, then the tangent at p is the line on $\{p, p\}$. The third point on the tangent is called the *tangential* of p. An inflexion point is a point at which the tangent contains only p equivalently, $\{p, p, p\}$ is a line.

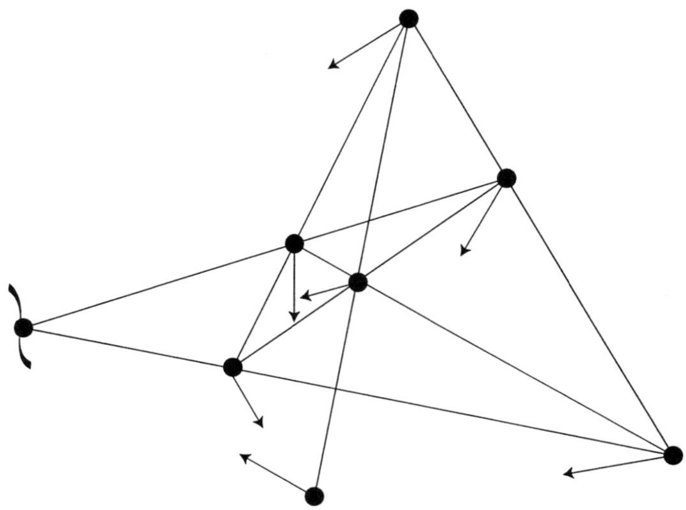

Figure 1. A cubic of 8 points

2.3. Totally symmetric quasigroup

It is clear that a GECC is formally the same object as a Totally Symmetric Quasigroup, a concept defined and studied in Bruck [5]. See also Bruck [6], Manin [23], Pflugfelder [24] and Bénéteau [1].

Let us recall a basic property, namely that a quasigroup Q is totally symmetric if and only if it satisfies the two identities $xy = yx$ and $x(xy) = y$ for all x, y in Q. Apparently, Bruck was not aware of the relationship to cubics. The first to observe, use and study the fact that every cubic has the structure of quasigroup was Etherington [14]. A word of warning: in the interesting article by Schwenk [26], 'abelian' is not a synonym of 'commutative' along with an old tradition.

It may also be observed that a GECC is formally the same object as an Extended Triple System as defined by Bénéteau [2].

2.4. Example

Every Steiner triple system is a GECC all of whose points are inflexion points. These are the oldest ancestors of GECC other than cubics.

2.5. Example

This is the smallest GECC which is not a cubic. It has 6 points, of which three are non-collinear inflexion points; see Figure 2.

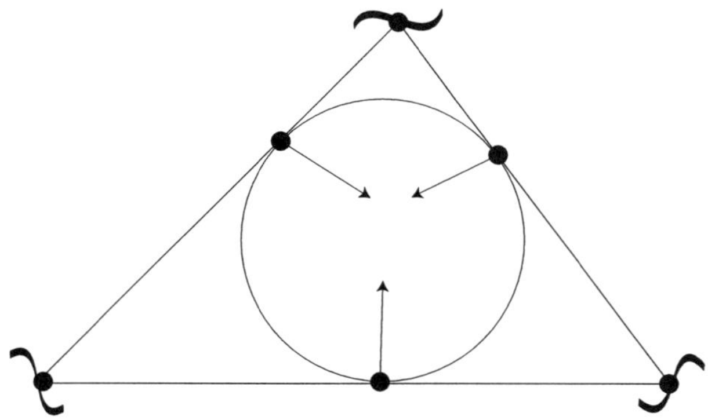

Figure 2. A GECC of 6 points which is not a cubic

2.6. Cayley table versus picture

The standard elementary way to present a GECC as a quasigroup is to give it by a Cayley table. The pictures used in a systematic way in Buekenhout [7] and in the two figures given so far illuminate the geometric structure. At a point p, the arrow points to the tangential of p unless p is an inflexion which is indicated by an S sign. Similar pictures appear in Hirschfeld and Thas [20] without the tangents, in order to describe some particular cubic curves.

2.7. Bruck's example

Bruck [5] gives a Cayley table for a GECC of 10 points. If we look for a picture we readily see that the GECC is an affine plane of order 3 completed by a tenth point which is the tangential of all points. As a matter of fact, every Steiner triple system can be completed to a GECC in the same way.

2.8. Sub-GECC

If C is a GECC it is immediately clear how to define a sub-GECC of C. It is a subset S of the set of points such that the line $\{x, y\}$ is contained in S for any x and y in S. It is the same concept as a sub-quasigroup of C.

2.9. Cycle

This concept is classical for cubics and immediately available for a general GECC. The idea is as follows. Let C be a GECC. For any point p of C let us write p' for the tangential of p. Then the 'derived series' of p is provided by $p, p', p'', p''', \dots$. It consists of p, its tangential, second tangential, etc. A *cycle of order n* is a derived series such that the nth tangential point of p is p itself and all earlier tangential points are distinct. As a matter of fact, it may be interesting to 'integrate' instead of 'deriving'. Thus we can consider a point, then all points admitting it as a tangential point and so on. We get the 'anti-derivative' of the point.

2.10. Switching

Assume that C is a GECC and Y is a cycle of C of at least three points. Then, we can get a new GECC from C, say $S(C, Y)$, by switching all tangentials within Y. This means that any point x in Y is now the tangential in $S(C, Y)$ of its tangential in C. By switching a cycle of length 3 at least, most cubics give an $S(C, Y)$ which is not a cubic anymore but this should be discussed in Part 2. Switching provides an interesting relationship on the class of GECC.

3. Embedding a GECC

3.1. Definition

Given a projective plane P and a GECC C, an embedding of C in P is a *monomorphism*, namely, an injective map from the pointset of C in the pointset of P such that every line of C is transformed into collinear points. Therefore, we may now consider an *embedded* GECC, namely, one for which the identity mapping is an embedding.

A classical cubic is a genuine embedded GECC. A definition and study of embedded GECC from the point of view of the characterization of cubic curves can be found in Hirschfeld and Voloch [18, 19] and in Ghinelli, Melone and Ott [15]. The latter paper is a model for a purely geometric approach to cubic curves.

3.2. Example

We see immediately that the GECC of Example 2.5 is not embeddable in the plane of order 2. Indeed, at an inflexion point, four distinct lines are required. It is easy to convince oneself that this GECC is embeddable in every plane of order 3 at least, infinite or not.

3.3. Problem

Is there any non-embeddable GECC? In order to embed a GECC it suffices to embed the underlying linear space with lines of three and two points. It is likely to be a difficult question within the class of finite planes.

3.4. Problem

Find all embeddings up to collineation of the small GECC namely those listed in Section 7. For the smallest GECC this is an exercise. Most non-obvious cases should not be too difficult.

3.5. Problem

Find all embedded GECC for any given known projective plane, especially the small ones. The GECC should become one of the distinguished class of subgeometries with ovals, unitals, etc.

3.6. Embedding of cubics

For a given GECC provided by a cubic there may exist an embedding as a cubic curve in various planes. This study should be made systematic. It may simplify the present views on cubics. It is not an easy task. For example, the GECC of order three with three inflexion points lives as an elliptic cubic curve over $GF(q)$ with $q = 2, 3, 4, 5, 7$. It lives as an embedded GECC in any plane.

3.7. To generate the plane

It may happen of course that an embedded GECC generates a non-trivial subplane. An embedding should be considered trivial if it works via the embedding of a non trivial subplane. What can be said for finite cubic curves in this respect? We could be demanding even more: namely, that the embedded GECC be not a proper sub-GECC of a larger one.

4. Small GECC

4.1. Enumeration

In [7], all GECCs of order at most 8 up to isomorphism are given. The table is a list of pictures, given in Section 7 without proof.

Table 1 gives the number of GECC for each order (number of points) up to 8 as well as the number of cubics that are embeddable as cubic curves in at least one Galois plane.

4.2. Problem

Classify the GECCs of 9 points. There is an interesting partial table in Hirschfeld and Thas [20] giving those cubics of 9 points that are cubics in more than one way. There are pictures without the tangents. It may

Table 1. GECCs and Cubics

Order :	1	2	3	4	5	6	7	8
Number of GECC :	1	1	2	2	1	3	3	13
Number of cubics :	1	1	2	2	1	2	1	2

be feasible computationally to classify the GECCs of order up to 15, according to Andries Brouwer.

4.3. Drawing

Let me make clear again how useful the pictures are. We cannot draw the plane of order 8 but we can draw each of its cubics.

It may be thought that it will be hard to draw all GECC of, say, 15 points. Indeed, it may be recalled that there are 80 Steiner triple systems; and the number of different Steiner triple systems explodes as the order increases. However, it is feasible to draw the cubics on 15 points, since there is only one commutative group of order 15.

5. Meaning of GECCs for the classification of cubics

5.1. Refining

In the classification of cubics given by Hirschfeld 1979, in particular in the final table, there are types of cubics that cannot be distinguished. It seems to me that we do not even know whether they are equivalent under collineation in fairly small cases. The structure of GECC offers a tool to refine the classification.

5.2. Refining by sub-cubics

The greatest effort in the classification is with respect to the number of points and the configuration of inflexion points and tangents. It looks like a matter of sub-cubics. Further configurations might be added and involved. Given a plane and a small GECC embedded in it, what are the cubics containing that GECC as a sub-GECC?

6. Final word

My views of 1962 on GECCs and other aspects of generalized cubics were shared with my friend and colleague Guy Valette. He adapted it and added topological structure on the set of points, proving some strong embeddability theorems. He did not quote my work, as it was then unpublished but he may confirm what I say. His work is in Valette [29], a paper finished in March 1964.

7. Table of small GECC

Here, the list of all GECCs with 8 points at most, up to isomorphism, is given. Each GECC is displayed by a picture. It uses standard conventions for the representation of points (little circles) and lines of three points. Each inflexion point can be recognized by a symbol S on that point in order to represent the tangent. Every other point p is the origin of an arrow pointing to the tangential of p. Every GECC in the list has a symbol $C_{i,j}$ where i is the number of points and j indicates that this GECC is the j-th one in the list among those with i points. Interesting papers on this subject and for beautiful pictures of small cubics are those of Keedwell [21, 22].

Acknowledgements

I thank James Hirschfeld for repeated help during the writing of this paper. During the conference, very helpful information on the abelian group structure of a finite cubic came from him and Andreas Enge; this will be crucial for Part II of my paper. Other thanks are due to Andries Brouwer for different interesting comments.

The generalized elliptic cubic curves with at most 8 points

C_0 is empty Not a cubic over $GF(q)$

C_1  Cubic: $X^3 + Y^3 + X^2Y + X^2Z + XZ^2 = 0$ over $GF(2)$

C_2 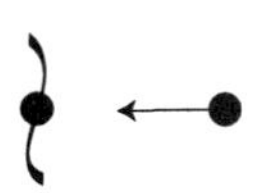 Cubic: $Z^3 + X^2Y + X^2Z + Y^2Z + XZ^2 + YZ^2 + XYZ = 0$ over $GF(2)$

$C_{3,1}$ 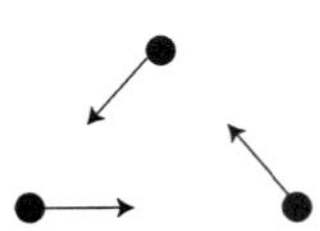 Cubic: $X^2Y + Y^2Z + Z^2X = 0$ over $GF(2)$

$C_{3,2}$ 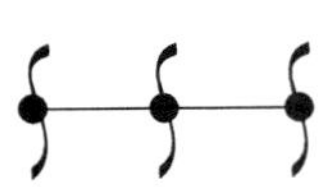 Cubic: $Z^3 + X^2Y + XY^2 = 0$ over $GF(2)$

$C_{4,1}$

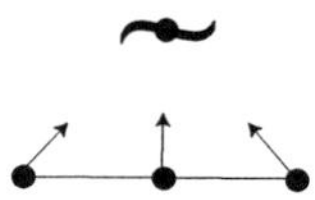

Cubic: $X^3 - Y^2Z - XZ^2 = 0$ over $GF(3)$

$C_{4,2}$

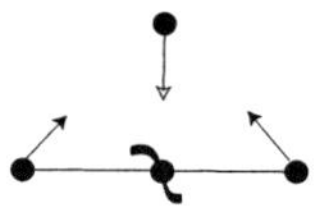

Cubic: $XY^2 + X^2Z + XZ^2 + YZ^2$
$+XYZ = 0$ over $GF(2)$

C_5

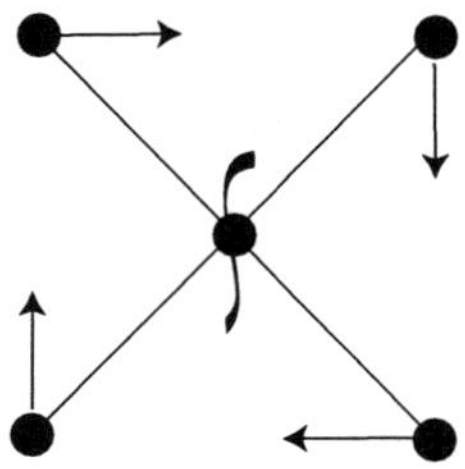

Cubic: $X^2Y + XYZ - X^2Z - Y^2Z$
$+XZ^2 + YZ^2 = 0$ over $GF(3)$

$C_{6,1}$

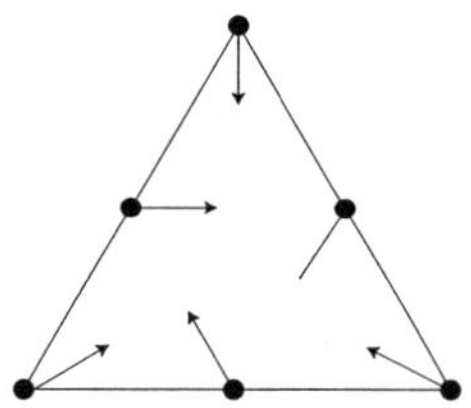

Cubic: $X^2Y - X^2Z - XY^2 + XZ^2$
$+Y^2Z - YZ^2 - XYZ = 0$ over $GF(3)$.

$C_{6,2}$

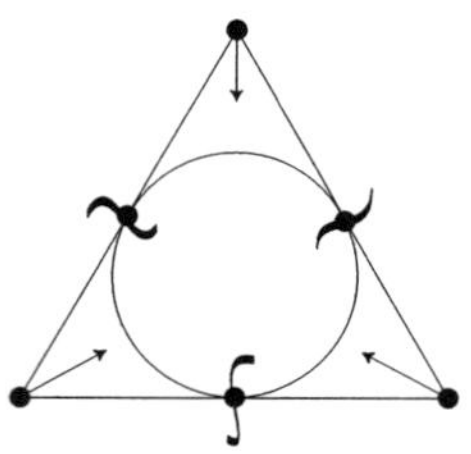

Cubic: $X^2Y - XY^2 + X^2Z - XZ^2$
$+Y^2Z - YZ^2 - XYZ = 0$ over $GF(3)$

$C_{6,3}$

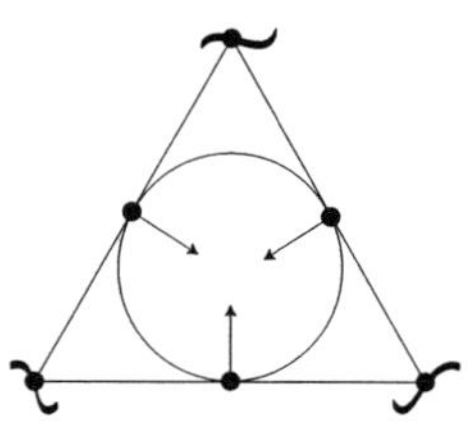

Not a cubic

$C_{7,1}$ 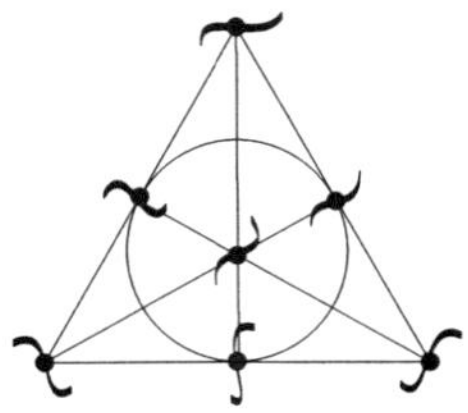 Not a cubic: a Steiner triple system

$C_{7,2}$ 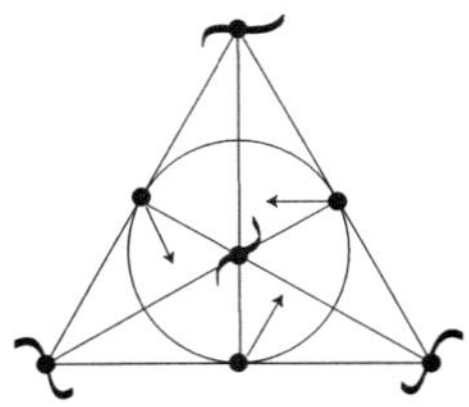 Not a cubic

$C_{7,3}$ 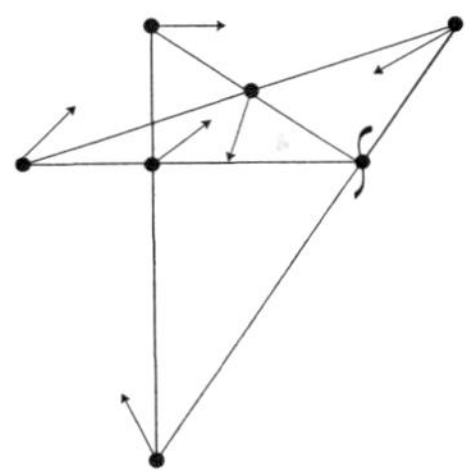 Cubic: $X^2Y - XY^2 - XYZ + X^2Z - XZ^2 + Y^2Z - YZ^2 = 0$ over $GF(3)$

$C_{8,1}$ 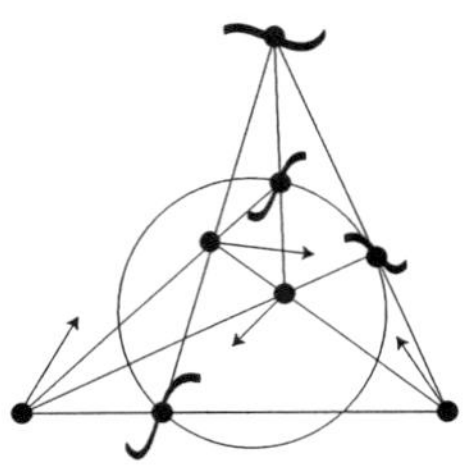 Not a cubic

$C_{8,2}$ 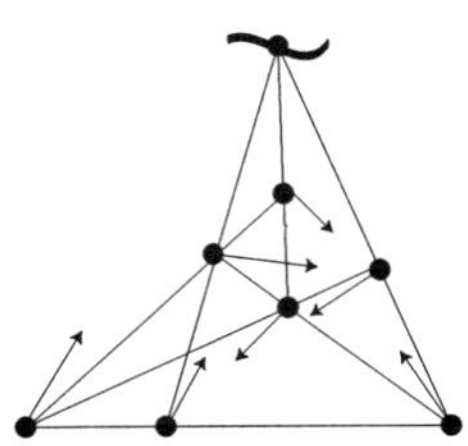 Not a cubic

$C_{8,3}$ 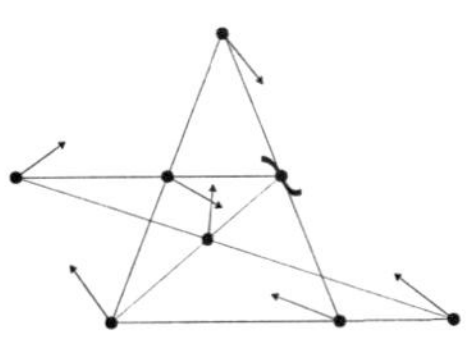 Not a cubic

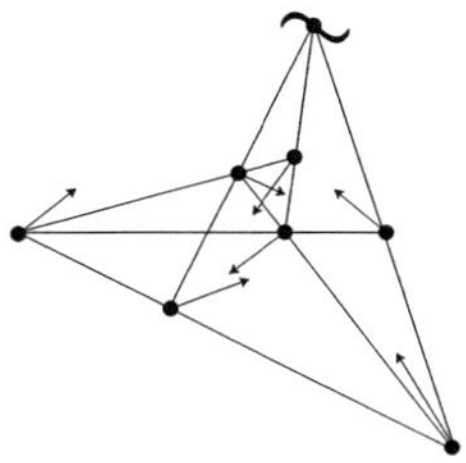

$C_{8,4}$ Not a cubic

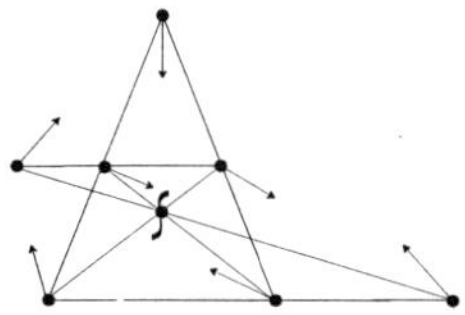

$C_{8,5}$ Cubic: $X^3 - XZ^2 - Y^2Z + Z^3 = 0$ over $GF(3)$

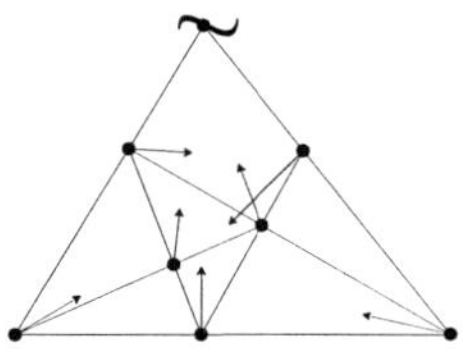

$C_{8,6}$ Not a cubic

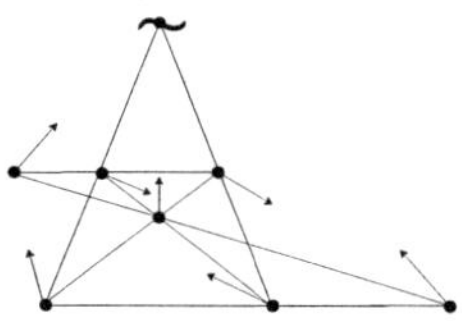

$C_{8,7}$ Not a cubic

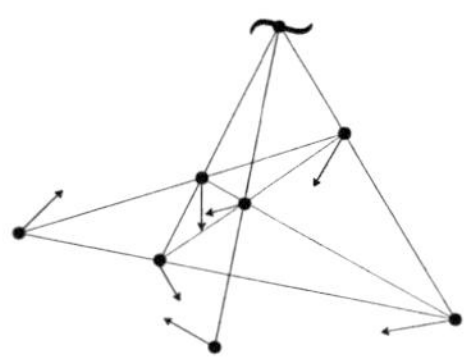

$C_{8,8}$ Not a cubic

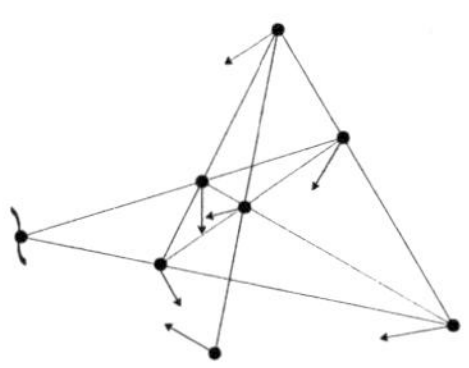

$C_{8,9}$ Cubic: $X^3 - XZ^2 - Y^2Z = 0$ over $GF(5)$

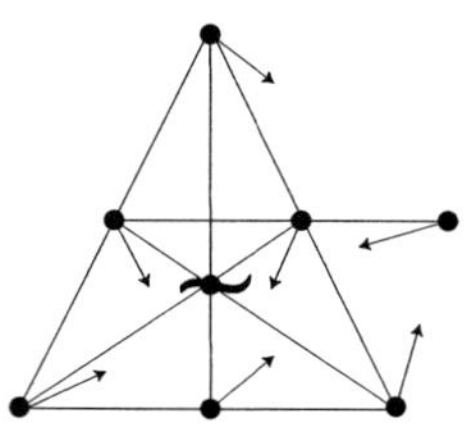

$C_{8,10}$ Not a cubic

$C_{8,11}$

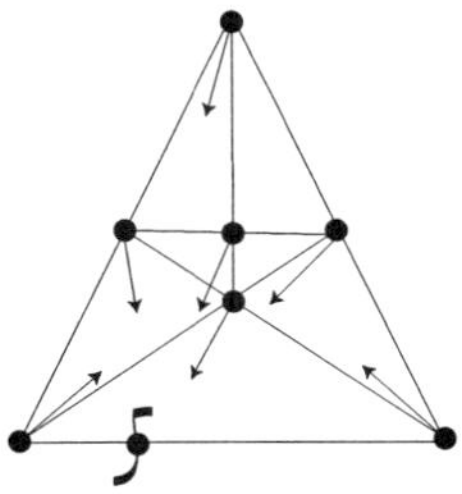

Not a cubic

$C_{8,12}$

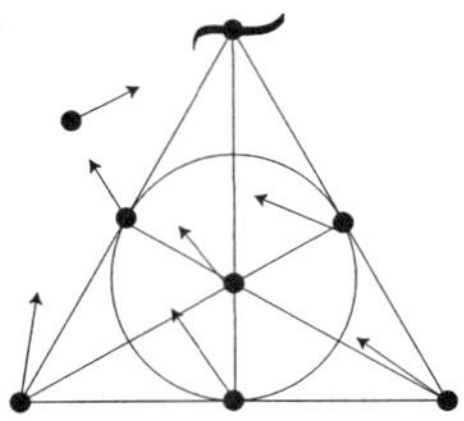

Not a cubic

$C_{8,13}$

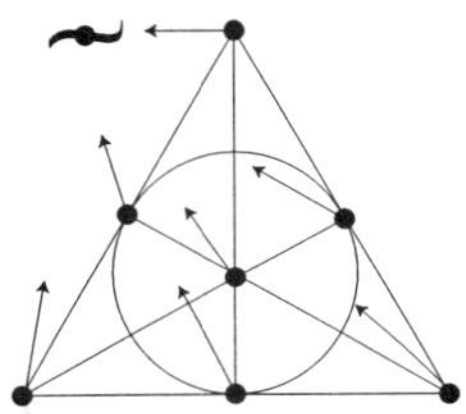

Not a cubic

References

[1] L. Bénéteau, Cubic hypersurface quasigroups, *Quasigroups and Loops: Theory and Applications*, Heldermann Verlag, Berlin, 1990, 143–150.

[2] L. Bénéteau, Extended triple systems: geometric motivations and algebraic constructions, *Discrete Math.* **208/209** (1999), 31–47.

[3] A. Beutelspacher and U. Rosenbaum, *Projective Geometry: from Foundations to Applications*, Cambridge University Press, Cambridge, 1998.

[4] R. Bix, *Conics and Cubics: a Concrete Introduction to Algebraic Curves*, Springer, New York, 1991.

[5] R.H. Bruck, Some results in the theory of quasigroups, *Trans. Amer. Math. Soc.* **55** (1944), 19–52.

[6] R.H. Bruck, *A Survey of Binary Systems*, Springer, Berlin, 1958.

[7] F. Buekenhout, Cubiques généralisées, unpublished manuscript, 1962, 18 pages.

[8] F. Buekenhout, Etude intrinsèque des ovales, *Rend. Mat.* **25** (1966) 1–61.

[9] F. Buekenhout, Ensembles quadratiques des espaces projectifs, *Math. Z.* **110** (1969), 306–318.

[10] F. Buekenhout, Non-singular cubic sets in projective planes, unpublished manuscript, 1987-1990, 14 pages.

[11] F. Buekenhout, *Handbook of Incidence Geometry. Buildings and Foundations*, North Holland-Elsevier, Amsterdam, 1995.

[12] M. Cicchese, Sulle cubiche di un piano di Galois, *Atti. Accad. Naz. Lincei Rend.*, **32** (1962), 38–42.

[13] F. Enriques and O. Chisini, *Lezioni sulle Teoria Geometrica delle Equazioni e delle Funzioni Algebriche*, Zanichelli, Bologna, 1915.

[14] I.M.H. Etherington, Quasigroups and cubic curves, *Proc. Edinburgh Math. Soc.* **14** (1964), 273–291.

[15] D. Ghinelli, N. Melone and U. Ott, On abelian cubic arcs, *Geom. Dedicata* **32** (1989), 31–52.

[16] J.W.P. Hirschfeld, Algebraic curves, arcs, and caps over finite fields, Quaderni del Dipart. Matemat. Univ. Lecce, 1986, 49 pages.

[17] J.W.P. Hirschfeld, *Projective Geometries over Finite Fields*, Oxford University Press, Oxford, 1979 (2nd edition in 1998).

[18] J. W. P. Hirschfeld and J.F. Voloch, The characterization of elliptic curves over finite fields, *J. Austral. Math. Soc. Ser. A* **45** (1988), 275–286.

[19] J.W.P. Hirschfeld and J.F. Voloch, Group-arcs of prime power order on cubic curves, *Finite Geometry and Combinatorics*, London Math. Soc. Lecture Note Series **191**, 1993, Cambridge University Press, Cambridge, 177–186.

[20] J.W.P. Hirschfeld and J.A. Thas, Sets with more than one representation as an algebraic curve of degree three, *Finite Geometries and Combinatorial Designs*, Contemp. Math. **111**, American Mathematical Society, Providence, 1990, 99–110.

[21] A.D. Keedwell, Simple constructions for elliptic cubic curves with specified small number of points, *European J. Combin.* **9** (1988), 463–481.

[22] A.D. Keedwell, More simple constructions for elliptic cubic curves with small numbers of points, *Pure Math. Appl. Ser. A* **3** (1992), 199–214.

[23] Y.I. Manin, *Cubic Forms: Algebra, Geometry, Arithmetic*, North-Holland, Amsterdam, 1974 (Translation of Russian version, 1970).

[24] H.O. Pflugfelder, *Quasigroups and Loops: Introduction*, Heldermann, Berlin, 1990.

[25] H.-G. Rück, A note on elliptic curves over finite fields, *Math. Comput.* **49** (1987), 301–304.

[26] J. Schwenk, A classification of abelian quasigroups, *Rend. Mat. Appl.* **15** (1995), 161–172.

[27] N. Schappacher, Développement de la loi de groupe sur une cubique, *Séminaire de Théorie des Nombres*, Paris, 1988–89, 159–184.

[28] R. Schoof, Nonsingular plane cubic curves over finite fields, *J. Combin. Theory Ser. A* **46** (1987), 183–211.

[29] G. Valette, Cubiques topologiques à 1 dimension, *Arch. Math.* **26** (1965), 265–273.

[30] J.F. Voloch, A note on elliptic curves over finite fields, *Bull. Soc. Math. France* **116** (1988), 455–458.

FIXED POINTS AND CYCLES

Peter J. Cameron
School of Mathematical Sciences
Queen Mary and Westfield College
London E1 4NS
United Kingdom
p.j.cameron@qmw.ac.uk

Abstract This paper presents a tapas of results on fixed points and cycles in permutation groups, arising in such disparate areas as matrix group recognition, Brauer groups, Latin squares, and relational databases.

1. Fixed points and orbits

The Orbit-Counting Lemma asserts that the number of orbits of a finite permutation group is equal to the average number of fixed points of its elements (or, in different language, the expected number of fixed points of a random element of the group). In this section, I discuss various extensions of this result.

1.1. A theorem of Boston *et al.*

A *derangement* in a permutation group G (a subgroup of the symmetric group S_n) is an element of G with no fixed points. It is well known that the probability that a random element of the symmetric group S_n is a derangement is very close to 1/e. This was generalised by Boston *et al.* [3] as follows.

Theorem 1.1 *Let G be a subgroup of S_n. For $0 \leq i \leq n$, let p_i be the probability that a random element of G fixes exactly i points, and F_i be the number of orbits of G acting on the set of i-tuples of distinct points. Then the sequences $(p_0, \ldots, p_n)$ and $(F_0, \ldots, F_n)$ determine each other: more precisely, if*

$$P(x) = \sum_{i=0}^{n} p_i x^i \qquad \textit{and} \qquad F(x) = \sum_{i=0}^{n} F_i x^i / i!\,,$$

A. Blokhuis et al. (eds.), Finite Geometries, 49–60.

then we have

$$F(x) = P(x+1).$$

In particular, the probability that a random element of G is a derangement is

$$F(-1) = \sum_{i=0}^{n} (-1)^i F_i / i!\,.$$

Proof Here is the simple proof of this result. The number of elements with i fixed points is $p_i|G|$, and each fixes $i(i-1)\cdots(i-j+1)$ ordered j-tuples of distinct elements. By the Orbit-Counting Lemma,

$$F_j = \sum_{i=0}^{n} p_i i(i-1)\cdots(i-j+1).$$

Multiply by $x^j/j!$, sum over j, reverse the order of summation, and use the Binomial Theorem to give the result. The last line follows from the fact that the probability that a random element is a derangement is $p_o = P(0) = F(-1)$. □

This result is useful because often it is easier to calculate the numbers F_i than the p_i. Boston *et al.* [3] give several examples. Most easily, if $G = S_n$, then $F_i = 1$ for $0 \leq i \leq n$, and so $F(x)$ is the truncated exponential series; this easily gives the classical formula for the number of derangements.

Independently of Theorem 1.1, Cameron and Cohen [7] have proved lower bounds for the proportion of derangements in a permutation group G in terms of its degree and rank.

1.2. Infinite permutation groups

The probability that an element of an infinite permutation group fixes i points has no meaning. We cannot just count, and the symmetric group is not compact in any natural topology so there is no finite Haar measure. However, sometimes the other side of the equation in Theorem 1.1 makes sense.

An infinite permutation group G on a set Ω is said to be *oligomorphic* if G has only finitely many orbits on Ω^n. For an oligomorphic group G, the numbers F_i are defined for all $i \geq 0$, and $F(x)$ exists as a formal power series (though it may not converge anywhere except the origin).

If $F(-1)$ exists, or can be defined by analytic continuation or some other summation method, then it is a parameter of the infinite group playing the role of the "proportion of derangements". It is not at all clear what this parameter means.

For example, if G is the infinite symmetric group, then $F_i = 1$ for all i, and $F(x) = \mathrm{e}^x$. So $F(-1) = 1/\mathrm{e}$, which happens to be the limiting proportion of derangements in finite symmetric groups.

More striking is the group of all order-preserving permutations of $\mathbb{Q}$. It is easy to see that $F_i = i!$ for this group, so that

$$F(-1) = 1 - 1 + 1 - 1 + 1 - \cdots$$

As Euler remarked, the sum of this series is $\frac{1}{2}$ (for example, by analytic continuation, since $F(x) = 1/(1-x)$ is defined everywhere except at $x = 1$). There is no obvious sense in which half of the elements of this group are derangements!

1.3. Linear groups

The problem of counting derangements in linear or other classical groups has been considered by Neumann and Praeger [20, 21] in connection with matrix group recognition algorithms. Many results about permutation groups have analogues for linear groups. This is the case for Theorem 1.1; so in principle we can calculate the proportion of derangements in any linear group.

Theorem 1.2 *Let G be a subgroup of the general linear group* $\mathrm{GL}(n,q)$. *For $0 \le i \le n$, let P_i be the probability that the fixed points of a random element of G form an i-dimensional subspace, and L_i be the number of orbits of G acting on the set of linearly independent i-tuples. Then the sequences $(P_0, \ldots, P_n)$ and $(L_0, \ldots, L_n)$ determine each other.*

Proof We consider G as a permutation group on the set of q^n vectors of the n-dimensional vector space over $\mathrm{GF}(q)$. Since the fixed points of any element form a subspace, the probabilities p_i of Theorem 1.1 are given by

$$p_i = \begin{cases} P_j & \text{if } i = q^j, \\ 0 & \text{otherwise.} \end{cases}$$

By the theorem, the numbers F_i of G-orbits on i-tuples are determined for all $i \le q^n$.

Let F_i^* be the number of orbits of G on all i-tuples (elements not necessarily distinct. Then (see Cameron and Taylor [9]),

$$F_i^* = \sum_{j=1}^{i} S(i,j)F_j,$$

where the $S(i,j)$ are *Stirling numbers of the second kind.* In terms of generating functions, if $F^*(x) = \sum F_i^* x^i/i!$, then we have

$$F^*(x) = F(\mathrm{e}^x - 1).$$

This statement has an analogue for linear groups, also found in [9]: if L_i is the number of orbits on linearly independent i-tuples, then

$$F_i^* = \sum_{j=0}^{i} \begin{bmatrix} i \\ j \end{bmatrix}_q L_j.$$

Note that the Gaussian coefficients play the role of q-analogues of the Stirling numbers here. This is because set partitions (counted by Stirling numbers) are the "kernels" of surjective set mappings; the kernel of a linear mapping is completely determined by a single subspace (the corresponding partition being into cosets of that subspace.)

So the numbers L_i determine the F_i^*, and hence the F_i, and hence the P_i; and conversely. □

After this roundabout argument, it is natural to ask if a more direct approach exists. First we require some notation. For any positive integer n and indeterminate q, we define the q-analogues of factorials and powers by

$$\begin{aligned} [n]!_q &:= (q^n - 1)(q^{n-1} - 1)\cdots(q-1), \\ [1+x]_q^{[n]} &:= (1+x)(1+qx)\cdots(1+q^{n-1}x). \end{aligned}$$

The *q-binomial theorem* asserts that

$$[1+x]_q^{[n]} = \sum_{i=0}^{n} q^{i(i-1)/2} \begin{bmatrix} n \\ i \end{bmatrix}_q x^i.$$

Theorem 1.3 *Let G be a subgroup of* $\mathrm{GL}(n,q)$, *and let the numbers P_i, L_i be as defined earlier. Then*

$$\sum_{i=0}^{n} \frac{L_i x^i}{[i]!_q} = \sum_{j=0}^{n} P_j [1+x]_q^{[j]}.$$

Proof The proof is the q-analogue of the argument of Boston *et al.* By the Orbit-Counting Lemma we have

$$L_i = \sum_{j=i}^{n} P_j (q^j - 1) \cdots (q^j - q^{i-1}).$$

Multiply both sides by $x^i/[i]!_q$, sum over i, and use the q-binomial theorem. □

This relation can be inverted, to give the P_j explicitly in terms of the L_i. The last displayed equation in the proof can be written as

$$\frac{L_i}{|\,\mathrm{GL}(i,q)|} = \sum_{j=i}^{n} \left[\begin{array}{c} j \\ i \end{array} \right]_q P_j,$$

and Möbius inversion over the lattice of subspaces of the projective space (see Rota [22]) gives

Theorem 1.4 *With the hypotheses of Theorem 1.3,*

$$P_i = \sum_{j=i}^{n} (-1)^{j-i} q^{(j-i)(j-i-1)/2} \left[\begin{array}{c} j \\ i \end{array} \right]_q \frac{L_j}{|\,\mathrm{GL}(j,q)|}.$$

(I am grateful to Thomas Müller and Shahn Majid for conversations about this.)

Note that, as in the permutation group case, the numbers L_i are often easier to calculate than the P_i. For example, if $G = \mathrm{GL}(n,q)$, then $L_i = 1$ for all $i \le n$.

The calculation of the L_i is also possible for the classical groups, but the arguments are a bit more complicated. I give details for the symplectic group $\mathrm{Sp}(2m,q)$ in the next subsection as an example.

Problem Calculate the numbers L_i for the "large" linear groups constituting the Aschbacher classes [1].

One could also ask for a similar analysis for projective groups. Note that a fixed point of an element of $\mathrm{PGL}(n,q)$ is spanned by an eigenvector of a correponding element of $\mathrm{GL}(n,q)$. There is a simple characterisation of the numbers of fixed points of elements of $\mathrm{PGL}(n,q)$: the integer m is the number of fixed points of an element of $\mathrm{PGL}(n,q)$ if and only if, when m is written in base q, the digits are non-decreasing and have sum at most n.

1.4. The symplectic group

As promised, in this section the number P_i of orbits on linearly independent i-tuples is computed for the symplectic group $\mathrm{Sp}(2m, q)$.

First note that, by Witt's Lemma [23, p.57],

> Two linearly independent i-tuples $(v_1, \ldots, v_i)$ and $(w_1, \ldots, w_i)$ lie in the same orbit of $G = \mathrm{Sp}(2m, q)$ if and only if the skew-symmetric matrices $(B(v_j, v_k))$ and $(B(w_j, w_k))$ are equal, where B is the form preserved by G.

So we must count the possible Gram matrices of linearly independent i-tuples. First, we decide which matrices are representable.

> An $i \times i$ skew-symmetric matrix of rank $2r$ is the Gram matrix of a linearly independent set in $V(2m, q)$ if and only if $\max\{0, i - m\} \leq r \leq \min\{\lfloor i/2 \rfloor, m\}$.

For clearly $r \geq 0$; and the span of the given vectors has a totally isotropic subspace of dimension $(i - 2r) + r = i - r$, and this cannot exceed m. Moreover, the rank cannot exceed either i or $2m$. For the converse, note first that the condition implies that $i \leq 2m$. It is enough to deal with skew-symmetric matrices in 'standard form', namely consisting of diagonal blocks $\begin{pmatrix} 0 & 1 \\ -1 & 0 \end{pmatrix}$ or 0. Let the vectors e_i and f_j form the standard basis, with $(e_i, f_i) = -(f_i, e_i) = 1$, $i = 1, \ldots, r$ and all other inner products zero. If $0 \leq r \leq m$ and $i - r \leq m$, then the vectors $e_1, f_1, \ldots, e_r, f_r, e_{r+1}, \ldots, e_{i-r}$ have the required Gram matrix.

Now to count the matrices:

(a) The number of $2r \times 2r$ skew-symmetric matrices of rank $2r$ is

$$N_r := q^{r(r-1)} \prod_{j=1}^{r} (q^{2j-1} - 1).$$

(b) The number of $i \times i$ skew-symmetric matrices of rank $2r$ is

$$M_{i,r} := \begin{bmatrix} i \\ 2r \end{bmatrix}_q N_r$$

for $2r \leq i$.

Here (a) is the number of symplectic forms, most easily computed as the index of $\mathrm{Sp}(2r, q)$ in $\mathrm{GL}(2r, q)$ (see [23, pp.19, 70]); for (b), we choose the null space in $\begin{bmatrix} i \\ i-2r \end{bmatrix}_q$ ways, and a symplectic form on the quotient in N_r ways.

We conclude that

$$L_i = \sum_{r=\max\{0,i-m\}}^{\min\{\lfloor i/2\rfloor,m\}} M_{i,r}.$$

In particular, $L_i = q^{i(i-1)/2}$ for $i \leq m$, and of course $L_i = 0$ for $i \geq 2m$.

1.5. Cycles

In fact Theorem 1.1 is a special case of a much older result. Suppose we are interested in not just the number of fixed points but the cycle structure (the number of cycles of each length) of a random element of G. The probability generating function for this is

$$\frac{1}{|G|}\sum_{g\in G} s_1^{c_1(g)} s_2^{c_2(g)} \cdots ,$$

where $s_1, s_2, \ldots$ are indeterminates, and $c_i(g)$ is the number of i-cycles of the element g. This is just the familiar Redfield–Pólya *cycle index*, whose role in orbit-counting is well-known.

In particular, it is known that $F(x)$ (the exponential generating function for orbits of G on ordered i-tuples) is obtained from the cycle index by the substitution $s_1 \leftarrow x+1$, $s_i \leftarrow 1$ for $i > 1$ (which yields Theorem 1.1). A related result is that the ordinary generating function for the number of orbits of G on unordered i-sets (that is, the polynomial $f(x) = \sum f_i x^i$, where f_i is the number of orbits of G on i-sets), is also a specialisation of $Z(G)$, obtained by the substitution $s_i \leftarrow x^i + 1$ for all i.

Some authors (see for example [17, 12]) have considered extensions of the cycle index to linear groups. This would have obvious applications to the material of the preceding section.

2. Cycle partitions

The cycles of a permutation form a partition. So there is a map C from S_n to the set P_n of partitions of $\{1, \ldots, n\}$, which is onto but not one-to-one. Alberto Leporati (personal communication), in connection with work on database query languages [2], asked: what do we know about a permutation group G if we are given the set $C(G)$ of cycle partitions of its elements? I will report briefly on some results, which are discussed more fully in [6].

Theorem 2.1 *The set $C(G)$ determines*

(a) *the order of G;*

(b) *the multiset* $C(G)$ *(that is, for each* $\pi \in P_n$*, the number of* $g \in G$ *with* $C(g) = \pi$ *is determined);*

(c) *the cycle index* $Z(G)$.

It does not determine G*, even up to isomorphism as an abstract group.*

The last assertion is due to Eamonn O'Brien (personal communication), who found two pairs of permutation groups with the same set of cycle partitions (all these groups have degree 16 and order 64).

There is also a recognition problem: given a subset X of P_n, is there a subgroup G of S_n with $C(G) = X$? It is not reasonable to ask whether this problem has a polynomial-time solution, since the input is potentially exponentially large. P. Kolaitis (personal communication) suggested the *polynomial delay model*, where we are allowed to perform a polynomial-time computation after reading each element of the input X. Now the analogues of asking if the problem is in P or NP in this model are as follows,, where 'polynomial' means 'polynomial in n':

(a) Is there an algorithm which decides, with polynomial delay, whether $X = C(G)$ for some group G?

(b) Is the a polynomial-length certificate for a set X, knowing which we can verify the existence of G such that $X = C(G)$ with a polynomial-delay computation?

The certificate in (b) might be a generating set for G. (It is known that any subgroup of S_n can be generated by $n/2$ elements if $n > 2$: see McIver and Neumann [19].)

The question posed in this section makes sense for infinite groups as well. One particular problem is worth stating. It follows from Theorem 2.1(a) that, if H is a subgroup of the finite permutation group G with $C(H) = C(G)$, then in fact $H = G$.

Problem Does the above assertion hold also for infinite permutation groups?

3. Derangements of prime order

The Orbit-counting Lemma immediately implies that a transitive finite permutation group of degree greater than 1 contains a derangement. (The average number of fixed points is equal to 1, and the identity fixes more than one point, so some element fixes fewer than one.) Much more difficult is the following result of Fein *et al.* [11], whose proof uses the Classification of Finite Simple Groups.

Theorem 3.1 *A finite transitive permutation group of degree greater than* 1 *contains a derangement of prime power order.* □

These authors also gave an example to show that the theorem is false if 'prime power' is replaced by 'prime'. Let $G = \mathrm{AGL}(1,p^2)$ and $H = \mathrm{AGL}(1,p)$, where p is a Mersenne prime. Consider G as a permutation group on the cosets of H (equivalently, $G \leq \mathrm{AGL}(2,p)$ acting on the set of lines of the affine plane). The degree is $p(p+1)$, so the only possible prime orders of derangements are 2 and p; and it is easily seen that any conjugacy class of elements of order 2 or p in G meets H, so that all such elements have fixed points.

A permutation group is called *elusive* if it is transitive and has no derangement of prime order. Not many such groups are known. The most surprisingly overlooked example is the Mathieu group M_{11} in its action of degree 12, found recently by Michael Giudici. A number of examples and constructions are presented in [8], including another sporadic example of structure $7^3 : (3^{1+2} : Q_8)$. (The translation complement, taken projectively, is the group of the nine inflexion points of a cubic in the plane of order 7. For us, this group acts as an elusive permutation group of degree 84.) In the direction of characterisations, Giudici [15] has shown the following.

Theorem 3.2 *A quasiprimitive elusive group is isomorphic to a wreath product $M_{11} \wr K$, for some transitive permutation group K of degree k, in its product action of degree 12^k.* □

(A permutation group G is *quasiprimitive* if every non-trivial normal subgroup of G is transitive.)

Part of the motivation for studying elusive groups is a conjecture first stated by Marušič and in the present form by Klin, asserting that there is no 2-closed elusive group. (A permutation group $G \leq S_n$ is 2-*closed* if every permutation in S_n which preserves all the orbits of G on ordered pairs belongs to G.)

The automorphism group of a graph, possibly directed and edge-coloured, is necessarily 2-closed. Moreover, if a graph Γ on n vertices has an automorphism g which is a derangement of prime order, then we can get information about Γ by studying a quotient $\Gamma/\langle g\rangle$ on n/p vertices. Our study has turned up no counterexamples to this conjecture.

The set of degrees of elusive groups is multiplicatively closed.

Problem Does the set of degrees of elusive groups have density zero?

Either this question or Klin's conjecture might be settled with less than a complete classification of elusive groups. The known degrees below 100 of elusive groups are 12, 24, 36, 48, 56, 72, 84 and 96.

4. Derangements and Latin squares

Dixon [10] proved that almost all pairs of elements of the symmetric group S_n generate either S_n or A_n (in the sense that the probability that a randomly-chosen pair of elements generates a smaller subgroup tends to zero as $n \to \infty$). Recently, Luczak and Pyber [18] proved a theorem from which Dixon's result follows. The interpretation of 'almost all' is as just described.

Theorem 4.1 *Almost all elements of S_n lie in no proper transitive subgroup of S_n except possibly A_n.* □

The easier part of the proof of Dixon's theorem is the fact that almost all pairs of elements generate a transitive subgroup; then the conclusion is immediate from Theorem 4.1.

The theorem has another application, to a new proof of the result of [4].

Theorem 4.2 *The rows of a random Latin square of order n, regarded as permutations, almost surely generate S_n.*

Proof The group generated by the rows of a random Latin square is obviously transitive, and the first row is a random permutation (that is, all permutations occur equally often as the first row of a Latin square). The alternating group is excluded by a result of Häggkvist and Janssen [16], according to which the proportion of Latin squares for which all rows are even permutations is exponentially small. □

Problem Does the conclusion of Theorem 4.2 also hold for random *normalised* Latin squares, those for which the first row and column are equal to the identity permutation?

The second row of a normalised Latin square is a derangement. It follows from Theorem 4.1 and the fact that derangements have non-zero probability that almost all derangements lie in no proper subgroup of S_n except possibly A_n. However, this does not settle the question, since derangements do not occur equally often as the second row of normalised Latin squares. Indeed, random normalised Latin squares define a probability measure on derangements, and the problem above asks whether the analogue of Theorem 4.1 holds for this measure.

5. Other topics

Two further topics will be considered here only briefly.

A permutation group G is *cycle closed* if it contains all the cycles of its elements. The finite cycle closed groups are direct products of symmetric groups and cyclic groups of prime order, and it is known that from any permutation group we reach a cycle-closed group in at most three steps (each step consisting of taking the group generated by the cycles of the given group). Interesting open problems remain for infinite groups. See [5] for further information.

The *Parker vector* of a permutation group $G \leq S_n$ is the sequence $(P_1, \ldots, P_n)$, where P_i is the number of orbits of G on the set $\mathcal{C}_i(G)$ of all i-cycles occurring in elements of G (the action is by conjugation). The Parker vector contains some information about the group, and is closely connected with the cycle index, from which it can be calculated. See Gewurz [13, 14] for further information on this topic.

References

[1] M. Aschbacher, On the maximal subgroups of the finite classical groups, *Invent. Math.* **76** (1984), 469-514.

[2] P. Bonizzoni, G. Della Vedova, A. Leporati, D. Lucarella and G. Mauri, A unifying framework to characterize the power of query languages, preprint.

[3] N. Boston, W. Dabrowski, T. Foguel, P. J. Gies, J. Leavitt, D. T. Ose and D.A. Jackson, The proportion of fixed-point-free elements of a transitive permutation group, *Commun. Algebra* **21** (1993), 3259–3275.

[4] P.J. Cameron, Almost all quasigroups have rank 2, *Discrete Math.* **106/107** (1992), 111–115.

[5] P.J. Cameron, Cycle-closed permutation groups, *J. Algebraic Combin.* **5** (1996), 315–322.

[6] P.J. Cameron, Partitions and permutations, preprint.

[7] P.J. Cameron and A.M. Cohen, On the number of fixed point free elements of a permutation group, *Discrete Math.* **106/107** (1992), 135–138.

[8] P.J. Cameron, M.R. Giudici, G.A. Jones, W.M. Kantor, M. Klin, D. Marušič and L.A. Nowitz, Transitive permutation groups without semiregular subgroups, in preparation.

[9] P.J. Cameron and D.E. Taylor, Stirling numbers and affine equivalence, *Ars Combin.* **20B** (1985), 3–14.

[10] J.D. Dixon, The probability of generating the symmetric group, *Math. Z.* **110** (1969), 199–205.

[11] B. Fein, W.M. Kantor and M. Schacher, Relative Brauer groups, II, *J. Reine Angew. Math.* **328** (1981), 39–57.

[12] H. Fripertinger, Cycle indices of linear, affine, and projective groups, *Linear Algebra Appl.* **263** (1997), 133–156.

[13] D.A. Gewurz, Reconstruction of permutation groups from their Parker vectors, *J. Group Theory* **3** (2000), 271–276.

[14] D.A. Gewurz, Parker vectors and cycle indices of permutation groups, in preparation.

[15] M.R. Giudici, Quasiprimitive permutation groups without semiregular subgroups, in preparation.

[16] R. Häggkvist and J.C.M. Janssen, All-even latin squares, *Discrete Math.* **157** (1996), 199–206

[17] J.P.S. Kung, The cycle structure of a linear transformation over a finite field, *Linear Algebra Appl.* **36** (1981), 141–155.

[18] T. Łuczak and L. Pyber, On random generation of the symmetric group, *Combin. Probab. Comput.* **2** (1993), 505–512.

[19] A. McIver and P.M. Neumann, Enumerating finite groups, *Quart. J. Math.* (2) **38** (1987), 473–488.

[20] P.M. Neumann and C.E. Praeger, A recognition algorithm for special linear groups, *Proc. London Math. Soc.* (3) **65** (1992), 555–603.

[21] P.M. Neumann and C.E. Praeger, Derangements and eigenvalue-free elements in finite classical groups, *J. London Math. Soc.* (2) **58** (1999), 564–586.

[22] G.-C. Rota, On the foundations of combinatorial theory, I: Theory of Möbius functions, *Z. Wahrsch. Verw. Gebiete* **2** (1964), 340–368.

[23] D.E. Taylor, *The Geometry of the Classical Groups*, Helderman, Berlin, 1992.

RWPRI GEOMETRIES FOR THE ALTERNATING GROUP A_8

Philippe Cara
Department of Mathematics
Vrije Universiteit Brussel
Pleinlaan 2
B-1050 Brussel
Belgium
pcara@vub.ac.be

Abstract We give an exhaustive list of the residually connected, flag-transitive geometries of the alternating group A_8 satisfying the $(IP)_2$ and RWPRI properties. The list was obtained with the aid of a computer and the software package MAGMA[1]. The programs used are based on the algorithms described in [15] and [16]. Many geometries were predicted by a theory of inductively minimal geometries. For the remaining geometries, we give an explicit construction proving their existence. These constructions rely on various geometrical objects resulting from the well-known isomorphisms $A_8 \cong PSL(4,2) \cong PO^+(6,2)$.

1. Introduction

In [3] Buekenhout explains the importance of experimentation in the quest for a theory generalizing Tits' Buildings (see [29]) to the class of all (almost) simple groups. Given such a group, the idea is to find interesting geometries for that group. By interesting, we mean that the geometry somehow helps to understand the group better, as is the case for the simple groups of Lie type where the buildings are the interesting geometries. In the latter case, the theory gives an explicit construction for the interesting geometry. To generalize this situation, we should first get an idea of the geometries associated to a given (almost) simple group, then select the most interesting ones and study these to understand which general properties they hide.

In [28], Tits describes an algorithm to construct what we now call *coset geometries* from a group. This algorithm can be implemented on

A. Blokhuis et al. (eds.), Finite Geometries, 61–97.

computer and experimentation can begin ... We can ask the computer to list all coset geometries for a given group. Soon one realizes that you get too many candidates to play the role of interesting geometry for a given group as small as $PGL(2,7)$. To reduce the number of geometries, axioms have to be imposed. In [3] a set of axioms is presented and discussed. It is emphasized that this set is not yet grown up and that it should be fine-tuned by further experimentation. Over the past five years many experiments have been done. These led to both new axioms and theoretical results. For an overview, we refer to [8].

In this paper, we use the set of axioms described in [3] and list all geometries satisfying these axioms for the alternating group of degree 8. The list was obtained with the aid of a computer and the software package MAGMA[1]. The programs used are based on the algorithms described in [15] and in [16]. The list was further analyzed using GAP [18]. The CPU time needed to compute the list was about 13 days on a Pentium computer at 400 MHz. In terms of number of conjugacy classes of subgroups, we can say that A_8 is the largest group we can tackle in reasonable time with the computer power available at the moment of writing of the present paper.

The list is useful for further experimentation and allows one to test conjectures concerning the group A_8. In this paper, we will also show that the list confirms theoretical work done earlier. For each of the 88 geometries of rank greater than one in the list, we determine the diagram and give a construction which proves its existence. Some of the constructions were known before, others are new. For the completeness of the list, that is, that there are no other geometries satisfying our axioms for the group A_8 than the ones appearing in the list, we rely on the computer. The constructions given are certainly not unique but can be divided into two groups: the rather combinatorial ones related to *inductively minimal geometries* and the others which use objects related to the projective space $PG(3,2)$ and the Klein quadric.

The paper is organised as follows. First we recall some basic definitions and describe the axioms used. We also state some observations concerning the geometries satisfying this set of axioms for the alternating group of degree 8.

After that we give an outline of the theory of inductively minimal geometries. We chose to formulate this theory using the connection with trees described in [13] and prove some further theorems on this connection. Then comes a first list of geometries related to the inductively minimal geometries. This list contains 60 geometries, showing that this theory explains a large number of geometries.

Section 5 recalls the concept of Petersen replacement and explains 8 more geometries. The next two sections give generalizations and variations on the Petersen theme which help to understand more geometries.

Finally we finish the list with constructions related to the projective space $PG(3,2)$.

2. Definitions and notation

We recall notation and definitions used for finite diagram geometries. More details can be found in [4]. Let I be a finite set of $n \geq 1$ elements called *types*. A *geometry over* I is a triple $\Gamma = (X, *, t)$ where X is a set whose members are called *elements* of Γ, the symbol $*$ denotes a symmetric, reflexive binary relation on X, called *incidence relation* and t is a mapping of X onto I called *type function*, such that $a * b$ and $t(a) = t(b)$ implies $a = b$. The *rank* of Γ is $|I| = n$. If $A \subseteq X$, the *type* of A is the set $t(A)$. A *flag* F of Γ is a set of elements of Γ which are pairwise incident. We assume that Γ is *firm*, that is, every flag F with $t(F) \neq I$ is contained in at least two flags of type I. A flag of type I will often be called a *chamber*.

The *residue* Γ_F of a flag F is the firm geometry over $I \setminus t(F)$ whose elements are all $a \in X \setminus F$ such that $\{a\} \cup F$ is also a flag, together with the restrictions of $*$ and t to these elements. For $J \subseteq I$, we define the *J-truncation* of Γ to be the geometry over J, whose element set is $t^{-1}(J)$, together with the restrictions of $*$ and t. A geometry is said to satisfy property $(IP)_2$ if for every residue of rank 2 we have that either every element of one type is incident with every element of the other type or that two different elements of the same type are incident with at most one element of the other type.

The *basic diagram* of Γ is the graph $(I, \sim)$ defined by $i \sim j$, for $i \neq j$ in I, if and only if there is a flag F of type $I \backslash \{i,j\}$ such that Γ_F is not a complete bipartite graph; this means that Γ_F has elements a, b with $t(a) = i$, $t(b) = j$ and $(a,b) \notin *$. We call Γ *thin* if every flag F, with $|t(F)| = n - 1$, has a residue containing exactly two elements. The geometry Γ is *residually connected* if, for every residue Γ_F with $|t(\Gamma_F)| \geq 2$, the graph induced by $*$ on the element set of Γ_F is connected.

An *automorphism* of a geometry Γ is an automorphism α of the graph $(X, *)$ mapping every element x of X onto an element $\alpha(x)$ such that $t(x) = t(\alpha(x))$. An automorphism of $(X, *)$, which does not necessarily preserve the type of elements but maps elements of the same type to elements of the same type, is called a *correlation*. Under composition, the automorphisms and correlations form groups which are denoted by $\mathrm{Aut}(\Gamma)$ and $\mathrm{Cor}(\Gamma)$ respectively.

Let G be a group of automorphisms of Γ. If G is transitive on the set of chambers of Γ, we say that G is *flag-transitive*. The stabilizer of a chamber is called the *Borel subgroup* of the geometry and is denoted by B.

The *diagram* of a geometry Γ with a flag-transitive group is the basic diagram together with additional information given as labels on the vertices and edges.

Every vertex i of the diagram carries an integer which is one less than the number of elements in the residue of a flag of type $I \setminus \{i\}$. This integer is called the *i-order* of Γ (or the *order* of Γ at i) and is denoted by s_i. Note that a geometry is thin if and only if the i-order is 1 for all $i \in I$.

An edge $i \sim j$ in the diagram of Γ is endowed with three labels which are integers denoted by d_{ij}, g_{ij} and d_{ji}. The middle parameter g_{ij} is half the girth of the incidence graph $\mathcal{G}_{ij}$ of a residue Γ_{ij} of type $\{i, j\}$ in Γ. To compute d_{ij}, we take an element x of type i in Γ_{ij} and look for a longest geodesic path in $\mathcal{G}_{ij}$ with first vertex x. The length of such a path is called the *i-diameter* of Γ_{ij} and is denoted by d_{ij}. Similarly the *j-diameter* is the maximum distance between an element of type j and any other vertex in $\mathcal{G}_{ij}$. The j-diameter is denoted by d_{ji}.

A vertex i of the diagram also carries two more labels which are the number N_i of elements of type i in Γ and G_i, the (structure of the) stabilizer of an element of type i. Figure 1 shows how the parameters are organised around the edge $i \sim j$.

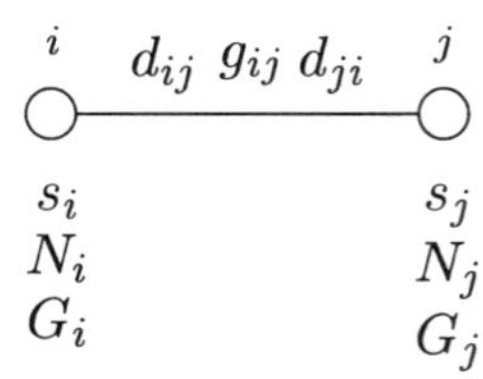

Figure 1. A generic edge $i \sim j$ in a diagram.

Some frequently appearing parameter sets (d_{ij}, g_{ij}, d_{ji}) are abbreviated as follows.

When $d_{ij} = g_{ij} = d_{ji}$, we do not write d_{ij} and d_{ji}. If Γ_{ij} has parameter set $(3, 3, 3)$, we draw it as a simple edge between i and j, without any label. For the parameter set $(4, 4, 4)$ we use a double line to join i and j and we omit the label 4 on the edge.

The vertex-edge geometry of the complete graph on n points has parameter set $(3, 3, 4)$. These are replaced by a "c" symbol oriented in such a way that the "open" side of the symbol is at the side where the parameter 4 would have been written. We remark that the point-order of this geometry on n points is 1 and that its line-order is $n-2$.

We now consider pairs (Γ, G) where Γ is a geometry and G is a flag-transitive group of automorphisms of Γ. A pair (Γ, G) is said to be *WPRI* if at least one stabilizer G_x of an element x of Γ is maximal in G. A geometry Γ is said to satisfy *RWPRI* if every nontrivial residue Γ_F, together with the automorphism group induced by the stabilizer G_F, is WPRI.

3. The axioms

In this paper we list all firm, residually connected geometries satisfying the $(IP)_2$ property on which the alternating group of degree 8 acts flag-transitively such that the RWPRI property is fulfilled. For a detailed motivation of the choice of these axioms, we refer to [3] and [8].

3.1. The influence of $(IP)_2$

The computer programs allow us to list all RWPRI geometries for A_8 without demanding the $(IP)_2$ property.

For each rank n, we give in Table 1 the number n_R of firm, residually connected, RWPRI geometries for A_8 and the number n_{RI} of these satisfying $(IP)_2$.

Table 1. Number of RWPRI geometries

n	1	2	3	4	5	6
n_R	6	21	44	62	52	48
n_{RI}	6	8	16	20	24	20

The $(IP)_2$ property severely restricts the number of geometries occurring. Hence it is certainly a good idea to consider $(IP)_2$ as an axiom in a future theory associating a small number of geometries to a given group.

3.2. Element stabilizers

In Table 2, we count the number of times each group appears as the stabilizer of an element. The champions are the intransitive maximal subgroups of A_8, that is, the stabilizers of bipartitions of a set of 8 elements. This reflects the fact that so many geometries are related to inductively minimal geometries; see Section 4.

Table 2. Stabilizers

A_7	134
S_6	106
$A_5 : S_3$	69
$(A_4 \times A_4) : 2^2$	12
$AGL(3,2)$	14
$(2 \times S_4) : 2$	10
$(A_4 \times A_4) : 2$	26
$2^3 : S_4$	7
$2^4 : D_{12}$	2
$2^4 : S_3$	2
$2 \times S_4$	2
$(S_3 \times S_3) : 2$	5
$S_3 \times S_3$	1

3.3. Rank 1

By the RWPRI requirement, the rank one geometries correspond to the maximal subgroups of A_8. These are A_7, $AGL(3,2)$ (two conjugacy classes), S_6, $S_4 \times S_4$ and $A_5 : S_3$. For a maximal subgroup H of A_8, the elements of the geometry are all left cosets of H in A_8, on which A_8 acts by left translation.

3.4. Rank 6

This is the highest possible rank of an RWPRI flag-transitive geometry for A_8. All these geometries are truncations of inductively minimal geometries. They are described in the next section.

4. Inductively minimal geometries and their truncations

Inductively minimal geometries were introduced in [6] and studied further in [5]. Summarising, we can say that we have an infinite family of geometries which are thin, residually connected and satisfy $(IP)_2$. If an inductively minimal geometry Γ has rank n, then the symmetric group of degree $n+1$ acts on it as a flag-transitive group of automorphisms. Together with this group, Γ satisfies the RWPRI property (see [5]). It is important to remark that this infinite family of geometries is the result of experimental work done some years ago on the symmetric and alternating groups of degree at most 5.

In [13], it is shown that there is a bijection between the set of non-isomorphic inductively minimal geometries of rank n and the set of non-isomorphic trees of $n+1$ vertices. We shall use this correspondence to give an alternative description of inductively minimal geometries and their truncations satisfying our properties.

Consider a tree $\mathcal{T}$ with $n+1$ vertices. We construct a geometry of rank n on a set S of cardinality $n+1$ as follows. The types are the n edges of the tree. Take a type i. Removing the edge i in $\mathcal{T}$ results in two trees with n_1 and n_2 vertices. The elements of type i are all bipartitions of S into two sets of cardinalities n_1 and n_2 where $n_1 + n_2 = n+1$. It is technically convenient to denote such a bipartition by $A \cup \complement A$ with the convention $|A| \leqslant |\complement A|$. Given two elements $A \cup \complement A$ and $B \cup \complement B$, we decide that they are incident according to the following rules:

- either $|A| = |B|$ and $A \subseteq \complement B$ or
- $|A| < |B|$ and $A \subseteq B$ or
- $|A| > |B|$ and $A \supseteq B$.

The geometry defined is called the *inductively minimal geometry* associated to the tree $\mathcal{T}$. We denote it by $\mathsf{IMG}(\mathcal{T})$.

We recall the the line graph $L\mathcal{G}$ of a graph $\mathcal{G}$ is the graph whose vertices are the edges of $\mathcal{G}$ and in which two vertices are adjacent if and only if they share a vertex when seen as edges of the graph $\mathcal{G}$.

Theorem 4.1 (see [13]) *The basic diagram of the inductively minimal geometry associated to $\mathcal{T}$ is the line graph $L\mathcal{T}$ of $\mathcal{T}$.*

From the basic diagram $(I, \sim) = L\mathcal{T}$ of an inductively minimal geometry $\mathsf{IMG}(\mathcal{T})$ we can reconstruct the tree $\mathcal{T}$ in the following way. The nodes of $\mathcal{T}$ are the maximal cliques of $L\mathcal{T}$ together with the set V_1 of vertices of $L\mathcal{T}$ that lie in just one maximal clique. Two cliques are adjacent if they have a nontrivial intersection and the adjacency between vertices in V_1 and maximal cliques is inclusion (thus, V_1 corresponds to the set of *leaves* of $\mathcal{T}$).

Theorem 4.2 (see [6, 5]) *For any tree $\mathcal{T}$, the inductively minimal geometry Γ associated to the tree $\mathcal{T}$ satisfies the following properties.*

1. *Γ is thin and residually connected.*
2. *The residue of an element whose type corresponds to a leaf f of $\mathcal{T}$ is an inductively minimal geometry associated to the tree obtained from $\mathcal{T}$ by removing f and the edge containing it.*
3. *The residue of any other flag is the direct sum of inductively minimal geometries corresponding to the connected components of $\mathcal{T}$*

after removing the set E of edges corresponding to the type of the flag together with the vertices which belong only to edges in E.

4. *The automorphism group of Γ is isomorphic to S_{n+1}, where $n+1$ denotes the number of vertices of $\mathcal{T}$. The action of S_{n+1} on Γ is the one induced by the natural action of S_{n+1} on the set of cardinality $n+1$ used to construct the geometry.*

Lemma 4.3 *Given a tree $\mathcal{T}$ and an automorphism α of it, there exists a correlation of the inductively minimal geometry associated to $\mathcal{T}$ whose permutation induced on the set of types corresponds to α acting on the edges of $\mathcal{T}$.*

Proof Consider the image $\alpha(i)$ of an edge i of $\mathcal{T}$ under an automorphism α. Removing the edge $\alpha(i)$ from $\mathcal{T}$ yields two connected components isomorphic to the components obtained by removing i. This shows that the elements of type $\alpha(i)$ are bipartitions of the same kind as the elements of type i. There is a natural bijection α_i between the elements of type i and the elements of type $\alpha(i)$ of $\mathsf{IMG}(\mathcal{T})$, since these are in fact two copies of the same set of bipartitions. Taking the union $\cup_{i\in I}\alpha_i$, where I denotes the set of edges of $\mathcal{T}$, we obtain a bijection $\overline{\alpha}$ from the set of elements of $\mathsf{IMG}(\mathcal{T})$ onto itself. This bijection maps elements with equal type onto elements with equal type and preserves incidence. □

Now consider a correlation $\overline{\alpha}$ of $\mathsf{IMG}(\mathcal{T})$. The bijection induced on the set of types I is an automorphism of the basic diagram of $\mathsf{IMG}(\mathcal{T})$. As seen above, we can reconstruct $\mathcal{T}$ from that basic diagram. This construction also shows that $\overline{\alpha}$ yields an automorphism α of the tree $\mathcal{T}$.

Theorem 4.4 *For any tree $\mathcal{T}$ with $n+1$ vertices, we have that the full correlation group $Cor(\mathsf{IMG}(\mathcal{T}))$ is isomorphic to the direct product $S_{n+1} \times Aut(\mathcal{T})$.*

Proof The automorphism group N of $\mathsf{IMG}(\mathcal{T})$ is a normal subgroup of the group $\mathrm{Cor}(\mathsf{IMG}(\mathcal{T}))$ which is isomorphic to S_{n+1} by Theorem 4.2. Lemma 4.3 also provides a subgroup K isomorphic to $\mathrm{Aut}(\mathcal{T})$. Consider a correlation γ of $\mathsf{IMG}(\mathcal{T})$. It yields an automorphism $\tilde{\gamma}$ of $\mathcal{T}$ which in turn provides a correlation $\overline{\tilde{\gamma}}$. The composition $\gamma(\overline{\tilde{\gamma}})^{-1}$ is a type preserving correlation of $\mathsf{IMG}(\mathcal{T})$. This shows that $\mathrm{Cor}(\mathsf{IMG}(\mathcal{T})) = NK$. Since an automorphism always induces the identity on the diagram of a geometry, we have $N \cap K = 1$. Now take $\alpha \in N$ and $\overline{\beta} \in K$ and consider an element $A \cup \complement A$ of type i in $\mathsf{IMG}(\mathcal{T})$. The image of $A \cup \complement A$ under $\overline{\beta}$ is simply $A \cup \complement A$, but now considered as an element of type j, say. Assume the automorphism α maps this element to $B \cup \complement B$. Since automorphisms of $\mathsf{IMG}(\mathcal{T})$ result from the action of S_{n+1} on the set on which the inductively minimal geometry was constructed, the element $A \cup \complement A$ of type i must be mapped to the element $B \cup \complement B$ of type i, which is mapped to the element $B \cup \complement B$ of type j by $\overline{\beta}$. This shows that the elements α and $\overline{\beta}$ commute with each other. □

By induction one easily establishes that the number of chambers in an inductively minimal geometry of rank n is exactly $(n+1)!$. This implies that the alternating group of degree $n+1$ cannot act flag-transitively on such a geometry. Truncations however have less maximal flags and allow flag-transitive action of an alternating group.

The truncations of inductively minimal geometries have been studied in [11]. We cite the main theorems.

Theorem 4.5 *A J-truncation of an inductively minimal geometry with diagram $L\mathcal{T}$ satisfies $(IP)_2$ if and only if the subgraph $\mathcal{S}$ of $\mathcal{T}$ generated by the set of edges J is a connected and for each vertex v in $\mathcal{S}$, the tree $\mathcal{S}$ contains either one or all neighbours of v (in $\mathcal{T}$).* □

Theorem 4.6 *Let Γ be an inductively minimal geometry. Any $(IP)_2$ truncation of Γ also satisfies RWPRI for the group S_{n+1}.* □

Theorem 4.7 *The Borel subgroup of a J-truncation of an inductively minimal geometry $\mathsf{IMG}(\mathcal{T})$ satisfying $(IP)_2$ is the direct product of symmetric groups whose degrees equal the number of vertices in the connected components of $\mathcal{T}$ after removing the edges corresponding to J together with the vertices which belong only to edges in J.*

Proof The stabilizer of a chamber in this J truncation corresponds to the stabilizer of a flag of type J in the geometry $\mathsf{IMG}(\mathcal{T})$. By Theorem 4.2, this is the automorphism group of a direct sum of inductively minimal geometries corresponding to the connected components of $\mathcal{T}$ after removing the edges corresponding to J together with the vertices which belong only to edges in J. □

Theorem 4.8 *Let Γ be an $(IP)_2$ truncation of rank $n-2$ of an inductively minimal geometry of rank $n-1$. The alternating group of degree n acts flag-transitively on Γ.*

Proof Let J be the set of types of Γ. By Theorem 4.5, the subtree $\mathcal{S}$ generated by J must be connected. This shows that $\mathcal{S}$ was obtained from $\mathcal{T}$ by removing one edge e containing a leaf. Consider two chambers C_1 and C_2 in Γ. These are contained in two chambers $\overline{C_1}$ and $\overline{C_2}$ of the inductively minimal geometry $\overline{\Gamma}$ on which S_n acts flag-transitively. Hence we can find $\sigma \in S_n$ with $\sigma(C_1) = C_2$. If σ is not an even permutation, use the fact that C_2 is contained in exactly two chambers (because $\overline{\Gamma}$ is thin, see Theorem 4.2) of $\overline{\Gamma}$ namely $\overline{C_2} = C_2 \cup \{x\}$ and $C_2 \cup \{x'\}$ where x and x' are of type e. The stabilizer of C_2 in $\overline{\Gamma}$ is isomorphic to the symmetric group of degree 2 acting faithfully on the elements x and x' which can be identified with two elements s and s' of the set S on which $\overline{\Gamma}$ is constructed. Hence there is a transposition $\tau \in S_n$ mapping x to x' and fixing C_2. Now $\tau \circ \sigma$ is an even permutation mapping C_1 to C_2. □

Since the alternating group of degree n acts flag-transitively on $(IP)_2$ truncations of rank $n-2$ of inductively minimal geometries of rank $n-1$, it is clear that it also acts flag-transitively on lower rank $(IP)_2$ truncations. This means that all proper $(IP)_2$ truncations of inductively minimal geometries of rank 7 will appear in the list of geometries satisfying our axioms for A_8.

Below we give the diagrams of all RWPRI and $(IP)_2$ geometries for the alternating group of degree 8 which are truncations of inductively minimal geometries. These diagrams can be completely described thanks to theoretical work done in [11].

The automorphism group of each J-truncation is S_8 and the correlation group is the direct product of S_8 with the group of automorphisms of $\mathcal{T}$ which induce an automorphism on the subtree $\mathcal{S}$ generated by the set J of edges. One proves this in the same way as in Theorem 4.4. The Borel subgroup of a truncation is the intersection with A_8 of the group given by Theorem 4.7.

Below we list all proper truncations of inductively minimal geometries of rank 7. As shown above these are geometries on which A_8 acts flag-transitively and which satisfy all of our axioms. For each of the 23 trees on 8 vertices, we first give the basic diagram of the corresponding inductively minimal geometry. After that, the diagrams of the truncations of that inductively minimal geometry are listed.

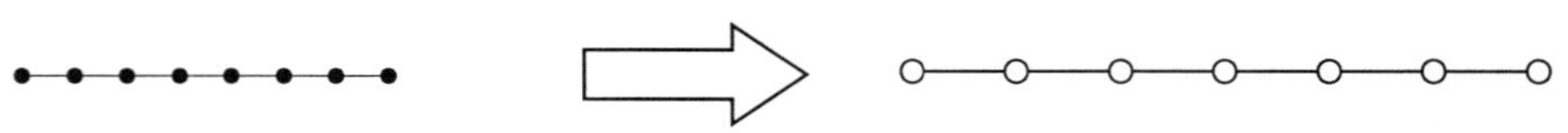

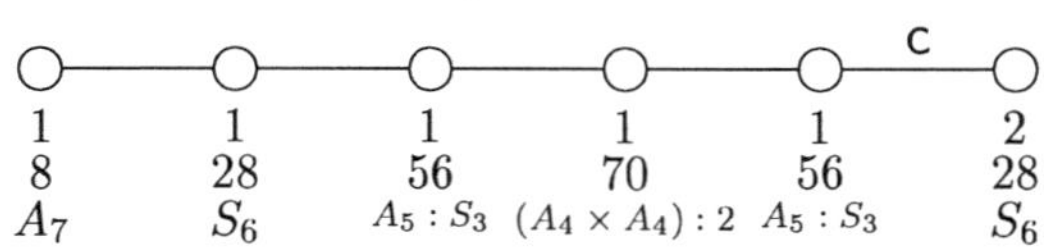

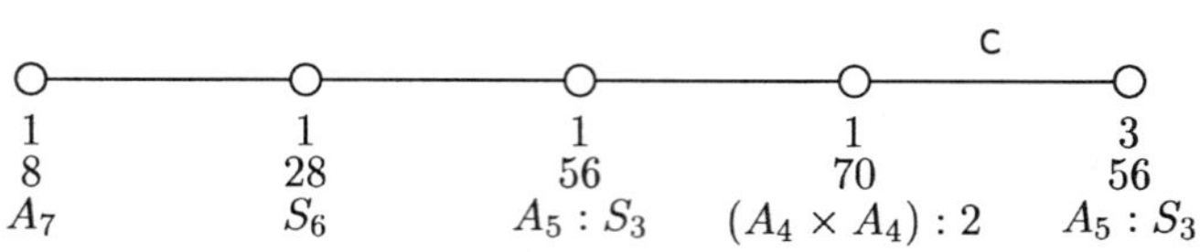

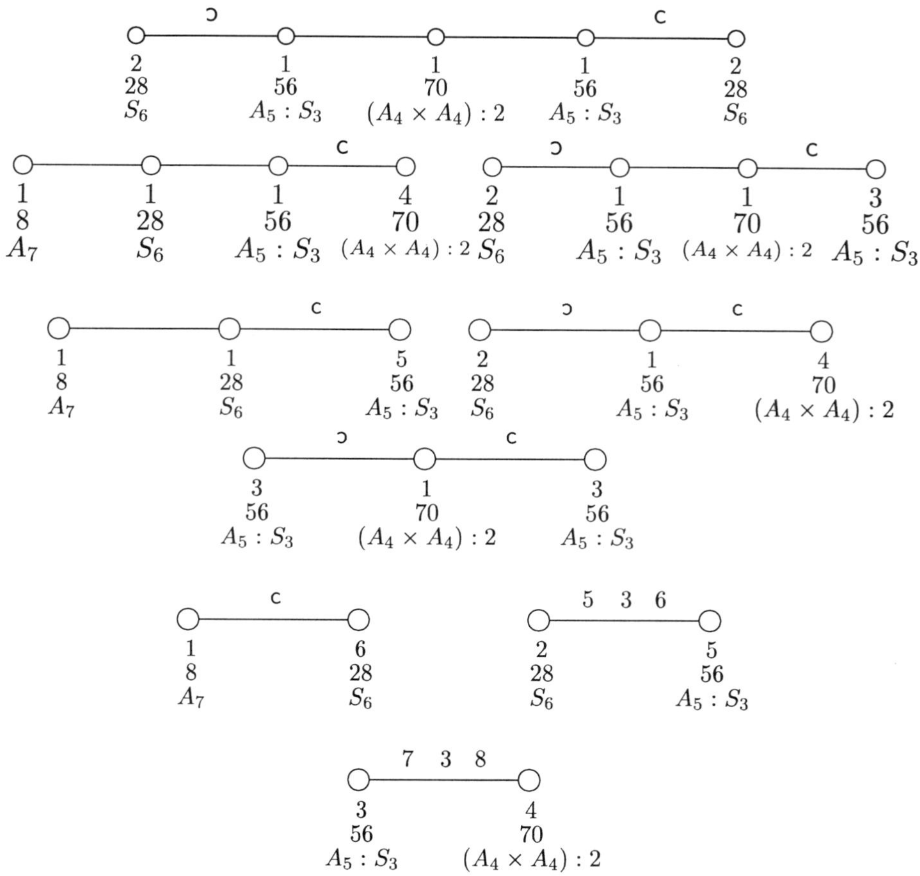

c
2 28 S6
1 56 A5 : S3
1 70 (A4 × A4) : 2
1 56 A5 : S3
2 28 S6
c
1 8 A7
1 28 S6
1 56 A5 : S3
4 70 (A4 × A4) : 2
c
2 28 S6
1 56 A5 : S3
1 70 (A4 × A4) : 2
3 56 A5 : S3
c
1 8 A7
1 28 S6
5 56 A5 : S3
c
2 28 S6
1 56 A5 : S3
4 70 (A4 × A4) : 2
c
3 56 A5 : S3
1 70 (A4 × A4) : 2
3 56 A5 : S3
c
1 8 A7
6 28 S6
5 3 6
2 28 S6
5 56 A5 : S3
7 3 8
3 56 A5 : S3
4 70 (A4 × A4) : 2

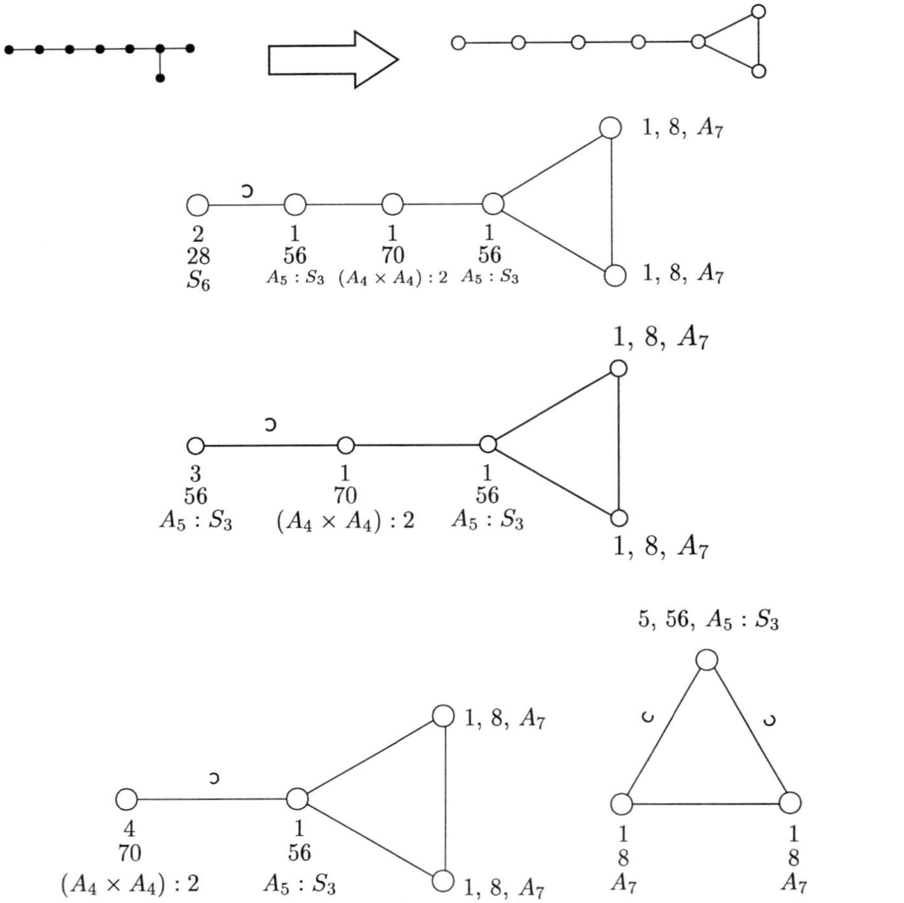
1, 8, A_7
2
28
S_6
1
56
$A_5 : S_3$
1
70
$(A_4 \times A_4) : 2$
1
56
$A_5 : S_3$
1, 8, A_7
1, 8, A_7
3
56
$A_5 : S_3$
1
70
$(A_4 \times A_4) : 2$
1
56
$A_5 : S_3$
1, 8, A_7
5, 56, $A_5 : S_3$
1, 8, A_7
4
70
$(A_4 \times A_4) : 2$
1
56
$A_5 : S_3$
1, 8, A_7
1
8
A_7
1
8
A_7

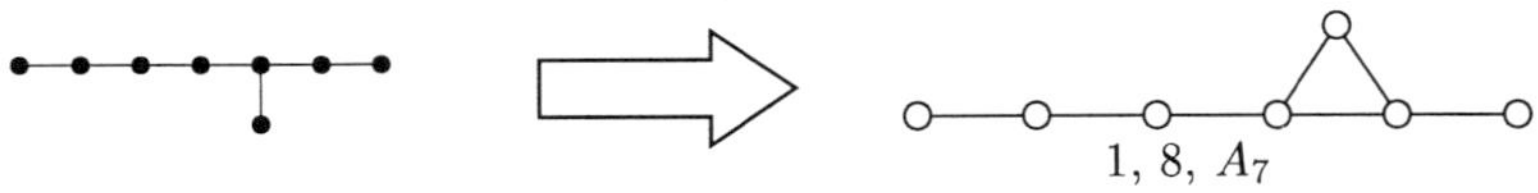
1, 8, A7

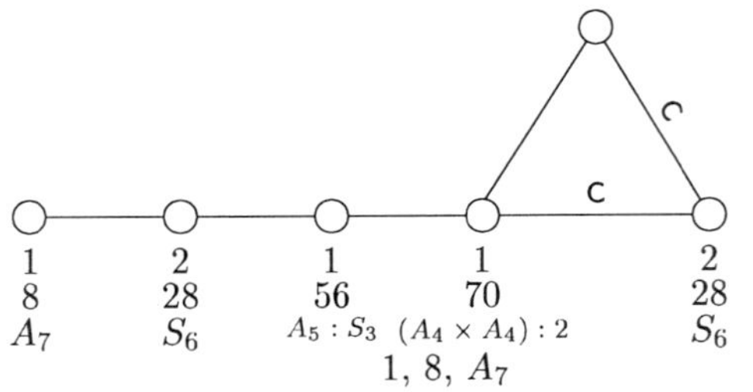
1, 8, A7
1 8 A7
2 28 S6
1 56 A5 : S3
1 70 (A4 × A4) : 2
2 28 S6

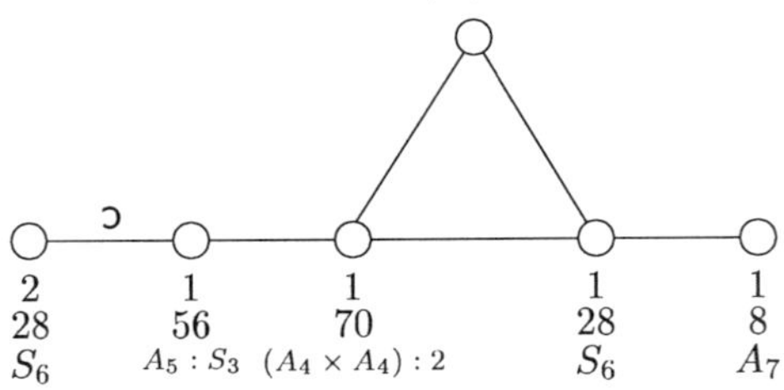
2 28 S6
1 56 A5 : S3
1 70 (A4 × A4) : 2
1 28 S6
1 8 A7

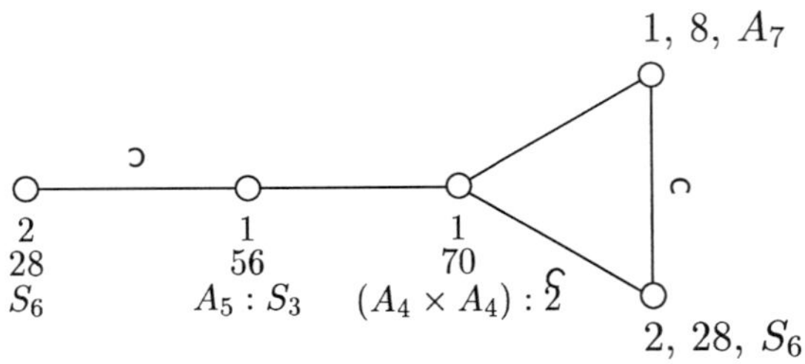
1, 8, A7
2 28 S6
1 56 A5 : S3
1 70 (A4 × A4) : 2
2, 28, S6

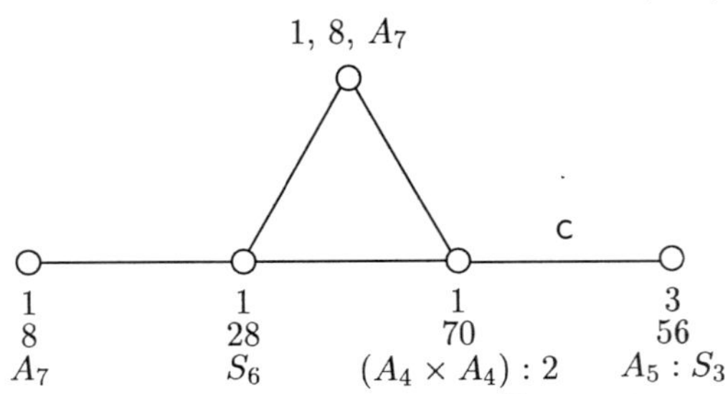
1, 8, A7
1 8 A7
1 28 S6
1 70 (A4 × A4) : 2
3 56 A5 : S3

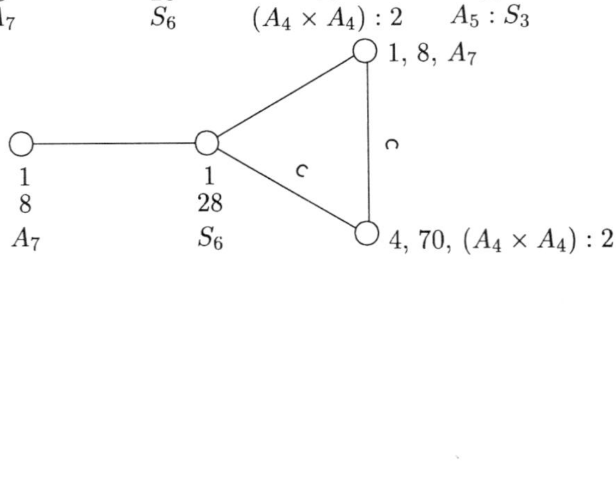
1, 8, A7
1 8 A7
1 28 S6
4, 70, (A4 × A4) : 2

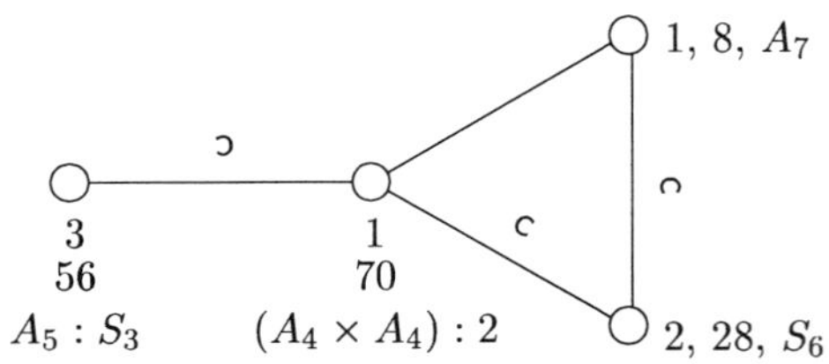
1, 8, A_7
3
56
$A_5 : S_3$
1
70
$(A_4 \times A_4) : 2$
2, 28, S_6

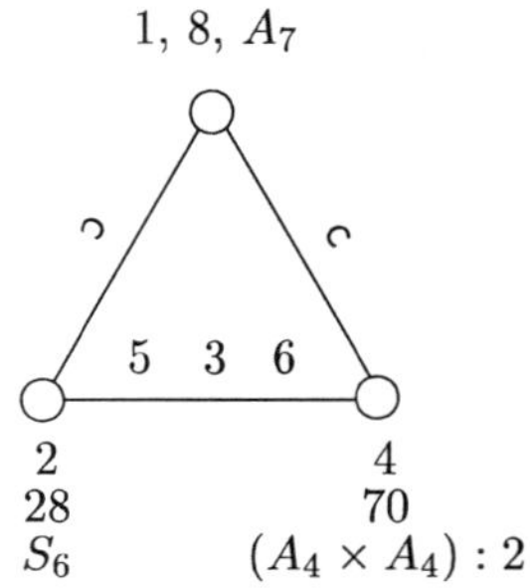
1, 8, A_7
5 3 6
2
28
S_6
4
70
$(A_4 \times A_4) : 2$

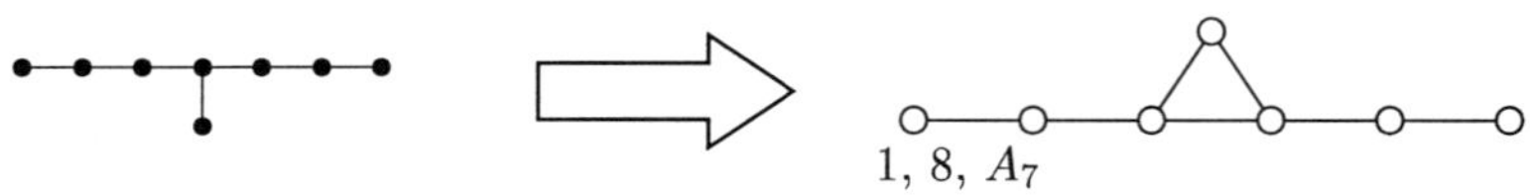
1, 8, A_7

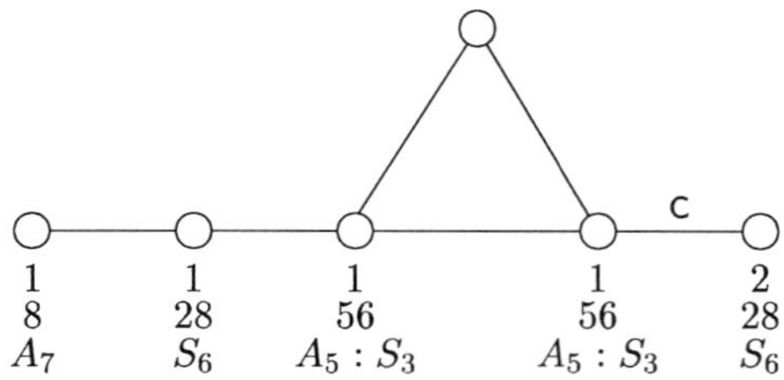
1
8
A_7
1
28
S_6
1
56
$A_5 : S_3$
1
56
$A_5 : S_3$
2
28
S_6

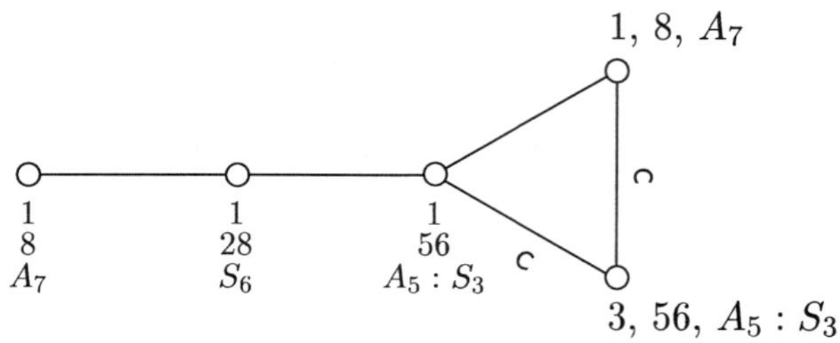
1, 8, A_7
1
8
A_7
1
28
S_6
1
56
$A_5 : S_3$
3, 56, $A_5 : S_3$

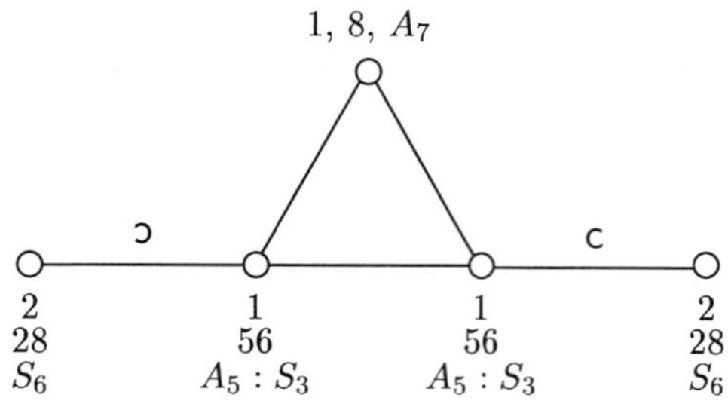
1, 8, A_7
2
28
S_6
1
56
$A_5 : S_3$
1
56
$A_5 : S_3$
2
28
S_6

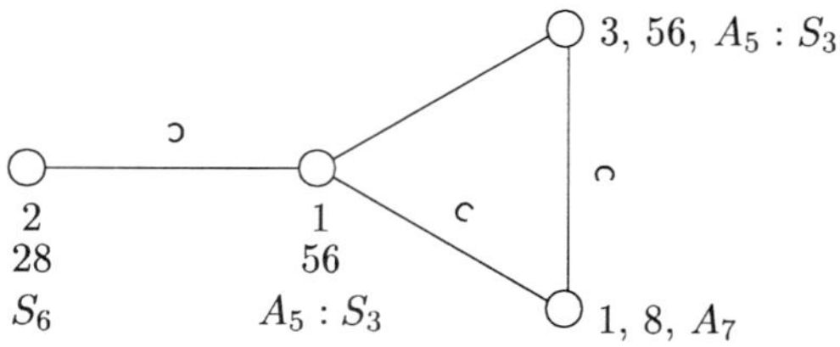
3, 56, $A_5 : S_3$
2
28
S_6
1
56
$A_5 : S_3$
1, 8, A_7

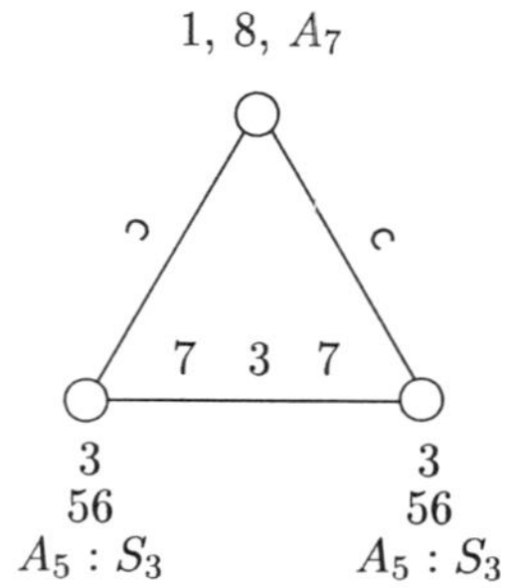
1, 8, A_7
7 3 7
3
56
$A_5 : S_3$
3
56
$A_5 : S_3$

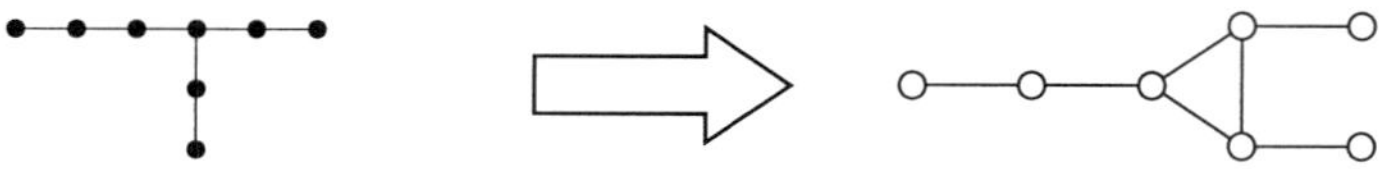

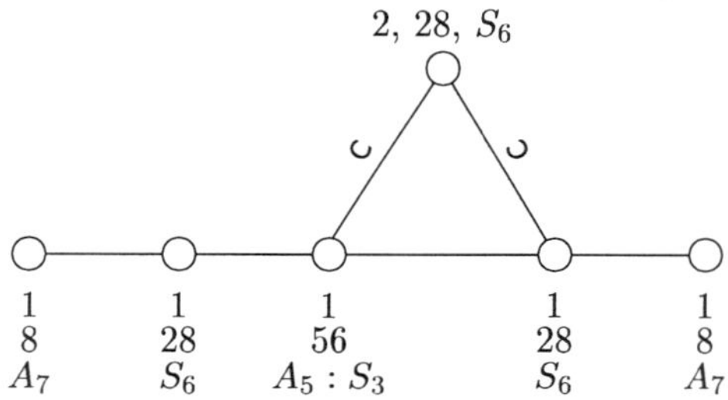
2, 28, S_6
1
8
A_7
1
28
S_6
1
56
$A_5 : S_3$
1
28
S_6
1
8
A_7

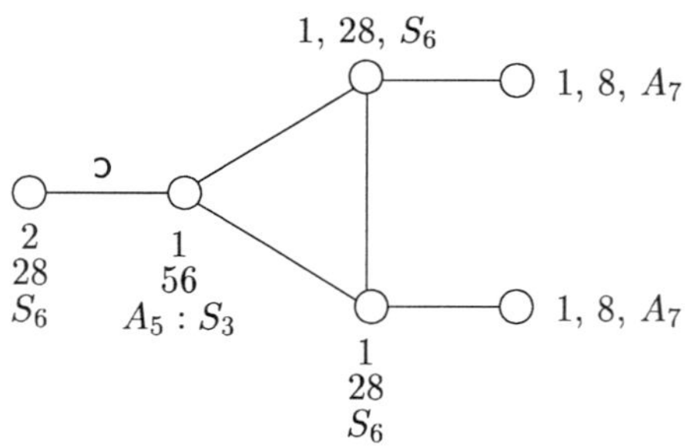
1, 28, S_6
1, 8, A_7
2
28
S_6
1
56
$A_5 : S_3$
1, 8, A_7
1
28
S_6

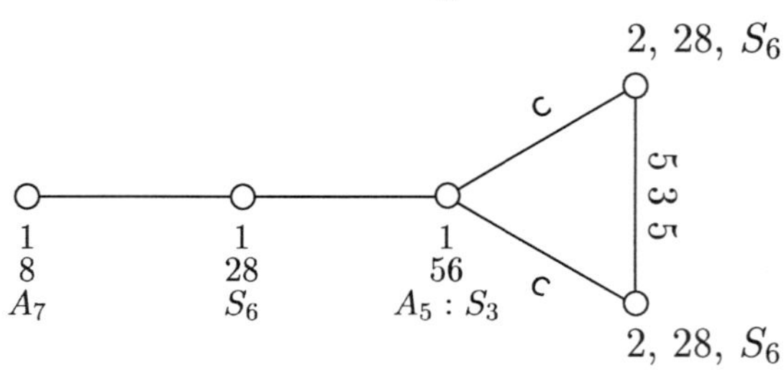
2, 28, S_6
5 3 5
1
8
A_7
1
28
S_6
1
56
$A_5 : S_3$
2, 28, S_6

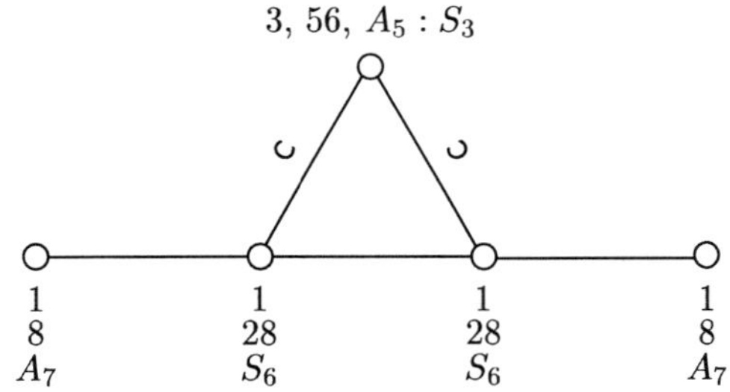
3, 56, $A_5 : S_3$
c
c
1
8
A_7
1
28
S_6
1
28
S_6
1
8
A_7

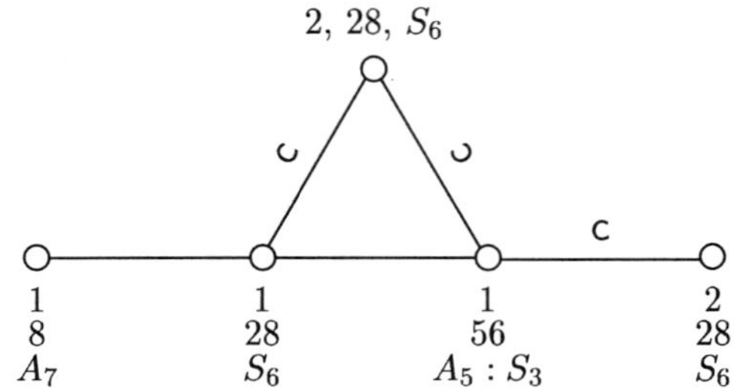
2, 28, S_6
c
c
c
1
8
A_7
1
28
S_6
1
56
$A_5 : S_3$
2
28
S_6

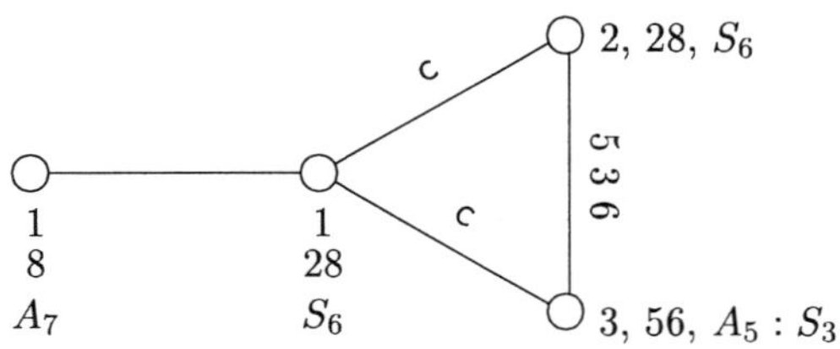
2, 28, S_6
c
c
5 3 6
1
8
A_7
1
28
S_6
3, 56, $A_5 : S_3$

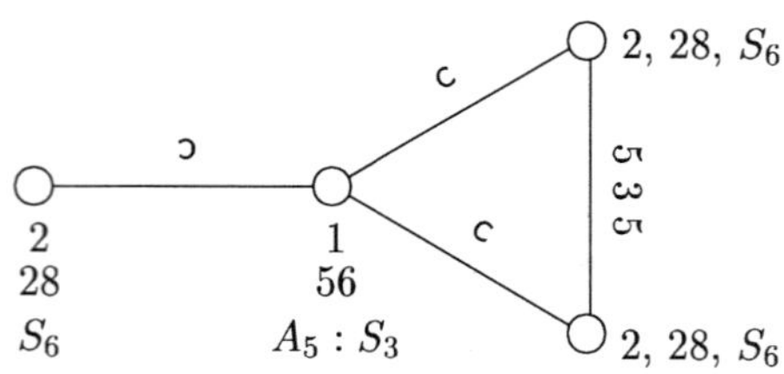
2, 28, S_6
c
c
c
5 3 5
2
28
S_6
1
56
$A_5 : S_3$
2, 28, S_6

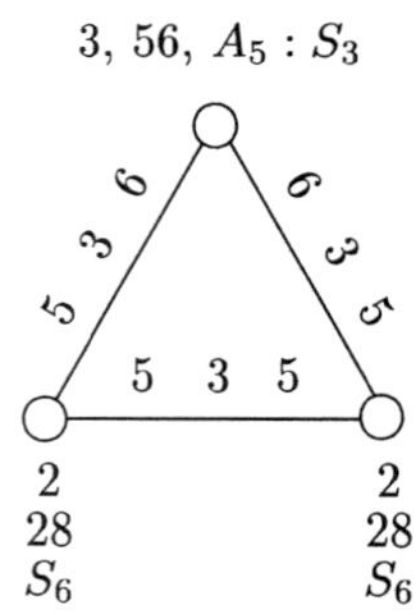
3, 56, $A_5 : S_3$
5 3 6
6 3 5
5 3 5
2
28
S_6
2
28
S_6

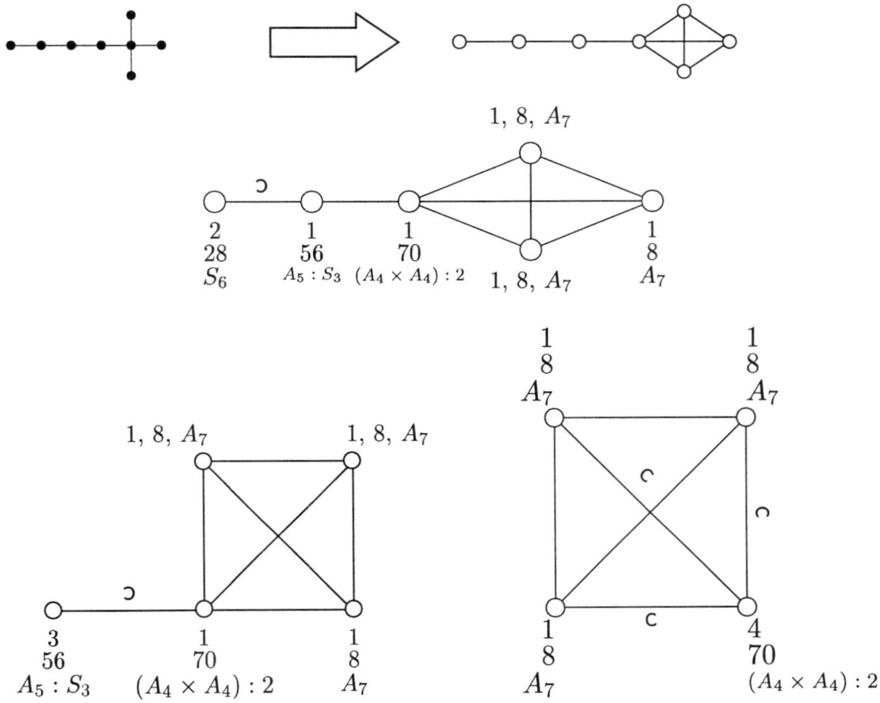
1, 8, A_7
2
28
S_6
1
56
$A_5 : S_3$
1
70
$(A_4 \times A_4) : 2$
1, 8, A_7
1
8
A_7
1, 8, A_7
1, 8, A_7
3
56
$A_5 : S_3$
1
70
$(A_4 \times A_4) : 2$
1
8
A_7
1
8
A_7
1
8
A_7
1
8
A_7
4
70
$(A_4 \times A_4) : 2$

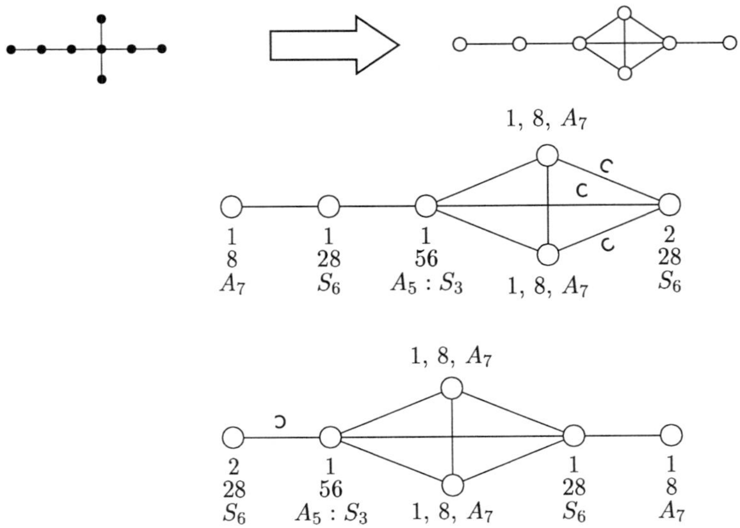
1, 8, A_7
1
8
A_7
1
28
S_6
1
56
$A_5 : S_3$
1, 8, A_7
2
28
S_6
1, 8, A_7
2
28
S_6
1
56
$A_5 : S_3$
1, 8, A_7
1
28
S_6
1
8
A_7

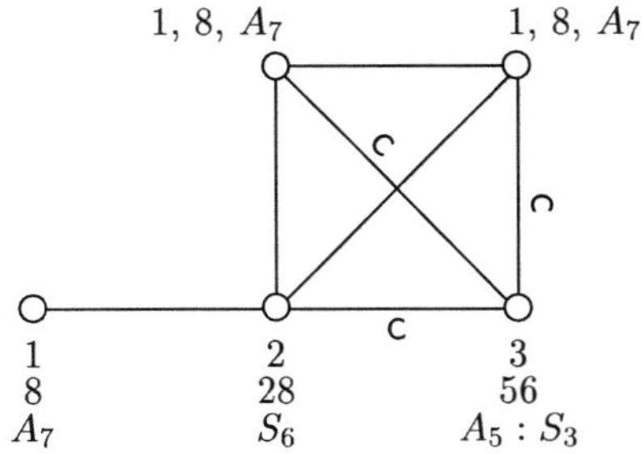
1, 8, A_7
1, 8, A_7
c
c
c
1
8
A_7
2
28
S_6
3
56
$A_5 : S_3$

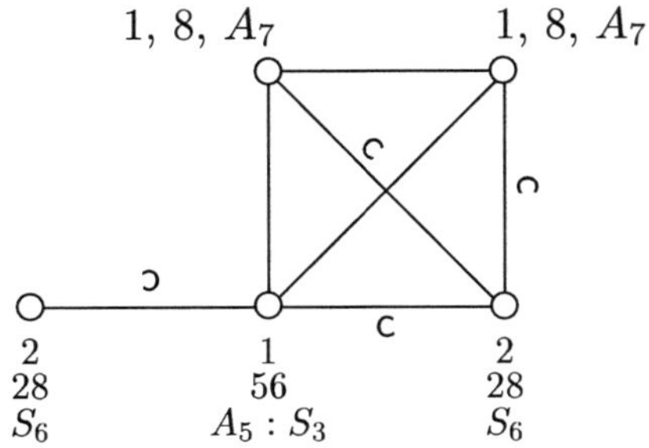
1, 8, A_7
1, 8, A_7
c
c
c
c
2
28
S_6
1
56
$A_5 : S_3$
2
28
S_6

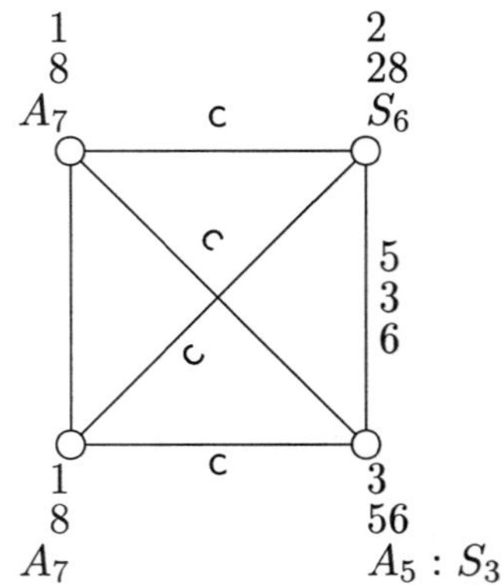
1
8
A_7
c
2
28
S_6
c
5
3
6
c
c
1
8
A_7
3
56
$A_5 : S_3$

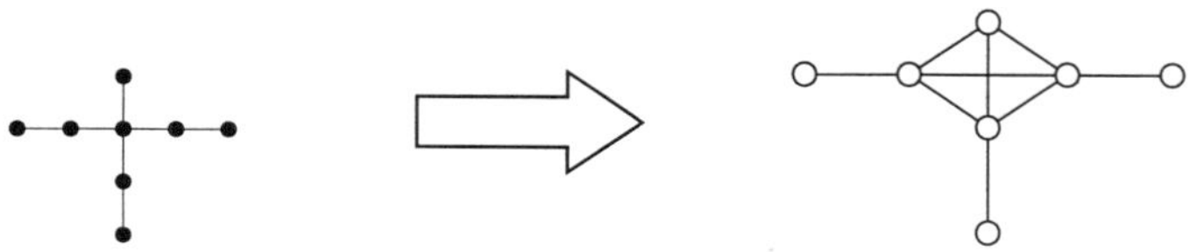

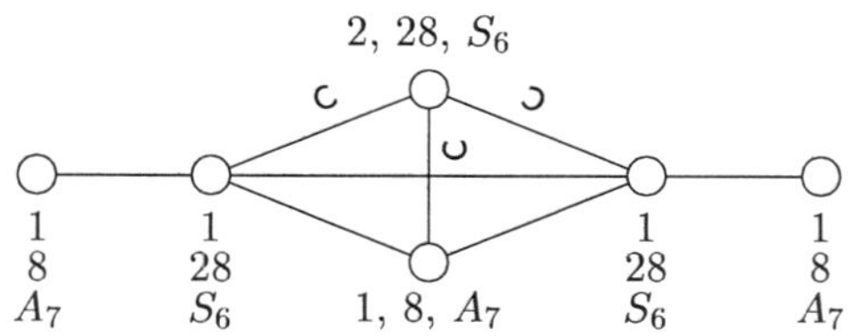
2, 28, S6
c
c
c
1
8
A7
1
28
S6
1, 8, A7
1
28
S6
1
8
A7

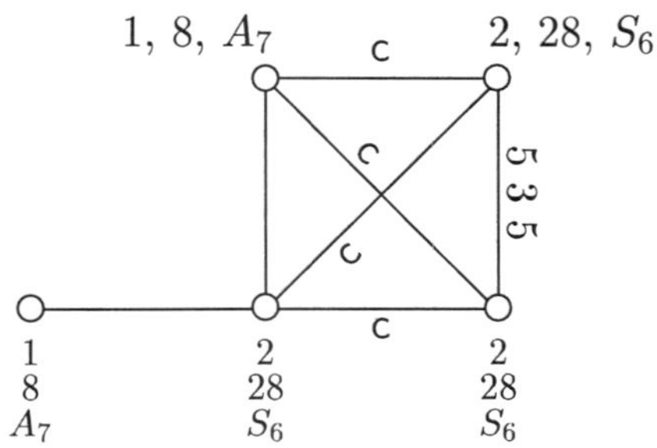
1, 8, A7
c
2, 28, S6
c
5 3 5
c
c
1
8
A7
2
28
S6
2
28
S6

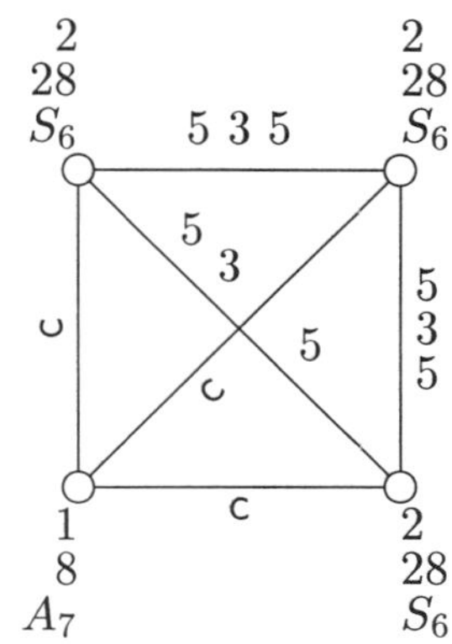
2
28
S6
5 3 5
2
28
S6
5
3
c
5
5
3
5
c
c
1
8
A7
2
28
S6

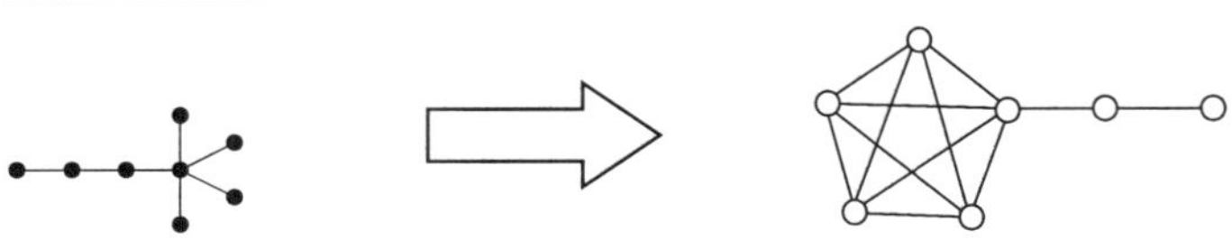

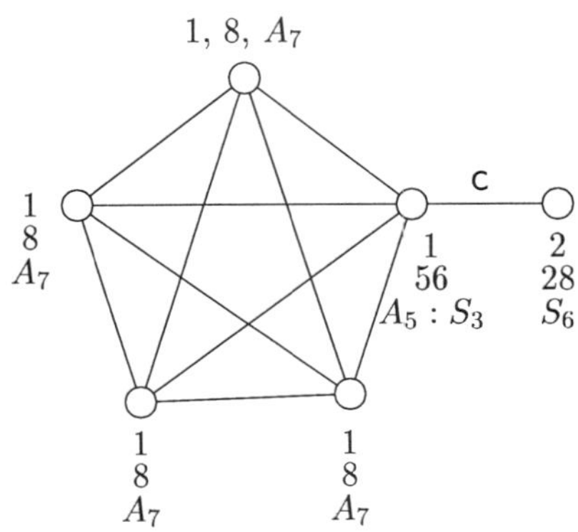
1, 8, A7
1
8
A7
c
1
56
A5 : S3
2
28
S6
1
8
A7
1
8
A7

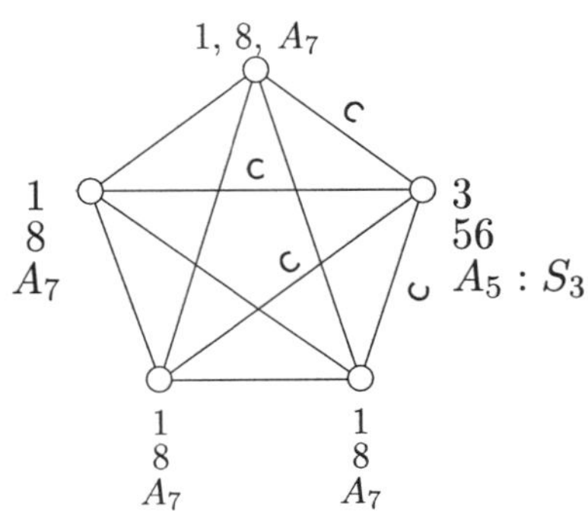
1, 8, A7
c
c
1
8
A7
3
56
A5 : S3
c
c
1
8
A7
1
8
A7

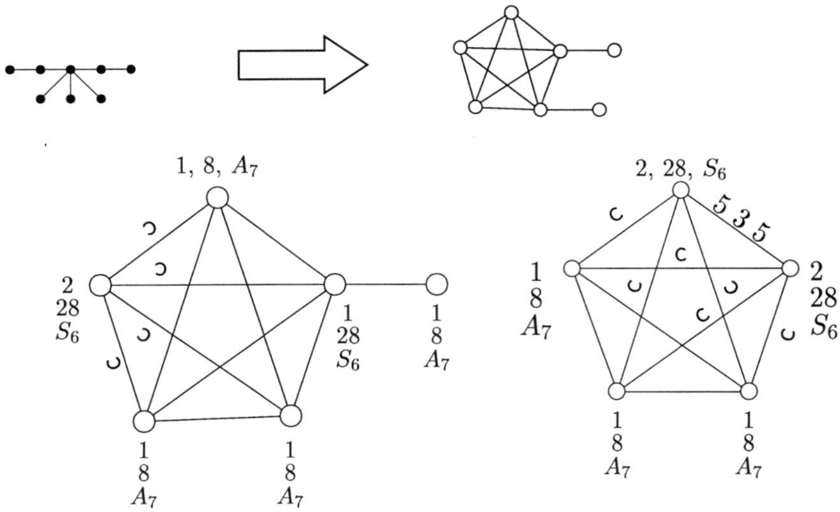
1, 8, A_7
2
28
S_6
1
28
S_6
1
8
A_7
1
8
A_7
1
8
A_7
2, 28, S_6
5 3 5
1
8
A_7
2
28
S_6
1
8
A_7
1
8
A_7

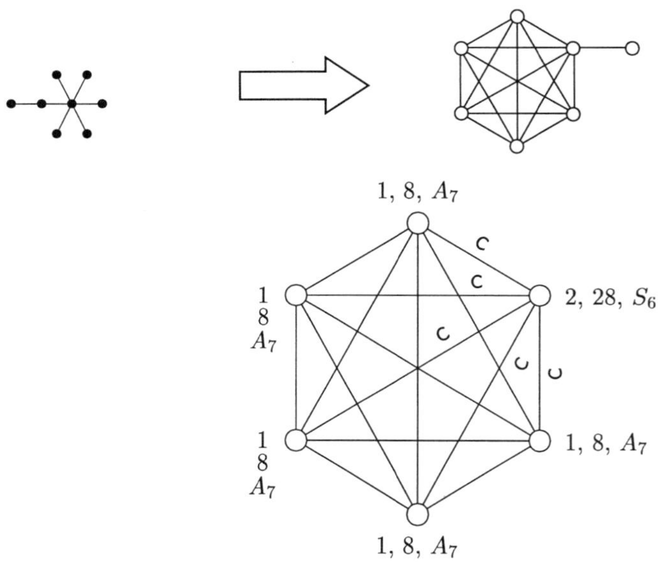
1, 8, A_7
1
8
A_7
2, 28, S_6
1
8
A_7
1, 8, A_7
1, 8, A_7

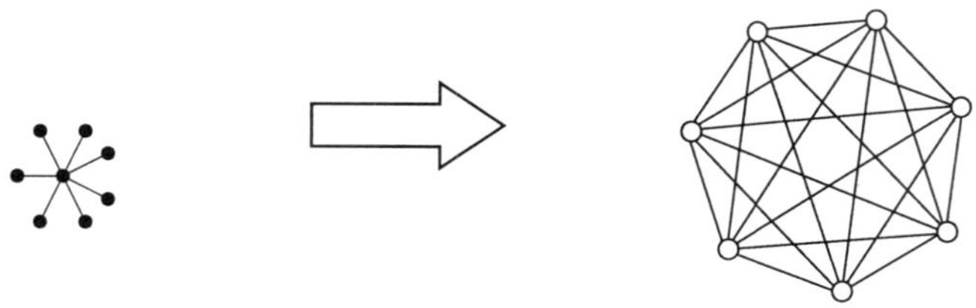

No $(IP)_2$ truncations.

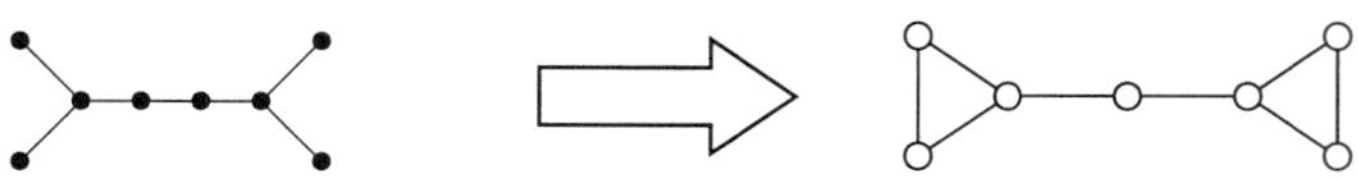

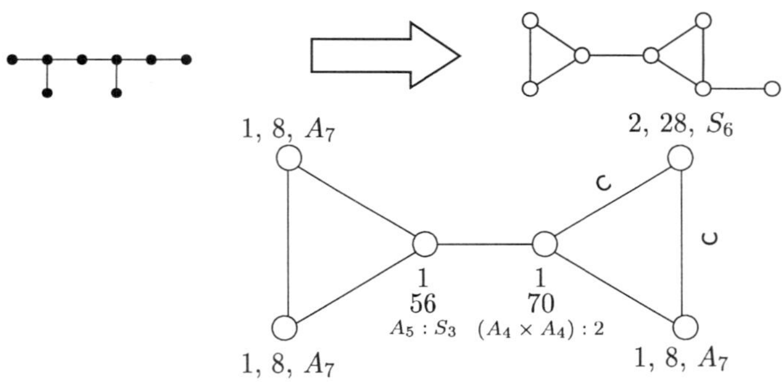

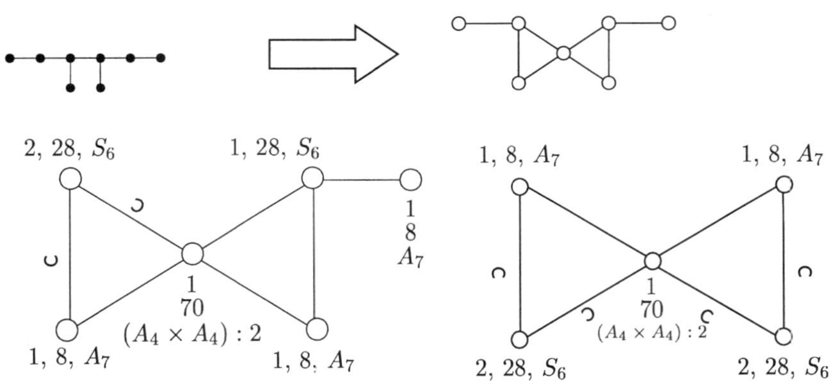

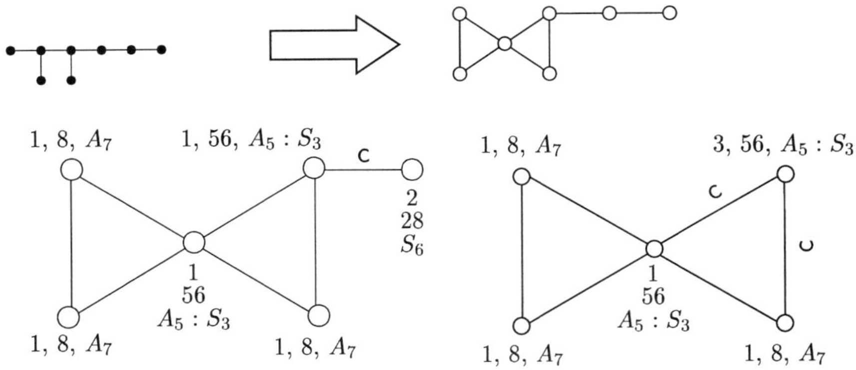
1, 8, A_7
1, 56, $A_5 : S_3$
c
2
28
S_6
1
56
$A_5 : S_3$
1, 8, A_7
1, 8, A_7
1, 8, A_7
3, 56, $A_5 : S_3$
c
c
1
56
$A_5 : S_3$
1, 8, A_7
1, 8, A_7

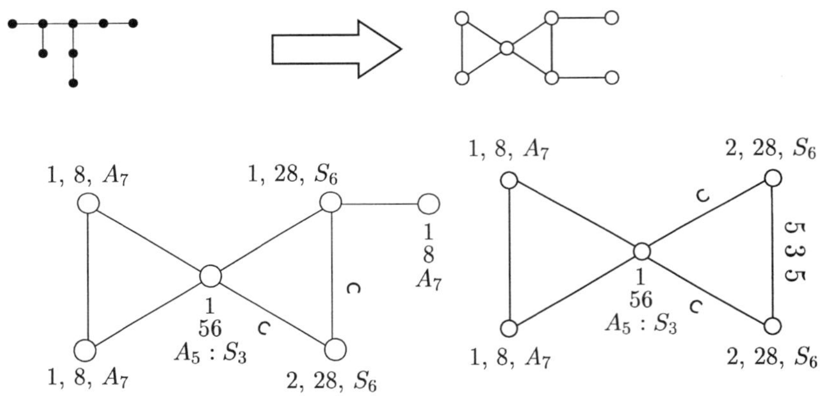
1, 8, A_7
1, 28, S_6
1
8
A_7
c
1
56
$A_5 : S_3$
c
1, 8, A_7
2, 28, S_6
1, 8, A_7
2, 28, S_6
c
535
1
56
$A_5 : S_3$
c
1, 8, A_7
2, 28, S_6

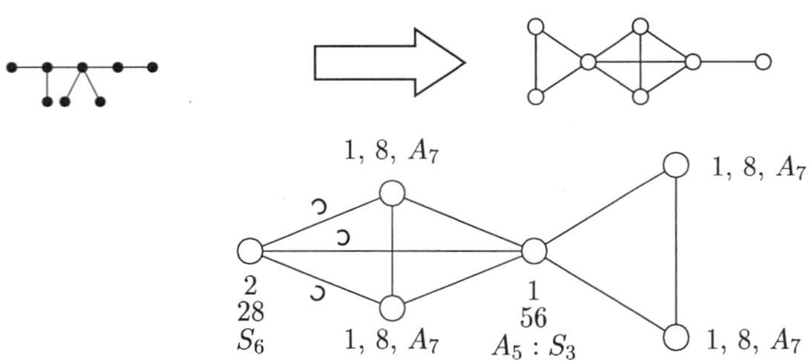
1, 8, A_7
1, 8, A_7
c
c
c
2
28
S_6
1, 8, A_7
1
56
$A_5 : S_3$
1, 8, A_7

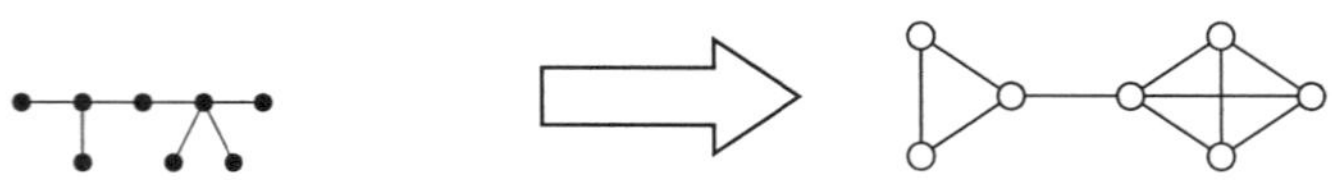

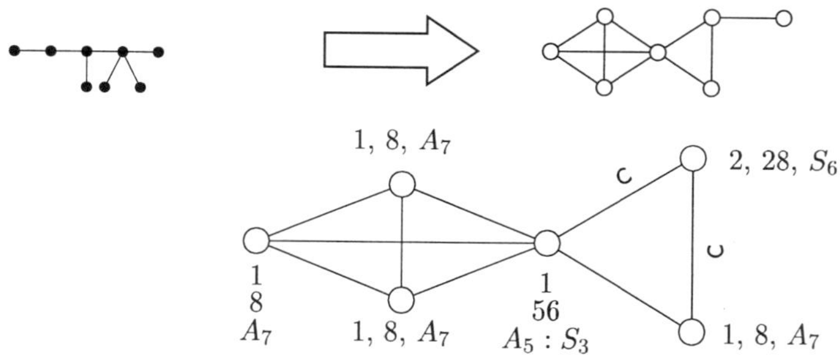

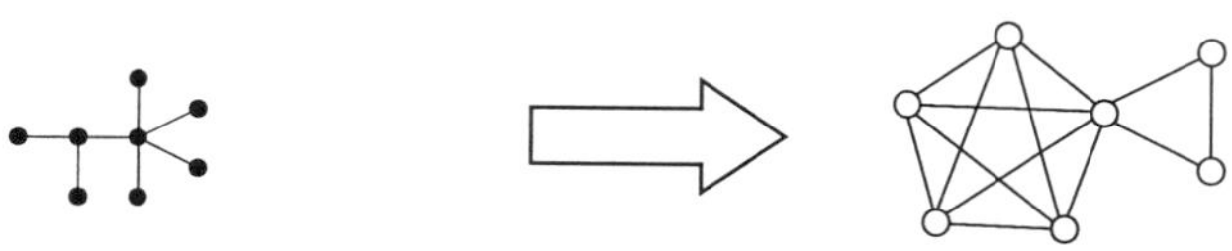

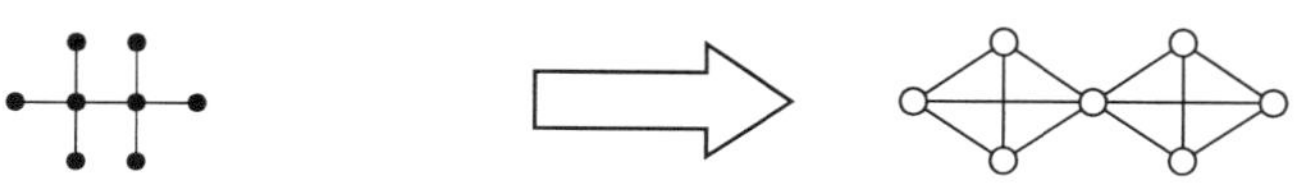

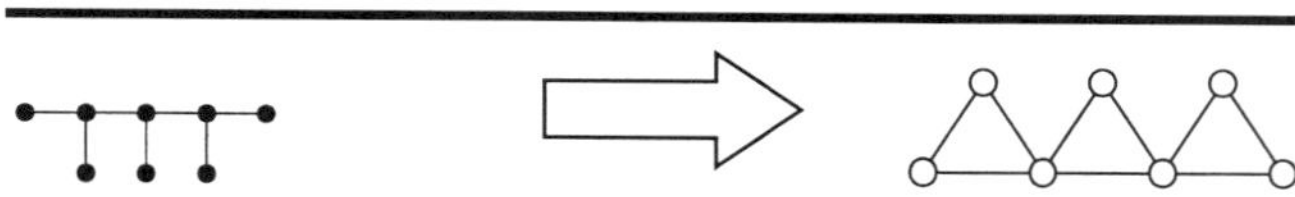

Remark 4.9 Some truncations can be obtained from different inductively minimal geometries. However they are mentioned only once in our list. That is why there are no truncations listed for the last inductively minimal geometries. All truncations of these geometries have already appeared earlier. The only inductively minimal geometry which has no proper $(IP)_2$ truncation is the one whose diagram is the complete graph on 7 vertices.

5. Petersen replacements

These geometries are generalizations of the vertex-edge geometry of the Petersen graph.

The familiar construction of the Petersen graph on a set of five elements can be generalized to a set S of $2k+1$ elements to define a geometry of rank k which we note $\mathsf{PeS}(k)$ and call *Petersen string* of length k. The elements of type $i \in \{1, 2, \ldots, k\}$ are the sets of i pairwise disjoint pairs of elements of S. Incidence is symmetrised inclusion and the vertex-edge geometry of the Petersen graph corresponds to $\mathsf{PeS}(2)$.

The full automorphism group of $\mathsf{PeS}(k)$ is clearly the symmetric group of degree $2k+1$ and the alternating group A_{2k+1} acts flag-transitively. It is easy to see that the diagram of $\mathsf{PeS}(k)$ is

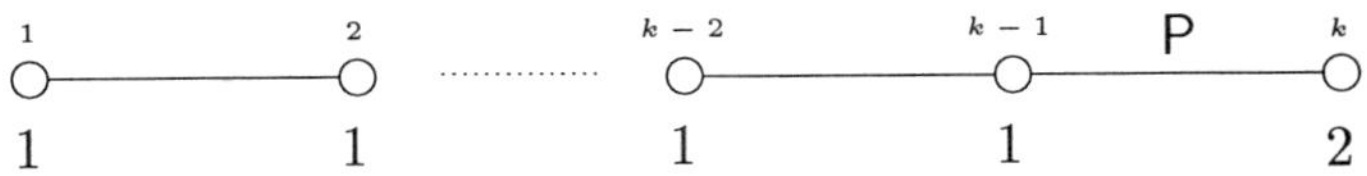

Where the label P above the last edge shows that the residue of type $\{k-1, k\}$ is isomorphic to the vertex-edge geometry of the Petersen graph.

The following theorem was proved in [12].

Theorem 5.1 *Given an inductively minimal geometry* Γ *and a non-cutting vertex* i *of its basic diagram* $(I, \sim)$ *together with a connected component* I_1 *of* $I \setminus \{i\}$ *such that* $|I \setminus I_1|$ *is an even integer* $2k$ *greater than* 2, *the Petersen replacement of* Γ *with respect to* i *and* I_1 *is a Petersen geometry whose diagram is obtained from* $(I, \sim)$ *by replacing the component* $I \setminus I_1$ *by a Petersen string diagram of length* k *and modifying some labels in* $I_1 \cup \{i\}$. *The symmetric and alternating group of degree* $|I|+1$ *both act flag-transitively on this new geometry which also satisfies the RWPRI and* $(IP)_2$ *properties. The Borel subgroup of this Petersen replacement in the alternating group is of isomorphism type* 2^{k-1}. □

From the family of inductively minimal geometries we get a new family of Petersen geometries. This family generalizes the family given in [2] as item (88) and which is in fact the family of Petersen strings. For the group A_8 we obtain the following Petersen geometries or truncations of them.

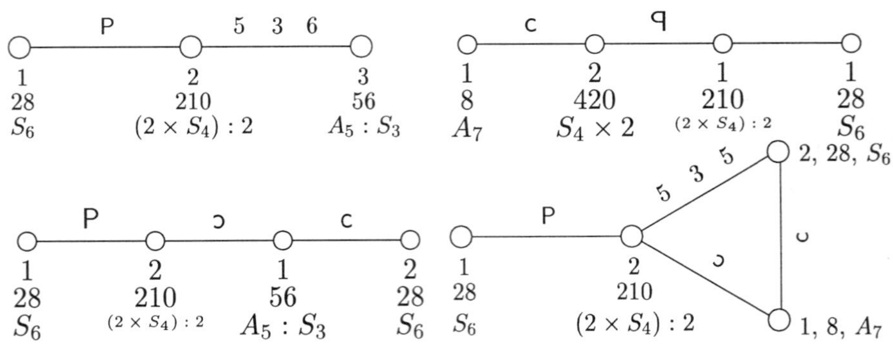

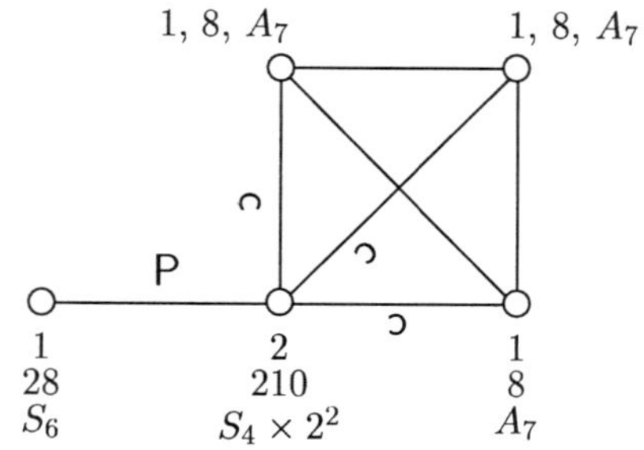

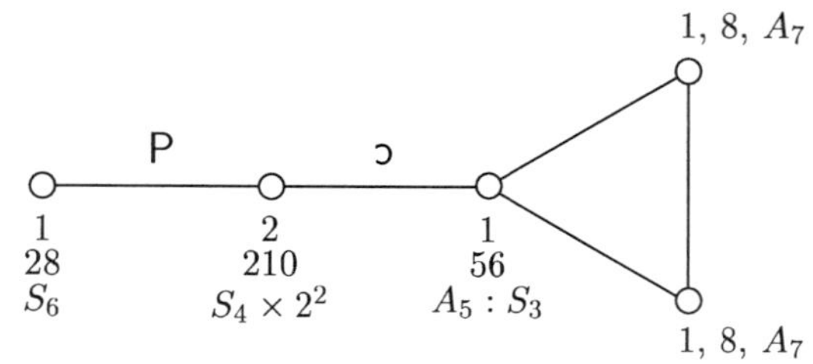

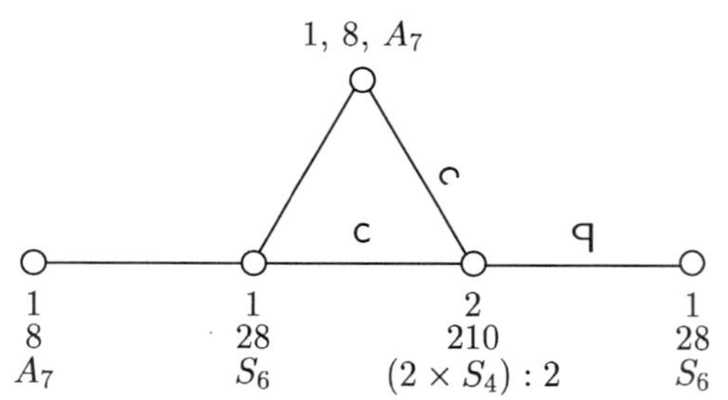

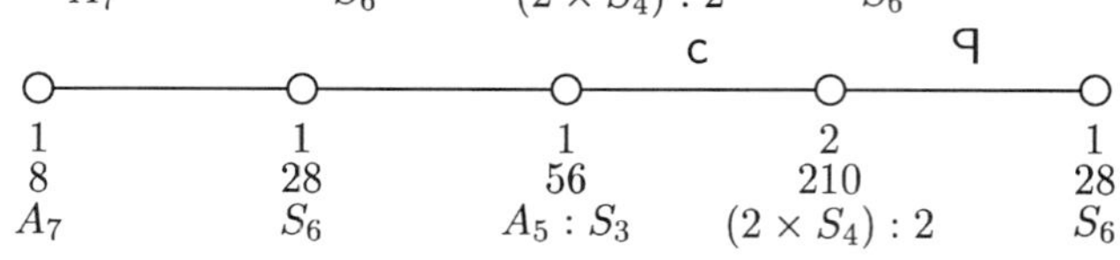

Before going to the next section, we explain the numbering of types in the constructions we give.

Remark 5.2 When giving a construction, we describe the elements of every type in the geometry and the incidence relation. The types are not written on the diagram but are assumed to be numbered 0, 1, 2, ... from left to right on the diagram. When an incidence is not mentioned, natural incidence (that is, containment, inclusion, ...) is assumed.

6. Generalizations of the Petersen geometry

In our list of geometries, we meet a generalization of a Petersen replacement. It has the following diagram.

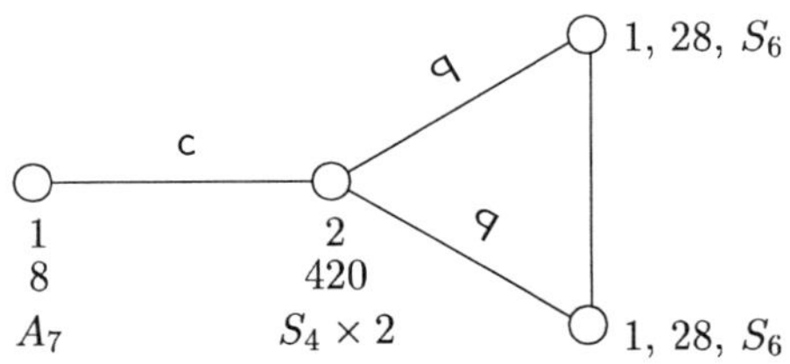

The geometry $(X, *, t)$ can be constructed on a set S of 8 elements as follows. The elements of type 0 are the elements of S, the elements of type 1 are the 3-subsets of pairwise disjoint pairs of S and the sets $t^{-1}(2)$ and $t^{-1}(3)$ are two disjoint copies of the set of pairs of S. An element of type 0 is incident with an element $\mathcal{A}$ of type 1 if and only if it is not contained in the union $\cup\mathcal{A}$. For incidence of an element of type 0 with an element of type 2 or 3 we also say that the element of type 0 must not belong to the element of type 2 or 3. An element P of type 2 or 3 is incident with an element $\mathcal{A}$ of type 1 if and only if $P \in \mathcal{A}$. Two elements P_1 and P_2 of types 2 and 3 respectively are incident if and only if they are disjoint.

The automorphism group of this geometry is S_8, its correlation group is $S_8 \times 2$ and the Borel subgroup is 2^2.

This geometry is also related to the second Petersen replacement Γ appearing in the previous section. The elements of types 0 and 1 are the same as in Γ, the sets of elements of types 2 and 3 are copies of the set of elements of type 3 in Γ. We remark that the elements of type 2 in Γ are pairs of disjoint 2-subsets of the set S. Incidences in our generalized Petersen geometry are the same as in Γ with the addition that two elements of respective type 2 and 3 are incident whenever there exists an element of type 2 in Γ which is incident to both, seen as elements of type 3 in Γ.

Remark 6.1 The construction just described is known as *doubling*. We say that our generalized Petersen geometry is a doubling of the second geometry of previous section. For more information on doublings, we refer to [25] and [22].

7. Polar towers

Let S be a set of $2k+2 \geqslant 6$ elements. We define a geometry $\mathsf{PT}(k)$ of rank k as follows. For $i \in \{1, 2, \dots, k-1\}$, the elements of type i are the sets of i pairwise disjoint 2-subsets of S, as for the Petersen string. The elements of type k are the sets of $k+1$ disjoint 2-subsets. It is easy to verify that this defines a geometry with the following diagram.

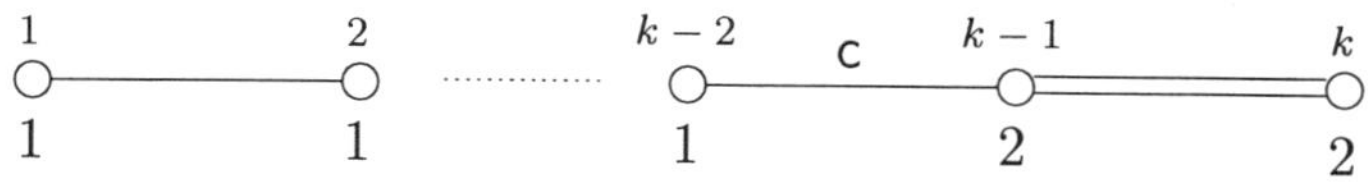

We remark that a residue of type $\{k-1, k\}$ in $\mathsf{PT}(k)$ is isomorphic to the *symplectic quadrangle* $W(2)$ corresponding to $\mathsf{PT}(2)$ (for details on generalized quadrangles, we refer to [26]). The geometries constructed here are members of a family of geometries called *polar towers*. Polar towers were classified by Meixner in [23] and in that classification, $\mathsf{PT}(k)$ corresponds to the polar tower of type $\mathbf{G}(T_{2l+6})$ with $l = k-2$. Polar towers satisfy the axioms used in this paper and our list contains one of them, $\mathsf{PT}(3)$, whose automorphism and correlation group are both isomorphic to S_8. Its Borel subgroup is $D_8 \times 2$.

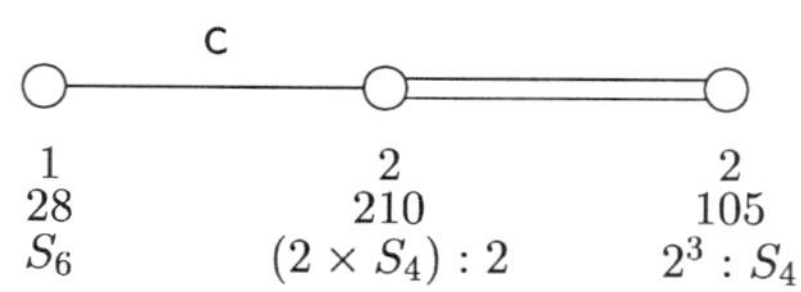

The construction of $\mathsf{PT}(k)$ is similar to the construction of Petersen geometries given in [12]. This similarity will be used in a forthcoming paper to construct "$W(2)$-geometries" with nonlinear diagram.

8. Geometrical objects related to A_8

The geometries of A_8 which are left now stem from the sporadic isomorphisms $A_8 \cong PSL(4,2) \cong PO^+(6,2)$. We first recall some important facts about these isomorphisms and gather information on the projective spaces $PG(3,2)$ and $PG(5,2)$.

A good reference on this subject is [20] but we must warn the reader that our notation differs from the one used in that reference.

8.1. The projective space $PG(3,2)$

The projective space $PG(3,2)$ of dimension 3 over the field with two elements has 15 points, 35 lines and 15 planes. The planes are isomorphic to the Fano plane $PG(2,2)$ and hence each one contains 7 points and 7 lines. By duality, there are 7 lines and 7 planes through any point of $PG(3,2)$. The lines have 3 points and that is also the number of planes containing a given line. From this information, we deduce that the number of flags of rank 2 is 105, no matter which two types of elements we choose. Given a line L, there are $3 \cdot 6 = 18$ lines intersecting it and hence we have 16 lines skew to L. It follows that the number of pairs of skew lines is $\frac{1}{2} \cdot 35 \cdot 16 = 280$. The number of *ordered* pairs of skew lines is twice as large. A line and a plane in $PG(3,2)$ have at least one point in common and two different planes meet in a line. A *spread* of $PG(3,2)$ is a partition of its point set into lines. Hence a spread of $PG(3,2)$ is a set of 5 mutually disjoint lines. There are 56 different spreads in $PG(3,2)$. The automorphism group of $PG(3,2)$ is denoted by $PSL(4,2)$ and is known to be isomorphic to A_8 since Jordan [21].

8.2. The Klein quadric

It is well known that the lines of $PG(3,2)$ correspond to the 35 points of a hyperbolic quadric $Q^+(5,2)$ with equation $X_0X_1+X_2X_3+X_4X_5=0$ in the five-dimensional projective space $PG(5,2)$. This correspondence is known as the *Klein correspondence* and the hyperbolic quadric is then called the *Klein quadric.* In what follows, the Klein quadric will be denoted by $\mathcal{Q}$. On $\mathcal{Q}$ there are two families of 15 planes each, which correspond to the 15 points and the 15 planes of $PG(3,2)$ respectively. Two different planes of the same family have one point in common and two planes of different families are either disjoint or share one line. A {point, plane}-flag $p * P$ of $PG(3,2)$ appears on $\mathcal{Q}$ as two planes from different families which intersect in a line, representing the three lines in $PG(3,2)$ which are included in P and contain the point p. The Klein correspondence for $PG(3,2)$ is described in detail in [14]. We quote some useful facts. Since $PG(5,2)$ has 63 points, there are 28 points not on $\mathcal{Q}$. These will be called *external points.* A line in $PG(5,2)$ not on $\mathcal{Q}$ can meet $\mathcal{Q}$ in either 0, 1 or 2 points. Such lines will be called *external, tangent* and *bisecant* respectively. The bisecants correspond to pairs of skew lines in $PG(3,2)$, implying that there are 280 of them. The number of external lines and tangents is 56 and 210 respectively. Since $PG(5,2)$ has 651 lines, there remain 105 lines which are contained in $\mathcal{Q}$. These lines correspond to pencils in $PG(3,2)$. A *pencil* in $PG(n,2)$ is a point together with three lines containing that point and lying in the same plane. Through an external point, there are 31 lines. These can be partitioned in 6 external lines, 15 tangents and 10 bisecants. A point on $\mathcal{Q}$ lies on 9 lines contained in $\mathcal{Q}$ in 3 pencils, 16 bisecants and 6 tangents in 2 pencils.

Consider an external line L and two points p and q on it. On each of these points, we have 5 external lines different from L. These 10 external lines intersect in 5 external points different from p and q. This yields a set H of 7 points off $\mathcal{Q}$. These points define 21 lines which are all external. If we choose any two points p' and q' in H, we can make the same construction and the set of 7 points that we obtain will again be H. Such a set of 7 points is called a *heptad* or *maximal exterior set.*

There are 8 different heptads and each external point belongs to exactly two heptads. Any two heptads have exactly one point in common.

Lemma 17.5.1 of [20] states that the 56 spreads of $PG(3,2)$ correspond to the 56 external lines in $PG(5,2)$. On an external line, there are 3 pairs of points which define 3 heptads. This shows that the spreads of $PG(3,2)$ correspond to the 3-subsets of the set of 8 heptads. In a construction given below, we will say that a spread corresponds to *five* heptads, the complement of the 3-subset of heptads corresponding to the spread.

It must be remarked that these 8 heptads are the key to an elegant proof of the sporadic isomorphisms between the alternating group of degree 8 and the group $PSL(4,2)$ via the orthogonal group $PO^+(6,2)$ which is the group of automorphisms of $PG(5,2)$ preserving the Klein quadric.

In the next subsection we study a geometry in detail and give a construction using some of the objects defined above.

Case study

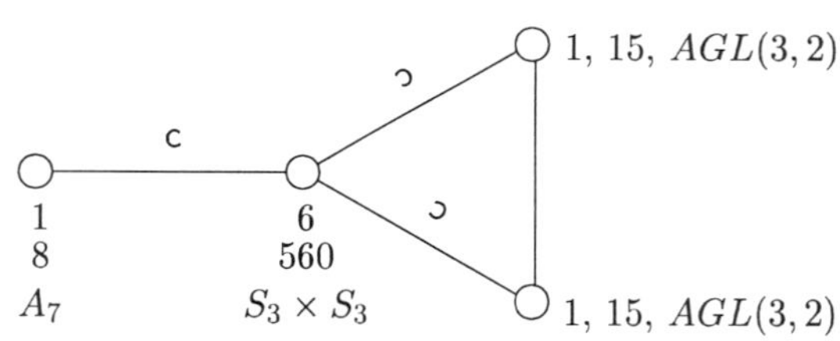

The geometry with the above diagram can be explained as follows. The elements of types 0, 1, 2, 3 are respectively the 8 heptads, the 560 ordered pairs of skew lines, the points and the planes of $PG(3,2)$. A point and a plane are defined to be incident if the point does not belong to the plane. A point and an ordered pair are declared incident if and only if the point belongs to the first line of that ordered pair. A plane is incident with an ordered pair if the second line of the pair lies in the plane. By definition all heptads are incident with all points and planes. We know that an ordered pair of skew lines corresponds to a pair of points on the Klein quadric (in fact an ordered pair) such that the line connecting them is a bisecant. This bisecant has one external point which belongs to two heptads. We declare that the ordered pair of skew lines is incident with exactly these two heptads.

Consider a point p and a plane P not containing it, that is, a {point, plane}-flag in the geometry just constructed. The residue of this flag contains all 8 heptads and 28 ordered pairs of skew lines, since there are 7 choices for the first line and then 4 possible lines of the plane which do not contain the intersection of the first line and the plane. As explained earlier, every pair of skew lines defines a bisecant line with respect to the

Klein quadric. Such a line has exactly one external point. The 28 pairs of skew lines in the residue hence define 28 external points. We claim that these are different. Suppose two bisecants from the residue define the same external point. These two bisecants define two points q_1 and q_2 in the plane corresponding to the point p and two points q_1' and q_2' in the plane corresponding to P. The Pasch property of the projective space $PG(5,2)$ then implies that the two lines q_1q_2 and $q_1'q_2'$ must intersect. This contradicts the fact the the planes corresponding to p and P on the Klein quadric are disjoint. Hence we have a bijective correspondence between the 28 external points and the 28 pairs of skew lines in the residue of $p * P$. Since two heptads define one external point and any two external points define an heptad, the residue of the flag considered is isomorphic to the vertex-edge geometry of the complete graph on 8 points. Now take a heptad and a point. This is a flag of type $\{0,2\}$. The residue contains the 8 planes not containing the points and $28 = 7 \cdot 2 \cdot 2$ ordered pairs of skew lines. Again the residue is isomorphic to the vertex-edge geometry of the complete graph on 8 points. A similar reasoning works for the residue of a flag of type $\{0,3\}$. If we take an ordered pair of skew lines and a plane containing the second line of the ordered pair, its residue contains two heptads and two points (the points of the first line which are not in the plane). This residue is clearly a complete bipartite graph. Finally we choose an ordered pair of lines and a heptad incident with it in the geometry. The points and planes in the residue are the three points of the first line and the three planes containing the second line of the ordered pair. Each of these planes contains one of the points of the residue and is hence incident with two points. We see that a residue of type $\{2,3\}$ is isomorphic to a triangle geometry. We have proved here that the geometry defined indeed has the given diagram. The automorphism group of this geometry is isomorphic to A_8 and the correlation group is S_8. The Borel subgroup of the geometry is a dihedral group D_{12}.

Below we give more geometries of A_8 which can be explained in $PG(3,2)$ or on the Klein quadric. Sometimes more than one construction is possible. The automorphism and correlation groups are mentioned as well as the Borel subgroup B.

Remark 8.1 In fact, each construction of this paper should be endowed with a similar proof to show that we really obtain the claimed diagram. Due to a lack of space, these rather lengthy proofs are omitted.

$B = PSL(3,2)$, $Aut = A_8 \times S_8$, $Cor = A_8 \times S_8$. The elements of the geometry are the points of $PG(3,2)$, which correspond to a family of planes on $\mathcal{Q}$, and the heptads. Incidence is empty intersection between planes and heptads.

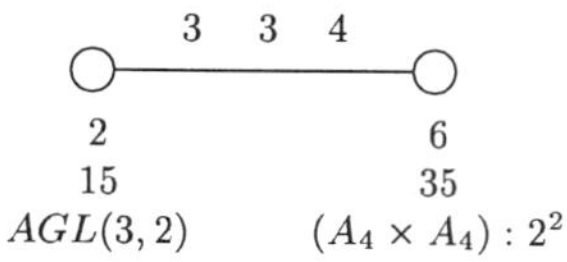

$B = 2^4 : D_{12}$, $Aut = A_8$, $Cor = A_8$.
This geometry is the {point, line}-truncation of $PG(3,2)$.

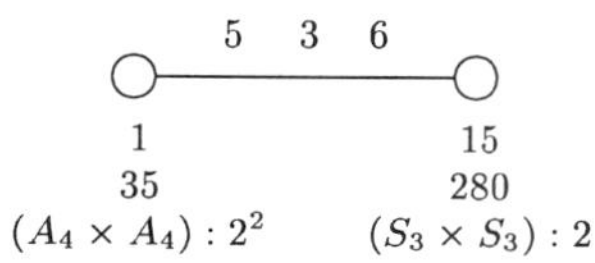

$B = S_3 \times S_3$, $Aut = S_8$, $Cor = S_8$. The elements of the geometry are the lines and pairs of skew lines in $PG(3,2)$. These can also be seen as the points of $\mathcal{Q}$ and the bisecants respectively.

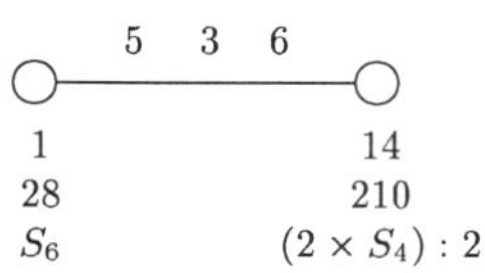

$B = 2 \times S_4$, $Aut = S_8$, $Cor = S_8$. The elements of the geometry are the external points and the tangents to $\mathcal{Q}$. This geometry corresponds to a graph on the external points: $x \sim y$ if and only if line xy is a tangent. This graph is also the complement of the triangular graph $T(8)$ (see [19]). Another way to construct the geometry is to take the pairs and pairs of pairs of a set of 8 elements. Hence it is a generalization of the Petersen geometry.

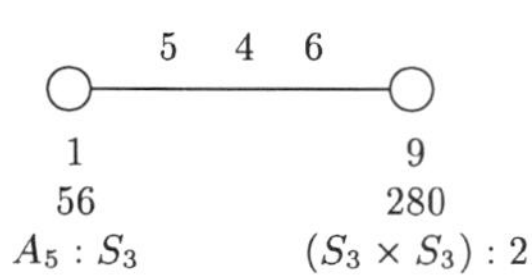

$B = S_3 \times S_3$, $Aut = S_8$, $Cor = S_8$. The elements are the spreads of $PG(3,2)$ and the pairs of skew lines. This geometry can also be constructed with the 3-subsets and pairs of disjoint 3-subsets of a set of 8 elements (another generalization of Petersen geometry!). In $PG(5,2)$ we can take the external lines of $\mathcal{Q}$ and the bisecants to construct the geometry. Two elements of different type are incident if and only if they intersect.

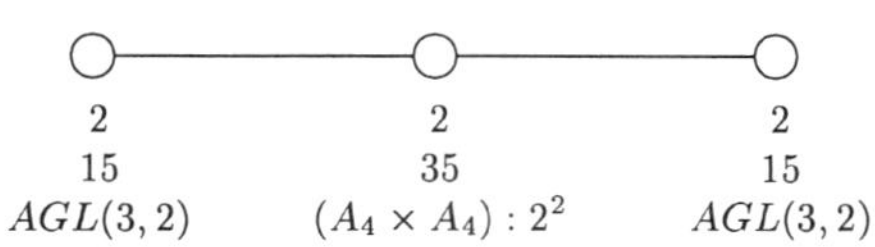

$B = 2^4 : 2^2$, $Aut = A_8$, $Cor = S_8$.
This is the projective space $PG(3,2)$!

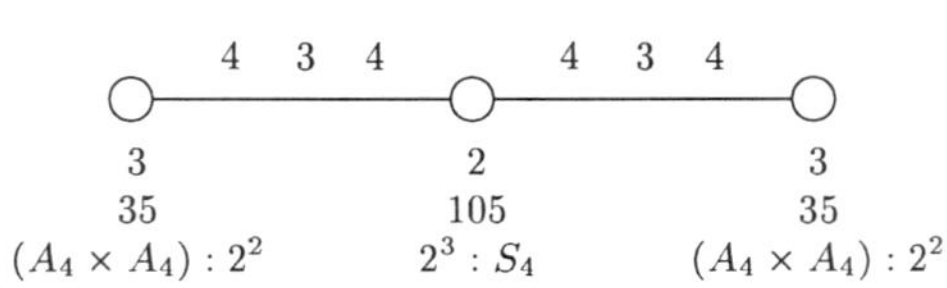

$B = D_{12}$, $Aut = A_8$, $Cor = S_8$. The elements of this geometry are the lines, the {point, plane}-flags and a copy of the lines of $PG(3,2)$. Lines of different type are incident if and only if they are skew. A line of type 0 is incident with a flag if and only if the line is in the plane and the point is not on the line. A line of type 2 is incident with a flag if and only if the line is not in the plane and the point belongs to the line.

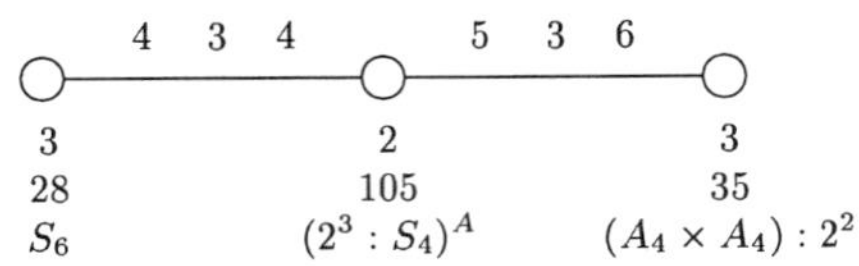

$B = D_{12}$, $Aut = A_8$, $Cor = A_8$. To construct this geometry in $PG(5,2)$, we take the points external to $\mathcal{Q}$, the planes intersecting $\mathcal{Q}$ in one line and the points on $\mathcal{Q}$. All points of $\mathcal{Q}$ are incident with all external points. A point of $\mathcal{Q}$ is incident with a plane P if and only if it is in the unique plane on $\mathcal{Q}$ containing the line $\mathcal{Q} \cap P$ and does not belong to that line.

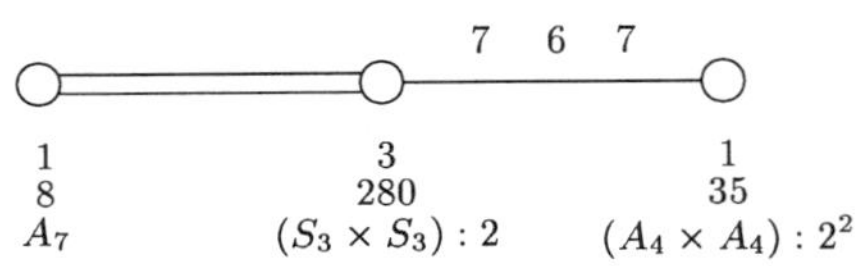

$B = 3^2 : 2$, $Aut = S_8$, $Cor = S_8$.
The elements of this geometry can be taken to be the heptads, the bisecants and the points on the Klein quadric $\mathcal{Q}$. Points and heptads are all incident with each other.

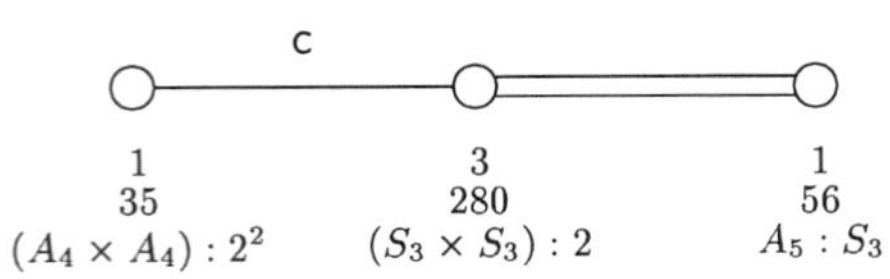

$B = 3^2 : 2$, $Aut = S_8$, $Cor = S_8$.
The elements of this geometry are th lines, the pairs of skew lines and the spreads of $PG(3,2)$. This geometry is also the halved Johnson geometry (see [17]).

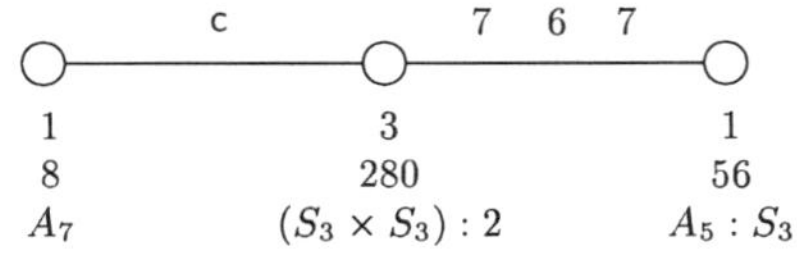

$B = 3^2 : 2$, $Aut = S_8$, $Cor = S_8$. In projective space we can construct this geometry with the heptads, the pairs of skew lines and the spreads. A spread defines 5 heptads (see Section 8.2) which are declared incident with the spread. Another construction uses heptads, bisecants and external lines in $PG(5,2)$.

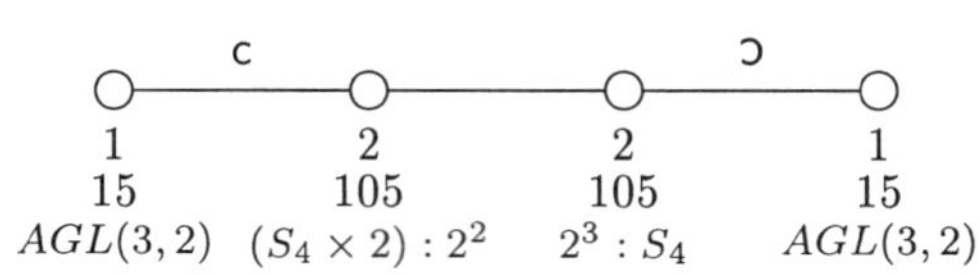

$B = D_8$, $Aut = A_8$, $Cor = S_8$. The elements of this geometry are the points, the {point, line}-flags, {line, plane}-flags and the planes in $PG(3,2)$. A point p and plane P are incident if and only if $p \notin P$. A point p is incident with a flag if it belongs to the plane or the line of the flag but is neither equal to the point nor on the line of the flag. A plane is incident with a flag if and only if the point or the line is in the plane of the flag. Flags are incident if the line of one flag is in the plane of the other.

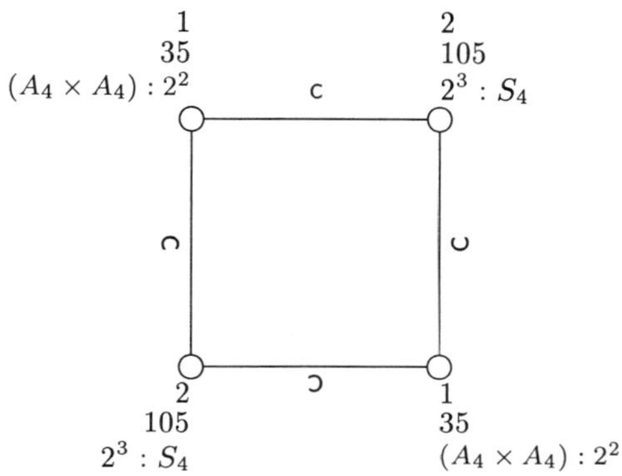

$B = 2^2$, $Aut = A_8$, $Cor = S_8 \times 2$. This geometry can be constructed with two copies of the lines of $PG(3,2)$ and two copies of the {point, plane}-flags. Two lines are incident if and only if they are skew. A line L_0 of type 0 is incident with a flag $p * P$ of type 1 if and only if $L_0 \subseteq P$ and $p \notin P$. It is incident with a flag $q * Q$ of type 3 if and only if $q \in L_0$ and $L_0 \not\subseteq Q$. For a line of type 2, we dualise. Two flags are incident if and only if $p \neq q$ and $p, q \in P \cap Q$.

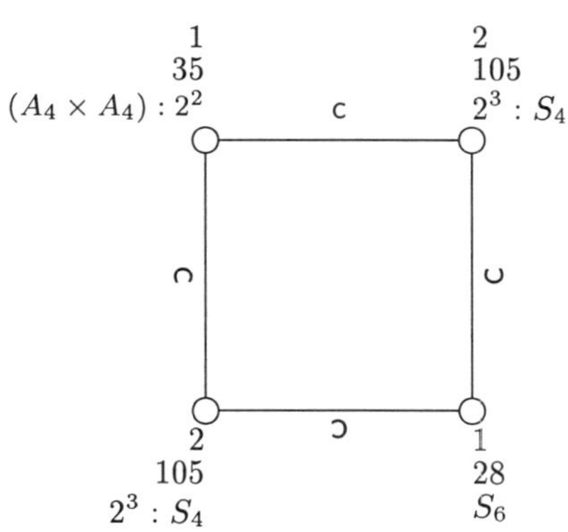

$B = 2^2$, $Aut = A_8$, $Cor = S_8$.
To construct this geometry, we take the points of $\mathcal{Q}$, two copies of the lines of $\mathcal{Q}$ and the external points. All elements of type 0 and 3 are incident with each other. Two lines are incident whenever they intersect. A line is incident with the external points of a plane intersecting $\mathcal{Q}$ in that line. Other incidences are natural.

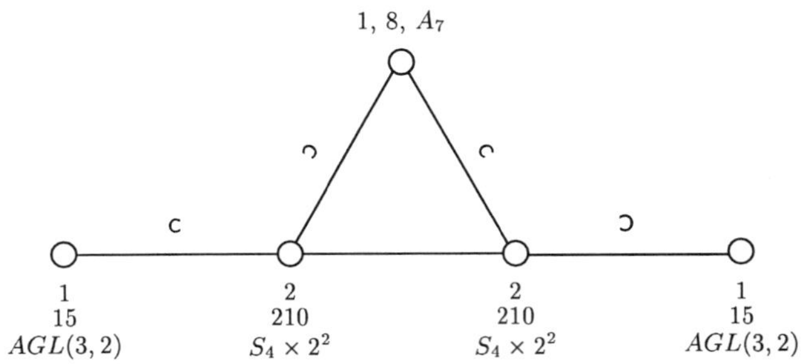

This geometry can be constructed on the Klein quadric as follows. The elements of type 0 are the planes of $\mathcal{Q}$ which correspond to the points of $PG(3,2)$. The elements of type 4 are the planes corresponding to the planes of $PG(3,2)$. The elements of types 1 and 3 are two copies of the lines tangent to $\mathcal{Q}$. The elements of type 2 are the heptads. Since a tangent line has two points off $\mathcal{Q}$, it defines two pairs of heptads. An element L of type 1 or 3 is incident with all elements of type 2 which are *not* the heptads intersecting L. An element of type 1 is incident with an element of type 3 if and only if they are tangent at the same point of $\mathcal{Q}$. An element of type 0 and an element of type 4 are incident if and only if they are disjoint. An element of type 1 or 3 is incident with an element of type 0 whenever they are disjoint. We declare all elements of type 2 to be incident with all elements of type 0 and 4. The automorphism group of this geometry is isomorphic to A_8 and its correlation group has isomorphism type S_8.

Remark 8.2 This is the only rank 5 geometry in our list which is not a truncation of an inductively minimal geometry.

8.3. The affine space $AG(3,2)$

The affine space of dimension 3 over the field with two elements is denoted by $AG(3,2)$. If we fix a plane P in $PG(3,2)$, the structure induced on the 8 points not in P by the lines and planes of $PG(3,2)$ is isomorphic to $AG(3,2)$. Since every two points of $AG(3,2)$ define a line, the collinearity graph of $AG(3,2)$, that is, the graph whose vertices are the points of $AG(3,2)$ and in which two vertices are adjacent whenever they are collinear, is the complete graph K_8 on 8 vertices. Given two copies of the complete graph on 8 vertices, one can label the vertices of each of them with the numbers from 1 to 8. These labellings are always equivalent when the two copies are seen as graphs, but not if they are understood as models of the affine space. The reason is that an affine space has parallel lines and to be affinely equivalent, the labellings must be such that edges which were parallel in the first labelling remain parallel in the second labelling. In fact there are 30 affinely inequivalent ways to label the vertices of K_8 on which A_8 has two orbits of size 15. The affine space has 14 planes of 4 points and there are 70 subsets of 4 elements in the vertex set of K_8. Each time we label K_8, there are 14 of the 70 sets of 4 elements which become planes of $AG(3,2)$. The remaining 4-subsets will be called *non-planes* for that labelling.

These concepts are useful to explain for example the geometry with the following diagram.

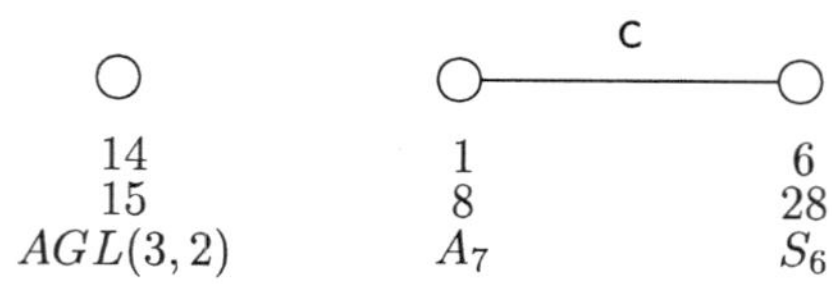

The elements of type 0 are the 15 inequivalent labellings of K_8 in one A_8-orbit and the elements of types 1 and 2 are the vertices and edges of K_8 respectively. Every labelling corresponds to a model of $AG(3,2)$ which contains all vertices and edges of K_8 and the residue of a labelling is simply the complete graph on 8 vertices. The automorphism and correlation group of this geometry are both isomorphic to $A_8 \times S_8$. The Borel subgroup is S_4.

Another geometry we can explain in this way is the geometry described by Neumaier in [24]. This geometry is quite important since its residue of cotype 0 is the famous A_7-geometry which is known to be the only flag-transitive finite thick C_3-geometry which is not a polar space (see [27] for details).

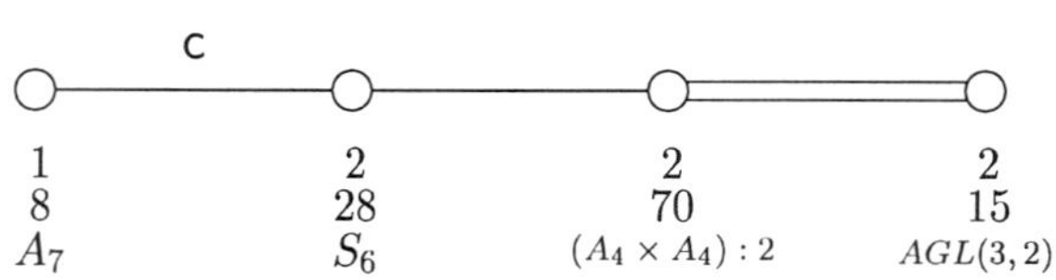

The Neumaier geometry can be explained as follows. The elements of types 0 and 1 are the vertices and edges of the complete graph K_8, the elements of type 2 are the 4-subsets of the vertex set of K_8 and the elements of type 3 are the 15 inequivalent labellings of K_8. Incidences are the same as in the previous geometry and an element of type 2 (this is a 4-subset) is naturally incident with the vertices and edges of the subgraph of K_8 induced on it. A 4-subset is incident with a labelling of K_8 if it is the set of points of a non-plane in the model of $AG(3,2)$ defined by that labelling. The automorphism and correlation group of the Neumaier geometry are both A_8 and the Borel subgroup is D_8.

Remark 8.3 In his original paper, Neumaier gives another construction and mentions only 35 elements of type 2 in his geometry. In [10], it is shown that this number must be 70.

Remark 8.4 J.W.P. Hirschfeld pointed out that the Neumaier geometry was already implicit in [14]. This was clarified with the help of J.A. Thas and yields an alternative construction of the Neumaier geometry. The elements of type 0, 1, 2 and 3 are respectively the 8 heptads, the 28 external points, the 70 planes of $PG(5,2)$ sharing one point with $\mathcal{Q}$ and the 15 planes of one family on $\mathcal{Q}$.

Acknowledgements

The author is a postdoctoral fellow of the Fund for Scientific Research-Flanders (Belgium) (F.W.O.-Vlaanderen) and thanks F. Buekenhout for the inspiring conversations concerning this work.

References

[1] W. Bosma, J. Cannon, and C. Playoust, The Magma algebra system I: The user language, *J. Symbolic Comput.* **24** (1997), 235–265.

[2] F. Buekenhout, Diagram geometries for sporadic groups, *Contemp. Math.* **45** (1985), 1–32.

[3] F. Buekenhout, Finite groups and geometries: a view on the present state and on the future, *Groups of Lie Type and their Geometries* (eds. W.M. Kantor and L.D. Martino), Cambridge University Press, Cambridge, 1995.

[4] F. Buekenhout, editor, *Handbook of Incidence Geometry: Buildings and Foundations*, Elsevier, Amsterdam, 1995.

[5] F. Buekenhout and P. Cara, Some properties of inductively minimal geometries, *Bull. Belg. Math. Soc. Simon Stevin* **5** (1998), 213–219.

[6] F. Buekenhout, P. Cara, and M. Dehon, Inductively minimal flag-transitive geometries, *Mostly Finite Geometries* (ed. N.L. Johnson), Marcel Dekker, New York, 1997, 185–190.

[7] F. Buekenhout, P. Cara, and M. Dehon, Geometries of small almost simple groups based on maximal subgroups, *Bull. Belg. Math. Soc. Simon Stevin*, supplement (128 pages), 1998.

[8] F. Buekenhout, P. Cara, M. Dehon and D. Leemans, Residually weakly primitive geometries for small almost simple groups: a synthesis, preprint.

[9] F. Buekenhout, M. Dehon, and D. Leemans, An atlas of residually weakly primitive geometries for small groups, *Mém. Acad. R. Belg., Cl. Sciences* **14** (1999).

[10] P. Cara, On the number of planes in Neumaier's A_8-geometry, *J. Combin. Theory Ser. A* **93** (2001), 199-200.

[11] P. Cara, Truncations of inductively minimal geometries, preprint.

[12] P. Cara, An infinite family of Petersen geometries with nonlinear diagram, *J. Geom.* **67** (2000), 73–88.

[13] P. Cara, S. Lehman, and D. Pasechnik, On the number of inductively minimal geometries, *Theoret. Comput. Sci.*, to appear.

[14] G. M. Conwell, The 3-space $PG(3,2)$ and its group, *Ann. of Math.* **11** (1910), 60–76.

[15] M. Dehon, Classifying geometries with CAYLEY, *J. Symbolic Comput.* **17** (1994), 259–276.

[16] M. Dehon and D. Leemans, Constructing coset geometries with Magma: an application to the sporadic groups M_{12} and J_1, preprint (1999).

[17] A. Del Fra, D. Ghinelli, T. Meixner, and A. Pasini, Flag-transitive extensions of C_n geometries, *Geom. Dedicata* **37** (1991), 253–273.

[18] The GAP Group, Aachen, St Andrews, *GAP – Groups, Algorithms, and Programming, Version 4.2*, 1999, (`http://www-gap.dcs.st-and.ac.uk/~gap`).

[19] J.I. Hall, Locally Petersen graphs, *J. Graph Theory* **4** (1980), 173–187.

[20] J.W.P. Hirschfeld, *Finite Projective Spaces of Three Dimensions*, Oxford University Press, Oxford, 1985.

[21] C. Jordan, *Traité des Substitutions et des Équations Algébriques*, Gauthier-Villars, Paris, 1870.

[22] C. Lefèvre-Percsy, N. Percsy, and D. Leemans, New geometries for finite groups and polytopes, *Bull. Belg. Math. Soc. Simon Stevin*, to appear.

[23] T. Meixner, Some polar towers, *European J. Combin.* **12** (1991), 397–415.

[24] A. Neumaier, Some sporadic geometries related to $PG(3,2)$, *Arch. Math. (Basel)* **42** (1984), 89–96.

[25] A. Pasini, A quarry of geometries, *Rend. Sem. Mat. Fis. Milano* **65** (1995), 179–247.

[26] S. E. Payne and J. A. Thas, *Finite Generalized Quadrangles*, Pitman, London, 1984.

[27] S. Yoshiara, The flag-transitive C_3-geometries of finite order, *J. Algebraic Combin.* **5** (1996), 251–284.

[28] J. Tits, Géométries polyédriques et groupes simples, *Atti 2ª Riunione Groupement de Mathématiciens d'Expression Latine*, Firenze, 1961, 66–88.

[29] J. Tits, *Buildings of Spherical Type and Finite BN-Pairs*, Lecture Notes in Math. **386**, Springer, Berlin, 1974.

A CHARACTERIZATION OF TRUNCATED $\mathbf{D}_n$-BUILDINGS AS FLAG-TRANSITIVE $\mathbf{PG.PG}^*$-GEOMETRIES

I. Cardinali

Dipartimento di Matematica
Università di Napoli I
Via Cintia
80126 Napoli
Italy
cardilar@mathna2.dma.unina.it

A. Pasini

Dipartimento di Matematica
Università di Siena
Via del Capitano 15
53100 Siena
Italy
pasini@unisi.it

Abstract We consider flag-transitive $L.L^*$-geometries where the residues of the planes are finite non-degenerate projective spaces of dimension at least 3 and the residues of the points are their duals. We prove that every geometry as above is a truncation of a building of type D_n, for a suitable n depending on the orders of that geometry.

1. Introduction

We recall that an $L.L^*$-geometry is a geometry belonging to a string diagram of rank 3 where the residues of the planes are linear spaces and the residues of the points are dual linear spaces.

The problem of classifying finite flag-transitive $L.L^*$-geometries is addressed by C. Huybrechts in her Ph.D. Thesis [12]. This problem splits in a number of subproblems, corresponding to the various ways of matching two families of linear spaces in an $L.L^*$-diagram. For instance, when

A. Blokhuis et al. (eds.), Finite Geometries, 99–119.

the residues of the planes are circular spaces, namely linear spaces where all lines have just two points, the we have the class of $c.L^*$-geometries. A classification has been obtained for this class, modulo two cases too hard to classify, namely the $c.c^*$-case, where the residue of a point is a dual circular space, and the 1-dimensional case, where the stabilizer of a point in the automorphism group of the geometry induces a 1-dimensional group in the residue of that point. We are not going to survey that classification here. We refer the interested reader to Huybrechts [12] (especially 5.4.6 and 6.5.17), Huybrechts and Pasini [16], [15] and [14] and Baumeister, Del Fra, Meixner and Pasini [3].

In this paper we make one more step in the classification of flag-transitive $L.L^*$-geometries. We consider geometries belonging to the following special case of the $L.L^*$-diagram,

$(PG.PG^*)$

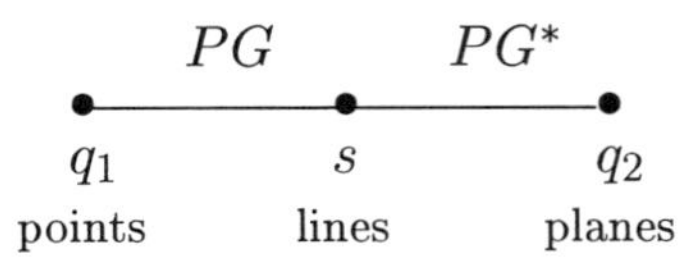

where the orders q_1, q_2 are prime powers, the residues of the planes are projective spaces of order q_1, the residues of the points are dual projective spaces of order q_2 and

$$s + 1 = \frac{q_1^{n_1} - 1}{q_1 - 1} = \frac{q_2^{n_2} - 1}{q_2 - 1}$$

for suitable integers $n_1, n_2 > 2$. We prove the following:

Theorem 1.1 (Main Theorem) *Let Γ be a flag-transitive geometry with diagram and orders as above. Then $q_1 = q_2$ ($= q$, say), $n_1 = n_2 = n$ and Γ is the following truncation of the building of type D_{n+1} defined over $GF(q)$:*

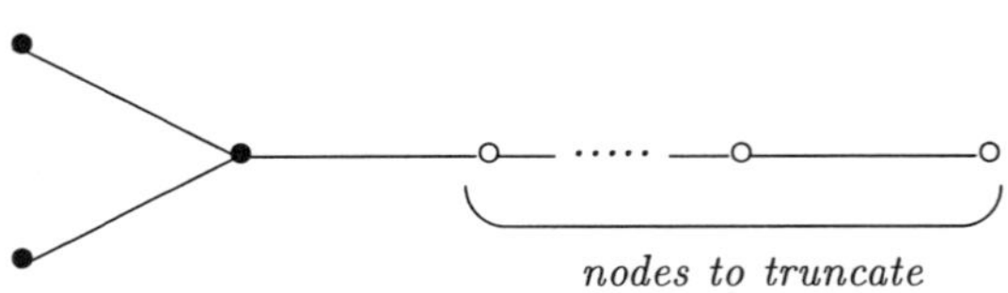

The rest of this paper is devoted to the proof of the above theorem. We first prove that $q_1 = q_2$ and $n_1 = n_2$ (Section 2). Then, in Sections 3 and 5, we prove that Γ satisfies the following:

(LL) no two distinct lines of Γ are incident with the same pair of distinct points;

(T) every triplet of pairwise collinear points belongs to the residue of some plane;

(T)* (dual of the above) every triplet of pairwise collinear planes belongs to the residue of some point.

Theorem 1.1 follows from this and the following theorem of Baumeister and Pasini [2, Theorem 30].

Theorem 1.2 *Let Γ be a thick $L.L^*$-geometry with finite orders r, s, t where $r, t < s$,*

$(L.L^*)$

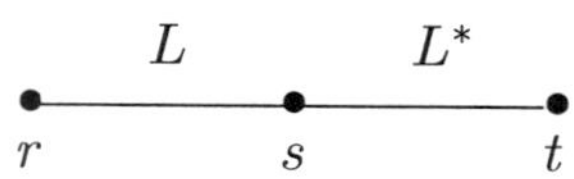

Suppose that (LL), (T) *and* (T)* *hold in Γ. Then $r = t = q$ for some prime power q, $s + 1 = (q^n - 1)/(q - 1)$ for some integer $n > 2$ and Γ is a truncation of the building of type D_{n+1} defined over $GF(q)$, as in Theorem* 1.1.

Remark 1.3 The following, contained in the main result of Huybrechts [13], can also be used instead of Theorem 1.2 to finish the proof of our Main Theorem.

Theorem 1.4 (Huybrechts) *Let Γ be a thick $PG.PG^*$-geometry with finite orders q_1, s, q_2 where $q_1, q_2 < s$. Suppose that both* (LL) *and* (T) *hold in Γ. Then Γ is a truncation of the building of type D_{n+1} defined over $GF(q)$, where $q = q_1 = q_2$ and $(q^n - 1)/(q - 1) = s + 1$.*

Remark 1.5 The flag-transitivity hypothesis is essential in Theorem 1.1. Indeed, some non-flag-transitive $PG.PG^*$-geometries exist that have nothing to do with truncated D_{n+1}-buildings. For instance, suppose that $PG(n, q)$ admits a packing, namely a partition of its set of lines in spreads. This is always the case when $n + 1 = 2^k$ for some $k \geq 2$ whatever q is, or when n is odd and $q = 2$; see Baker [1], Beutelspacher [6] and Denniston [9]. A packing of $PG(n, q)$ is parallelism in the meaning of Buekenhout, Huybrechts and Pasini [4]. Thus, if $PG(n, q)$ admits a packing, we can glue two copies of $PG(n, q)$ to obtain a (flat) $PG.PG^*$-geometry Γ with orders q, s, q where $s+1 = (q^n - 1)/(q - 1)$ (Buekenhout, Huybrechts and Pasini [4]). This geometry, being flat, is not a truncated D_{n+1}-building. It is not even a quotient of a truncated D_{n+1}-building, as finite thick truncated D_{n+1}-buildings do not admit any proper quotient; see the proof of Theorem 30 of Baumeister and Pasini [2]. In view of our Main Theorem, Γ is not flag-transitive.

2. Preliminaries

Throughout this paper Γ is a flag-transitive $PG.PG^*$-geometry with orders q_1, s, q_2 where $q_1, q_2 > 1$ are prime powers,

$$s+1 = \frac{q_1^{n_1}-1}{q_1-1} = \frac{q_2^{n_2}-1}{q_2-1},$$

and $n_1, n_2 > 2$. Also, G is a given flag-transitive subgroup of $\mathrm{Aut}(\Gamma)$.

2.1. Terminology and notation

Given an element x of Γ, we denote its residue by $\mathrm{Res}(x)$. According to the custom to give string diagrams an orientation from left to right, when x is a point we distinguish $\mathrm{Res}(x)$ from its dual, henceforth denoted $\mathrm{Res}(x)^*$. Thus, in our case, $\mathrm{Res}(x)^*$ is a projective space whereas $\mathrm{Res}(x)$ is a dual projective space. Accordingly, we call *points of* $\mathrm{Res}(x)^*$ the planes of Γ incident to x.

We say that two distinct points x, y of Γ are *collinear* and we write $x \perp y$, if there is a line incident with both x and y. Given a point x, we denote by $x^\perp$ the set of points collinear with x or equal to x. The same terminology and notation will be used for planes.

Given a point or a plane x of Γ, we denote by $\sigma(x)$ the set of planes or points incident to x. Given a line l, we denote by $\sigma^-(l)$ and $\sigma^+(l)$ the set of points and the set of planes incident to l.

Given an element x of Γ, G_x is its stabilizer in G and K_x is the element-wise stabilizer of $\mathrm{Res}(x)$. Thus, $\overline{G}_x := G_x/K_x$ is the group induced by G_x on $\mathrm{Res}(x)$.

Given a chamber $\{p, l, u\}$ of Γ, with p a point, l a line and u a plane, we also denote by G_0, G_1 and G_2 the stabilizers of p, l and u, respectively. Similarly, K_0, K_1, K_2 will stand for K_p, K_l and K_u and $\overline{G}_0$, $\overline{G}_1$, $\overline{G}_2$ stand for $\overline{G}_p$, $\overline{G}_l$ and $\overline{G}_u$ respectively.

2.2. Orders

By Higman [10], for $i = 0$ or 2 we have either $L_{n+1}(q) \leq \overline{G}_i \leq P\Gamma L_{n+1}(q)$ or $n = 3, q = 2$ and $\overline{G}_i = A_7$, where $(n, q) = (n_1, q_1)$ when $i = 0$ and $(n, q) = (n_2, q_2)$ when $i = 2$. In any case, $PGL_n(q) \leq \overline{G}_{0,2} \leq P\Gamma L_n(q)$ and, as $n > 2$ by assumption, we have $\overline{G}_{0,2}^{(\infty)} = L_n(q)$.

Proposition 2.1 *We have $q_1 = q_2$ and $n_1 = n_2$.*

Proof As noted above, $\overline{G}_{0,2}^{(\infty)} = L_{n_1}(q_1) = L_{n_2}(q_2)$. However, when $n > 2$, the number $q^{n(n-1)/2}$ is the largest prime power dividing $|L_n(q)|$.

Hence $q_1^{n_1(n_1-1)/2} = q_2^{n_2(n_2-1)/2}$. Therefore, q_1 and q_2 are powers of the same prime. Consequently, a subgroup of $\overline{G}_1^{(\infty)}$ is a Borel subgroup of $L_{n_1}(q_1)$ if and only if it is a Borel subgroup of $L_{n_2}(q_2)$. A Borel subgroup of $L_n(q)$ has index $(q^n - 1)/(q-1)$. Hence

$$\frac{q_1^{n_1}-1}{q_1-1} = \frac{q_2^{n_2}-1}{q_2-1}.$$

The above, combined with the fact that q_1 and q_2 are powers of the same prime, implies that $q_1 = q_2$. Finally, the equalities $q_1 = q_2$ and $q_1^{n_1(n_1-1)/2} = q_2^{n_2(n_2-1)/2}$ imply that $n_1 = n_2$. □

Henceforth we write q for q_1 $(= q_2)$ and n for n_1 $(= n_2)$.

3. The property (LL)

This section is devoted to the proof of the following.

Proposition 3.1 *The property* (LL) *holds in* Γ.

By this proposition and Lemma 7.25 of [18],

Corollary 3.2 *The Intersection Property holds in* Γ.

In particular, the dual (LL)* of (LL) holds. Also, if a line l and a plane u are incident with the same pair of distinct points, then l is incident to u. Dually, if a line l and a point a are incident with the same pair of distinct planes, then l is incident to a. If two distinct points a, b are incident with the same pair of distinct planes u, v, then $a \perp b$ and the line incident to a and b, which is unique by (LL), is incident to both u and v.

3.1. Proof of Proposition 3.1. The generic case

As noted in Section 2, either $L_{n+1}(q) \leq \overline{G}_0$ or $n = 3$, $q = 2$ and $\overline{G}_0 = A_7$. The same holds for $\overline{G}_2$. We call the case of $L_{n+1}(q) \leq \overline{G}_0$ the *generic case* and the case of $\overline{G}_0 = A_7$ the *exceptional case.*

Assume $L_{n+1}(q) \leq \overline{G}_i$ for $i = 0$ or 2 and suppose that (LL) fails to hold in Γ, by way of contradiction. Hence (LL)* neither holds in Γ, as (LL) and (LL)* are equivalent by [18, Lemma 7.25]. Thus, we may assume without loss that $L_{n+1}(q) \leq \overline{G}_0$.

Let a, b be distinct points incident with the same pair of distinct lines l, m and let u be a plane on m. If $l \in$Res(u) then the projective space Res(u) has two distinct lines l and m meeting in two distinct points a and b, which is impossible. Therefore, l and m are skew in the projective space $\mathrm{Res}(a)^*$. That space contains $q^{n-1}+q^{n-2}+...+q^2$ lines on u skew to l and, as $L_{n+1}(q) \leq \overline{G}_a$, those lines form an orbit O of $G_{a,l,u}$ in $\mathrm{Res}(a,u)$. As $m \in O$, all lines of O meet l in a point of Γ different from a. Furthermore, if $m_1, m_2 \in O$, then $\sigma^-(m_1)\cap\sigma^-(m_2)\cap\sigma^-(l) = \{a\}$ for, if otherwise, we get distinct lines m_1, m_2 of $\mathrm{Res}(u)$ meeting in at least two distinct points, which cannot be. Therefore, $\sigma^-(l) \geq q^{n-1}+q^{n-2}+ ...+q^2+1$. This forces $q^{n-1}+q^{n-2}+...+q^2 \leq q$; this is a contradiction. Thus, (LL) holds in the generic case. □

3.2. Proof of Proposition 3.1. The exceptional case

Assume now that $n = 3$, $q = 2$ and $\overline{G}_0 = A_7$ and, as above, suppose that (LL) fails to hold in Γ. We exploit the model Π of $PG(3,2)$ described in Section 6.

Given a, b, l, m and u as in Subsection 3.1, let U be the point of Π corresponding to u in a given isomorphism from $\mathrm{Res}(a)^*$ to Π. The lines of $\mathrm{Res}(a)^*$ skew to m appear as conics in U, with U regarded as a copy of $PG(2,2)$. Four of these conics do not meet the M corresponding to m and twelve of them meet M in two points. The line l corresponds to one of those conics and $G_{a,l,u}$, regarded as a group of automorphisms of the projective plane U, has two orbits on the set of lines of U. One of those orbits is a singleton, formed by the unique line of U exterior to the conic L corresponding to l. The other orbit, say O, contains the three lines of U that meet L in two points. If $M \in O$, then an argument similar to that used in the generic case implies that l has at least four points; contradiction. Hence $M \notin O$, namely M is exterior to L and $G_{a,l,u}$ fixes m. Therefore, $G_{a,l,u} \leq G_{a,m,u}$. However, $G_{a,m,u}$ transitively permutes the four conics of U exterior to M. As L is one of them, we get four lines $l_1(=l), l_2, l_3, l_4$ of Γ such that $\sigma^-(l_i)\cap\sigma^-(m)\setminus\{a\} \neq \varnothing$ for $i = 1,2,3,4$. For $i = 1,2,3,4$, let $b_i \in \sigma^-(l_i)\cap\sigma^-(m)\setminus\{a\}$. As m has 3 points, we have $b_i = b_j$ for at least one choice of distinct indices i, j. Suppose $b_1 = b_2 = b$, to fix ideas. The conics L_1, L_2 of U corresponding to the lines l_1 and l_2 meet in two points of U. Thus, we can replace m with l_2 and we are led back to the case of $M \in O$, which ended with a contradiction. Proposition 3.1 is proved. □

Remark 3.3 With a little of additional work one can prove that (LL) also holds in any flag-transitive $L.PG^*$-geometry, where point-residues are dually isomorphic to projective spaces of dimension $n > 2$ over $GF(q)$ and planes are linear spaces of orders r, s with $r < s$ and $s + 1 = (q^n - 1)/(q - 1)$. Indeed, the argument used for the generic case in the proof of Proposition 3.1 implies that, if (LL) fails to hold, then

$$1 + q^2 + ... + q^{n-1} \leq r + 1 < s + 1 = 1 + q + q^2 + ... + q^{n-1}.$$

However, no flag-transitive linear space exists with orders r, s satisfying the above (compare Buekenhout et al. [5]).

4. A few elementary lemmas

The following lemmas will be used in Section 5. We state them in the most general form, considering an arbitrary $L.L^*$-geometry Σ with finite orders r, s, t,

$(L.L^*)$

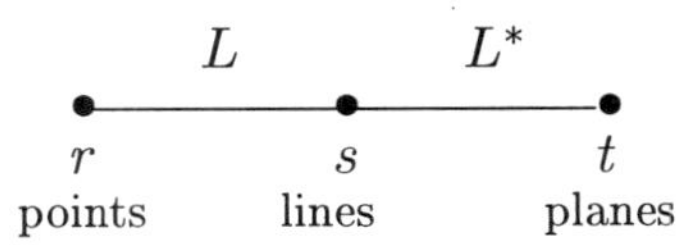

4.1. A lemma on (T) and (T)*

Lemma 4.1 *Suppose that Σ satisfies* (LL) *and* (T) *but not* (T)*. *Then $r < t$.*

Proof As (LL) holds in Σ, the Intersection Property (IP) also holds, by [18, Lemma 7.25]. In particular, Σ satisfies the dual (LL)* of (LL): no two distinct lines are incident with the same pair of distinct planes.

By assumption, (T)* fails to hold in Σ. Hence there are three distinct mutually collinear planes u_1, u_2, u_3 such that $\sigma(u_1) \cap \sigma(u_2) \cap \sigma(u_3) = \emptyset$. For $\{i, j, k\} = \{1, 2, 3\}$, let l_k be the line incident to u_i and u_j (unique, since (LL)* holds in Σ). If $a \in \sigma^-(l_i) \cap \sigma^-(l_j)$ for $i \neq j$, then $a \in \sigma(u_1) \cap \sigma(u_2) \cap \sigma(u_3)$ by (IP), contrary to the choice of u_1, u_2 and u_3. Hence $\sigma^-(l_i) \cap \sigma^-(l_j) = \emptyset$ for any $i \neq j$.

For $i = 1, 2$, pick $a_i \in \sigma^-(l_i)$. For any $a \in \sigma^-(l_3)$ let $l_1(a)$ and $l_2(a)$ be the lines of $\mathrm{Res}(u_2)$ and $\mathrm{Res}(u_1)$ joining a with a_1 and a_2 respectively and let l be the line of $\mathrm{Res}(u_3)$ through a_1 and a_2. By (T), there is a plane $u(a)$ incident with all of a_1, a_2 and a. By (IP), $u(a)$ is also incident with $l, l_1(a)$ and $l_2(a)$ and it is the unique plane with these properties. We have $u(a) \neq u_3$. Indeed, if otherwise, $a \in \sigma(u_1) \cap \sigma(u_2) \cap \sigma(u_3)$, contrary to the choice of u_1, u_2, u_3. Also, $u(a) \neq u_1, u_2$, as one can see by replacing a with a_1 or a_2 in the above argument.

Suppose $u(a) = u(b)$ for distinct points a, b of l_3. As $l_1(a)$ and $l_1(b)$ are incident with both u_2 and $u := u(a) = u(b)$ and since $u \neq u_2$, we have $l_1(a) = l_1(b)$, as we can see in $\mathrm{Res}(a_1)$. Hence $l_1(a) = l_1(b) = l_3$ by (LL). Thus, a_1 belongs to l_3; this is a contradiction. Therefore, if

$a \neq b$ then $u(a) \neq u(b)$. That is, the mapping sending $a \in \sigma^-(l_3)$ to $u(a) \in \sigma^+(l) \setminus \{u_3\}$ is injective. Hence $r+1 \leq t$. □

4.2. Lemmas on flag-transitive $L.L^*$-geometries

Suppose now Σ is flag-transitive, let G be a flag-transitive subgroup of $\mathrm{Aut}(\Sigma)$ and let G_x, K_x, G_0 and G_2 be defined as in Subsection 2.1, but for replacing Γ with Σ. Every $L.L^*$-geometry with finite orders is finite (Del Fra and Ghinelli [8]). Hence Σ is finite. Accordingly, G is finite, too.

Lemma 4.2 *Let $r = t$. Then $|G_0| = |G_2|$.*

Proof Let N_0 and N_2 be the number of points and the number of planes of Σ and let $N_{0,2}$ be the number of point-plane flags. Clearly,

$$N_{0,2} = ((s+1)t+1)N_0 = ((s+1)r+1)N_2.$$

Hence $N_0 = N_2$, as $r = t$. That is, $|G : G_0| = |G : G_2|$. Consequently, $|G_0| = |G_2|$. □

Lemma 4.3 *Suppose $r, t > 1$ and, given a line l of Σ, let a, b distinct points of l and u, v distinct planes on l. Then $K_a \cap K_b = K_u \cap K_v$.*

Proof Given $c \in \sigma(u) \setminus \sigma^-(l)$, let l_a and l_b be the lines of $\mathrm{Res}(u)$ joining c with a and b. As $K_a \cap K_b$ fixes u, l_a and l_b, it also fixes c. Now, let $d \in \sigma^-(l) \setminus \{a, b\}$ and let m be the line of $\mathrm{Res}(u)$ through c and d. By the above, $K_a \cap K_b$ fixes all points of $\sigma^-(m) \setminus \{d\}$. As $|\sigma^-(m) \setminus \{d\}| = r > 1$, $K_a \cap K_b$ also fixes m and, consequently, it fixes d. Thus, $K_a \cap K_b$ fixes all points of u. Therefore, $K_a \cap K_b \leq K_u$. Similarly, $K_a \cap K_b \leq K_v$. Hence $K_a \cap K_b \leq K_u \cap K_v$. Dually, $K_u \cap K_v \leq K_a \cap K_b$. Therefore, $K_u \cap K_v = K_a \cap K_b$. □

Lemma 4.4 *Let $r, t > 1$ and suppose that $K_u \leq K_a$ for some point a and some plane u incident with a. Then $K_u = 1$ and $K_b \cap K_c = 1$ for any pair of distinct collinear points b, c.*

Proof Suppose $K_u \leq K_a$. Then $K_u \leq K_b$ for every point $b \in \sigma(u)$, as G_u is transitive on $\sigma(u)$ and normalizes K_u. Therefore, $K_u = K_b \cap K_c = K_v$ for any two distinct points $b, c \in \sigma(u)$ and for any plane v incident with b and c, by Lemma 4.3. This implies that $K_u = K_v$ for any plane $v \perp u$. Hence $K_v = K_w$ for any two collinear planes u, v of Γ, by the transitivity of G on the set of planes of Γ. Finally, $K_u = K_w$ for any plane w of Γ, by the strong connectedness of Γ. This forces $K_u = 1$. As $K_u = K_b \cap K_c$ for any plane u and any two distinct points $b, c \in \sigma(u)$, we also have $K_b \cap K_c = 1$ for any two distinct collinear points b, c. □

5. The property (T)

We now turn to the flag-transitive $PG.PG^*$-geometry Γ of our Main Theorem. We shall prove the following, thus finishing the proof of that theorem.

Proposition 5.1 *Both* (T) *and its dual* (T)* *hold in* Γ.

Note first that the following holds, by Lemma 4.1 and since (LL) holds in Γ.

Corollary 5.2 *The property* (T) *holds in* Γ *if and only if* (T)* *holds.*

We will prove Proposition 5.1 arguing by contradiction. Assuming that (T) fails to hold, we shall prove first that $q \leq 3$. Then we will show that q cannot be equal to 3 and that, if $q = 2$, then $n = 3$. Finally, we shall rule out the case of $q = 2$ and $n = 3$.

5.1. Reduction to the case of $q \leq 3$

Suppose that (T) fails to hold in Γ and that $L_{n+1}(q) \leq \overline{G}_i$ for $i = 0$ and 2 (which is always the case when $(q, n) \neq (2, 3)$, by Higman [10]).

Henceforth p stands for the prime of which q is a power, namely $q = p^f$ for a given positive integer f. Also, if l is a line, we denote by K_l^- and K_l^+ the stabilizers of all points and of all planes incident to l. Clearly, $K_l = K_l^- \cap K_l^+$.

Lemma 5.3 *One of the following properties holds.*

(1) $K_0 = K_2 = 1$.

(2) *For* $\{i, j\} = \{0, 2\}$, *we have* $K_i/(K_j \cap K_i) \cong V_i.C_i$ *with* V_i *elementary abelian of order* p^{nf} *and* C_i *cyclic of order a divisor of* $q - 1$. *Furthermore,* V_i *can be regarded as a copy of* $V(n, q)$ *and* C_i *acts as a group of scalar matrices on* $V_i = V(n, q)$.

Proof Given a point-plane flag (a, u), let L_u be the stabilizer of u in $L := \overline{G}_a$ and U_u the element-wise stabilizer of $\mathrm{Res}(a, u)$ in L_u. As $L_{n+1}(q) \leq L \leq P\Gamma L_{n+1}(q)$, the group U_u is a split extension $V_2{:}Z$ where $V_2 := O_p(L_u)$ is elementary abelian of order p^{nf} and Z is cyclic of order z, with z a divisor of $q - 1$. Furthermore, V_2 can be regarded as a copy of $V(n, q)$ and Z acts as a group of scalar matrices on $V_2 = V(n, q)$.

The group K_aK_u/K_a is contained in U_u and is normalized by L_u. It is easy to see that all non-trivial subgroups of U_u normalized by L_u contain V_2. Therefore, either we have case (2) of the lemma for $(i, j) = (2, 0)$ or $K_u \leq K_a$.

Assume the latter. Then $K_u = 1$ by Lemma 4.4. Also, $|G_a| = |G_u|$ by Lemma 4.2. As $|G_a| = |\overline{G}_a| \cdot |K_a|$, we have $|K_a| = |G_a|/|\overline{G}_a|$. As $K_u = 1$, $|G_u|$ is a divisor of $|P\Gamma L_{n+1}(q)|$. However, $L_{n+1}(q) \leq \overline{G}_a$. Hence

$$|K_a| \text{ divides } \frac{|P\Gamma L_{n+1}(q)|}{|L_{n+1}(q)|} = (q-1, n+1)f.$$

On the other hand, the same conclusion obtained for K_u holds for K_a as well: either $K_a = 1$ or $K_a K_u/K_u \cong V_0.C_0$ with V_0 elementary abelian of order p^{nf} and C_0 cyclic of order a divisor of $q-1$. However, $p^{nf} = q^n$ does not divide $(q-1, n+1)f$. Hence p^{nf} does not divide $|K_a|$, as the latter divides $(q-1, n+1)f$. Therefore $K_a = 1$ and we have case (1). □

Lemma 5.4 *Let l be a line. For $\epsilon \in \{+, -\}$, the group $X = G_l/K_l^\epsilon$ induced by G_l on $\sigma^\epsilon(l)$ admits a subgroup L containing $PGL_2(q)$ and acting on $\sigma^\epsilon(l)$ as a subgroup of $P\Gamma L_2(q)$ in its natural 3-transitive action of degree $q+1$. Furthermore, $|X : L| = d$ for some divisor d of $q+1$.*

Proof Given a point-line flag (a, l) and regarded $\text{Res}(a)^*$ as the projective space of 1- and 2-dimensional subspaces of $V(n+1, q)$, let U be the 2-dimensional subspace of $V(n+1, q)$ corresponding in $\text{Res}(a)^*$ to the line l. Then $G_{a,l}/K_a$ is the stabilizer of U in $\overline{G}_a$. The planes of $\sigma^+(l)$ are the 1-dimensional subspaces of U. Hence the group induced by $G_{l,a}$ on $\sigma^+(l)$ is contained in $P\Gamma L_2(q)$ and contains $PGL_2(q)$, in its natural 3-transitive action on the $q+1$ 1-dimensional subspaces of U. Similarly, given $u \in \sigma^+(l)$, the group L induced by $G_{l,u}$ on $\sigma^-(l)$ contains $PGL_2(q)$ and is contained in $P\Gamma L_2(q)$. Therefore, the group A induced by $G_{a,l,u}$ on $\sigma^-(l) \setminus \{a\}$ lies between $AGL_1(q)$ and $A\Gamma L_1(q)$.

Let X and Y be the groups induced by G_l and $G_{a,l}$ on $\sigma^-(l)$. Then $L \leq X$, $A \leq Y$ and $|X : Y| = |L : A| = q+1$. Hence $|X : L| = |Y : A| = d$, say. As there are $q+1$ planes on l and G_l transitively permutes them, d is a divisor of $q+1$. □

Lemma 5.5 *Suppose $K_0 = K_2 = 1$. Then $q \leq 3$.*

Proof Given a line l, let L and d be as in Lemma 5.4. Given $a \in \sigma^-(l)$ let $A = L_a$ be its stabilizer in L. Then A lies between $AGL_1(q)$ and $A\Gamma L_1(q)$ and, if $X = G_l/K_l^-$, the stabilizer X_a of a in X contains A as a subgroup of index d. The group X_a is a quotient of $G_{a,l}$. However, $K_a = 1$ by assumption. Hence $\overline{G}_a = G_a$ and $G_{a,l}$ is isomorphic to the stabilizer of l in $\overline{G}_a$. The latter is the stabilizer of a 2-dimensional subspace U of $V(n+1, q)$. By assumption, $L_{n+1}(q) \leq \overline{G}_a \leq P\Gamma L_{n+1}(q)$. Hence, when $q > 3$, the quotients of $G_{l,a}$ either have order a divisor of $(q-1)^2 f$ or involve a copy of $L_2(q)$ or $L_{n-1}(q)$. Therefore X_a is a quotient of $G_{l,a}$ only if $q \leq 3$. □

Note that so far we have not used the hypothesis that (T) fails to hold. We will exploit it in the next lemma.

Lemma 5.6 *We have $K_a \cap K_b = 1$ for any two distinct collinear points a, b.*

Proof Let a, b, c be a triple of mutually collinear points not in the same plane. Let l and m be the lines joining a with b and c respectively, and let u be a plane on m. As l, m are non-coplanar in Γ, they are skew in the projective space $\mathrm{Res}(a)^*$. On the other hand, as $\overline{G}_a \geq L_{n+1}(q)$, the stabilizer $G_{a,l,u}/K_a$ of l and u in $\overline{G}_a$ is transitive on the set of lines of $\mathrm{Res}(a)^*$ containing u and skew to l. The number of these lines is $q^{n-1}+...+q^3+q^2$. On the other hand, the stabilizer of b in $G_{a,l,u}$ has index at most q in $G_{a,l,u}$. Therefore, there are at least $q^{n-2}+...+q^2+q$ lines of $\mathrm{Res}(a, u)$ non coplanar with l in Γ but containing a point collinear with b. Clearly, m is one of those lines. Let X be the set of lines of $\mathrm{Res}(a, u)$ containing points collinear with b. As the $q+1$ lines of $\mathrm{Res}(a, u)$ coplanar with l in Γ also contain points collinear with b, we have

$$|X| \geq q^{n-2} + q^{n-1} + ... + q^2 + 2q + 1.$$

Therefore X is not contained in any hyperplane of the projective space $\mathrm{Res}(u)$ and we can pick n lines $l_1, l_2, ..., l_n$ of X in such a way that $\cup_{i=1}^{n} l_i$ spans $\mathrm{Res}(u)$. We can also assume to have picked l_1 from the set of lines of $\mathrm{Res}(a, u)$ coplanar with l in Γ.

For every line $t \in X$ and every point $x \in \sigma^-(t) \setminus \{a\}$ collinear with b, let t_x be the line of Γ through x and b. Then $K_a \cap K_b$ fixes t and t_x, as $K_a \cap K_b \leq K_a$ and $K_a \cap K_b \leq K_b$. Hence $K_a \cap K_b$ fixes all lines of X and at least one point different from a in each of them. In particular, for every $i = 1, 2, ..., n$, $K_a \cap K_b$ fixes a point $a_i \in \sigma^-(l_i) \setminus \{a\}$.

The set $\{a, a_1, ..., a_n\}$ is a basis of the projective space $\mathrm{Res}(u)$. Furthermore, as we have chosen l_1 coplanar with l in Γ, all points of l_1 are collinear with b, whence $K_a \cap K_b$ fixes each of them. $K_a \cap K_b$ also fixes every line of $\mathrm{Res}(u)$ through a. It is now clear that $K_a \cap K_b$ acts trivially in $\mathrm{Res}(u)$. Therefore, $K_a \cap K_b \leq K_u$. The same holds for any other plane v on m. Hence $K_a \cap K_b \leq K_u \cap K_v$. However, $K_u \cap K_v = K_a \cap K_c$ by Lemma 4.3. Hence $K_a \cap K_b \leq K_a \cap K_c$. By symmetry, $K_a \cap K_c \leq K_a \cap K_b$. Hence $K_a \cap K_b = K_a \cap K_c$. Furthermore, with u and v as above, we have $K_a \cap K_y = K_u \cap K_v = K_a \cap K_c$ for any $y \in \sigma^-(m) \setminus \{a\}$. Therefore, $K_a \cap K_x = K_a \cap K_y$ for any $x \in \sigma^-(l) \setminus \{a\}$ and $y \in \sigma^-(l) \setminus \{a\}$. The lines l and m are skew in $\mathrm{Res}(a)^*$ and G_a is transitive on the set of ordered pairs of skew lines of $\mathrm{Res}(a)^*$. Therefore, $K_a \cap K_x = K_a \cap K_y$ for any two points x, y collinear with a but such that no plane of Γ contains $\{x, y, a\}$.

Now, let t_1, t_2 be coplanar lines on a and $x \in \sigma^-(t_1) \setminus \{a\}$, $y \in \sigma^-(t_2) \setminus \{a\}$. The projective space $\mathrm{Res}(a)^*$ contains lines skew to t_1 and t_2. Let t be one of them and $z \in \sigma^-(t) \setminus \{a\}$. Then $K_a \cap K_x = K_a \cap K_z = K_a \cap K_y$, by the above. Hence the equality $K_a \cap K_x = K_a \cap K_y$ holds for any choice of $x, y \in a^\perp \setminus \{a\}$. By transitivity, $K_x \cap K_y = K_y \cap K_z$ for any choice of distinct points x, y, z with $x \perp y \perp z$. By the connectedness of the collinearity graph of Γ, $K_a \cap K_b = K_x \cap K_y$ for any two distinct collinear points x, y. Hence $K_a \cap K_b \leq K_x$ for every point x of Γ. That is, $K_a \cap K_b = 1$. □

Lemma 5.7 $K_0 = K_2 = 1$.

Proof Suppose the contrary. Then, for any point a the group $O_p(K_a)$ is elementary abelian of order p^{nf}, by Lemma 5.3.

Let b be a point of $a^\perp \setminus \{a\}$. Then $O_p(K_a) \cap K_b = 1$ by Lemma 5.6. Therefore, $O_p(K_a) \cap G_b \cong (O_p(K_a) \cap G_b)K_b/K_b$. Furthermore, $O_p(K_a) \cap G_b$ is normal in $G_{a,b}$ and has order at least $p^{f(n-1)}$, as the line l through a and b has exactly q points different from a.

Let $L := G_{b,l}/K_b$. The group $K := (O_p(K_a) \cap G_b)K_b/K_b$ is contained in $O_p(L)$ and normalized by $X := G_{a,b}K_b/K_b$. The latter has index d in L, with d a divisor of q. L contains a product $S = S_1 \times S_2$ where $S \cap O_p(L) = 1$, $S_1 \cong SL_2(q)$ and $S_2 \cong SL_{n-1}(q)$. The quotient $L/O_p(L)$ is contained in the central product of $GL_2(q)$ and $GL_{n-1}(q)$, possibly extended by field automorphisms. Furthermore,

$$O_p(L) = U_1 \oplus U_2 \oplus ... \oplus U_{n-1} = W_1 \oplus W_2$$

with $U_1, U_2, ..., U_{n-1}$ elementary abelian of order p^{2f}, W_1, W_2 elementary abelian of order $p^{(n-1)f}$ and $|U_i \cap W_j| = q$ for every $i = 1, 2, ..., n-1$ and $j = 1, 2$. S_1 stabilizes each of the summands $U_1, U_2, ..., U_{n-1}$, acting on each of them as $SL_2(q)$ on $V(2, q)$, whereas S_2 stabilizes W_1 and W_2 acting on each of them as $SL_{n-1}(q)$ on $V(n-1, q)$. This implies that no non-trivial proper subgroup of $O_p(L)$ is normalized by $S_1 \times S_2$. The index of X in L is a divisor of q. It easily follows from the above and Lemma 6.1 of Section 6 that one of the following holds:

(1) $O_p(L)(S_1 \times S_2) \leq X$;

(2) $q = 2, 3, 5$ or 11, $O_p(L)S_2 \leq X$ and $|S_1 : S_1 \cap X| = q$;

(3) $q = 2, 3, 5$ or 11, $n = 3$, $O_p(L)S_1 \leq X$ and $|S_2 : S_2 \cap X| = q$.

When $q = 2, 3, 5$ or 11, the subgroups of $SL_2(q)$ of index q are known (see Huppert [11]). They act irreducibly on $V(2, q)$. Therefore, in any case, $X/O_p(L)$ acts irreducibly on $O_p(L)$. Accordingly, X cannot normalize any non-trivial proper subgroup of $O_p(L)$, contrary to the fact that X normalizes K. □

Corollary 5.8 $q \leq 3$.

Proof This is clear, by Lemmas 5.5 and 5.7. □

5.2. The case of $q \leq 3$

We still assume $L_{n+1}(q) \leq \overline{G}_i$ for $i = 0, 2$. We have $q \leq 3$ by Corollary 5.8 and $K_0 = K_2 = 1$ by Lemma 5.7. Hence $\overline{G}_i = G_i$ for $i = 0, 2$.

Lemma 5.9 *Given a chamber (a, l, u) of Γ and a line m on a, skew to l in* Res$(a)^*$, *the group $K_l^- \cap G_{m,u}$ acts transitively on $\sigma^-(m) \setminus \{a\}$.*

Proof In G_u we see that $K_l^- \cap G_u = VL$ where V is the direct sum $V_1 \oplus V_2$ of two copies of $V(n-1, q)$ and L lies between $SL_{n-1}(q)$ and $GL_{n-1}(q)$ and stabilizes both V_1 and V_2, acting on each of them according to the natural action of $GL_{n-1}(q)$ on $V(n-1, q)$. On the other hand, $K_l^- \cap G_u$ is a normal subgroup of $G_{a,l,u}$. The latter can bee seen inside G_a. Regarding Res$(a)^*$ as the projective space of 1- and 2-dimensional linear subspaces of $V(n+1, q)$ and having chosen a basis $\{e_1, e_2, ..., e_{n+1}\}$ of $V(n+1, q)$ in such a way that $\langle e_1 \rangle = u$ and $\langle e_1, e_2 \rangle = l$, the elements of $G_{a,l,u}$ are represented by the matrices of the following form

$$\begin{bmatrix} A & V \\ O & M \end{bmatrix},$$

where O is the $(n-1) \times 2$ null matrix, V is any $2 \times (n-1)$ matrix, M is a non-singular square matrix of order $n-1$, and

$$A = \begin{bmatrix} a & b \\ 0 & c \end{bmatrix},$$

with $ac \neq 0$ and, perhaps additionally, $\det(A)\det(M) = 1$. It is not difficult to see that a normal subgroup of $G_{a,l,u}$ with the same structure as $K_l^- \cap G_u$ can only be formed by elements represented by matrices as follows (modulo scalars), where I is the 2×2 identity matrix, O is the null $(n-1) \times 2$ matrix, V is any $2 \times (n-1)$ matrix and M is any non-singular square matrix of order $n-1$:

$$\begin{bmatrix} I & V \\ O & M \end{bmatrix}.$$

Now, let $m = \langle e_3, e_4 \rangle$ and $v = \langle e_3 \rangle$. The matrices representing elements of $K_m^- \cap G_v$ are obtained from those representing elements of $K_l^- \cap G_u$ by permuting rows and columns according to the permutation $(1,3)(2,4)$. The following element $g \in K_l^- \cap G_u$ has order q, stabilizes m and v but does not belong to $K_m^- \cap G_v$:

$$\begin{bmatrix} J & O \\ O & I \end{bmatrix} \quad \text{where} \quad J = \begin{bmatrix} 1 & 0 & 0 & 0 \\ 0 & 1 & 0 & 0 \\ 0 & 0 & 1 & 1 \\ 0 & 0 & 0 & 1 \end{bmatrix}$$

As $q = 2$ or 3, $\langle g \rangle$ is transitive on $\sigma^-(m) \setminus \{a\}$. □

End of the proof in this case. As (T) fails to hold in Γ by assumption, there is a triple a, b, c of collinear points of Γ not contained in the same plane. Let l and m be the lines joining a with b and c, respectively. By Lemma 5.9, there are elements of G fixing a and b but sending c to any other point of $\sigma^-(m) \setminus \{a\}$. Hence b is collinear with all points of m. By the same argument, all points of m are collinear with all points of l. However, G_a contains $L_{n+1}(q)$ (by assumption when $q = 2$ and $n = 3$). Hence it is transitive on the pairs of skew lines of $\mathrm{Res}(a)^*$. Therefore, any two points of $a^\perp$ are mutually collinear. This forces the collinearity graph of Γ to be a complete graph. Thus, the point-line system Λ of Γ is a linear space with orders q, t where $q = 2$ or 3 and

$$t + 1 = \frac{(q^n + ... + q^2 + q + 1)(q^{n-1} + ... + q^2 + q + 1)}{q + 1}$$

(and $n \geq 3$). Furthermore, Λ is flag-transitive. However, no flag-transitive linear space exists with orders as above (Buekenhout et al. [5]). We have reached a contradiction. □

5.3. The case of $q = 2$ and $n = 3$

The case of $q = 2$ and $n = 3$ remains to be examined, where at least one of $\overline{G}_0$ and $\overline{G}_2$ is isomorphic to A_7. We still assume that (T) fails to hold in Γ.

Lemma 5.10 $\overline{G}_0 \cong \overline{G}_2 \cong A_7$.

Proof By way of contradiction, suppose that either $\overline{G}_0$ or $\overline{G}_2$ is isomorphic to $L_4(2)$. By Corollary 5.2, we may assume without loss that $\overline{G}_0 = L_4(2)$ and $\overline{G}_2 = A_7$.

Given a point-plane flag (a, u), $K_a K_u / K_u$ is normal in $G_{a,u}/K_u$. However, $G_{a,u}/K_u \cong L_3(2)$, which is simple. Hence $K_a K_u / K_u = 1$, namely

$K_a \leq K_u$. Therefore $K_a = 1$ and $K_u \cap K_v = 1$ for any plane v collinear with u, by the dual of Lemma 4.4.

As $K_a = 1$, we have $G_a = \overline{G}_a$ ($\cong L_4(2)$). Hence $G_{a,u} = V{:}L$ with $L = L_3(2)$ and V the natural module for L. As K_u is a subgroup of V normal in $G_{a,u}$, either $K_u = V$ or $K_u = 1$. Suppose $K_u = 1$. Then $G_u = A_7$ and $G_{a,u}$, regarded as the stabilizer of a in G_u, is isomorphic to $L_3(2)$. However, if we regard $G_{a,u}$ as the stabilizer of u in G_a, then we get $G_{a,u} = 2^3{:}L_3(2)$, a contradiction. Therefore, $K_u = V$.

Clearly, if l is a line incident to a and u, then $K_u \leq K_l^-$. Also, $K_v \leq K_l^-$ for any other plane v on l. Thus, regarding $\mathrm{Res}(a)^*$ as the projective space of 1- and 2-dimensional linear subspaces of $V(4,2)$ and given a basis $\{e_1, e_2, e_3, e_4\}$ of $V(4,2)$ in such a way that $\langle e_1 \rangle = u$ and $\langle e_1, e_2 \rangle = l$, K_l^- contains all matrices

$$\begin{bmatrix} 1 & x & y & z \\ 0 & 1 & 0 & 0 \\ 0 & 0 & 1 & 0 \\ 0 & 0 & 0 & 1 \end{bmatrix} \quad \text{and} \quad \begin{bmatrix} 1 & 0 & 0 & 0 \\ x & 1 & y & z \\ 0 & 0 & 1 & 0 \\ 0 & 0 & 0 & 1 \end{bmatrix}.$$

Hence K_l^- also contains all non-singular matrices

$$\begin{bmatrix} A & B \\ O & I \end{bmatrix},$$

where O is the null 2×2 matrix, I is the identity matrix of order 2 and $\det(A) = 1$. Furthermore, K_l^- has index 2 in $G_{a,l}$ and the latter consists of all non-singular matrices

$$\begin{bmatrix} A & B \\ O & C \end{bmatrix},$$

where O is the null 2×2 matrix. Therefore, the elements of K_l^- are the matrices as above, with $C \in \langle C_0 \rangle$ for a given 2×2 matrix C_0 of exponent 3. Similarly, with $m = \langle e_3, e_4 \rangle$, the group K_m^- consists of the non-singular matrices as follows, with C as above:

$$\begin{bmatrix} C & O \\ B & A \end{bmatrix}.$$

Hence the elements of $K_l^- \cap K_m^-$ are the matrices

$$\begin{bmatrix} C & O \\ O & D \end{bmatrix},$$

with $C, D \in \langle C_0 \rangle$ and C_0 as above. This shows that K_l^- contains elements permuting the two points of $\sigma^-(m) \setminus \{a\}$. We have reached the same conclusion as in Lemma 5.9 and we can obtain a contradiction as in the final part of Subsection 5.2. □

Lemma 5.11 $K_0 = K_2 = 1$.

Proof Given a point-plane flag (a, u), we have $G_{a,u}/K_a = L_3(2)$, by Lemma 5.10. Hence $K_uK_a/K_a = 1$, namely $K_u \leq K_a$. Therefore $K_u = 1$, by Lemma 4.4. Dually, $K_a = 1$. □

By Lemma 5.11, $G_i = \overline{G}_i = A_7$ for $i = 0, 2$. We will freely use this fact below.

Lemma 5.12 *Given a point a of Γ, let l be a line on a and u a plane on a but not on l. Then $G_{a,l,u} = S_3$, it stabilizes one line m of* $\mathrm{Res}(a, u)$ *and permutes the remaining six lines of* $\mathrm{Res}(a, u)$ *with two orbits O and O' of size 3, one of which contains the three lines of* $\mathrm{Res}(a, u)$ *that are coplanar with l in the projective space* $\mathrm{Res}(a)^*$.

Proof Let Π be the model of $PG(3, 2)$ described in Section 6 and let P be the point of Π corresponding to u in a given isomorphism from $\mathrm{Res}(a)^*$ to Π. Let L be the line of Π corresponding to l. Then, regarding P as a copy of $PG(2, 2)$, L is a conic of P and is contained in a unique hyperoval H of P. The group $G_{a,l,u}$, regarded as a group of automorphisms of the projective plane P, stabilizes H. Hence $G_{a,l,u} = S_3$ and $G_{a,l,u}$ stabilizes the set $M := P \setminus H$, which is a line of P. Furthermore, $G_{a,l,u}$ has two orbits of size 3 on the remaining six lines of P, formed by the lines of P that are tangent or secant to the conic L. □

Lemma 5.13 *Given a, l and u as in Lemma 5.12, the group $G_{a,l,u}$ permutes the two points of $\sigma^-(l) \setminus \{a\}$.*

Proof Suppose the contrary. Then $G_{a,l,u} \leq K_l^-$. Hence K_l^-, which has index 2 in $G_{a,l}$, contains a copy of S_3. However, $G_{a,l} = (3 \times A_4)2$, with $A_42 = S_4$. The direct product $3 \times A_4$ is the unique subgroup of $G_{a,l}$ of index 2 and it does not contain any copy of S_3, a contradiction. □

Corollary 5.14 *Given a plane u of Γ, the group G_u is transitive on the set of pairs of points (a, b), with $a \in \sigma(u)$ and $b \perp a$.*

Proof G_u is transitive on $\sigma(u)$ and, given $a \in \sigma(u)$, we see in G_a that $G_{u,a}$ is transitive on the set of lines incident to a but not to u. If l is one of those lines, $G_{a,l,u}$ permutes the two points of $\sigma^-(l) \setminus \{a\}$, by Lemma 5.13. □

Given a plane u of Γ and a point $b \notin \sigma(u)$ with $b^\perp \cap \sigma(u) \neq \emptyset$, we define $k := |b^\perp \cap \sigma(u)|$. By Corollary 5.14, the number k does not depend on the particular choice of u and b as above.

Lemma 5.15 $k < 15$.

Proof Suppose $k = 15$, by way of contradiction. Suppose furthermore that $c^\perp \cap \sigma(u) = \emptyset$ for some plane u and some point c. We may assume to have chosen c in such a way that c has distance 2 from $\sigma(u)$ in the collinearity graph of Γ. Pick a point $b \in c^\perp$ with $b^\perp \cap \sigma(u) \neq \emptyset$. As $k = 15$, we have $b^\perp \supseteq \sigma(u)$. Given a point $a \in \sigma(u)$, let l be the line on a and b and let v be a plane on l. Then $b \in c^\perp \cap \sigma(v)$. Hence $\sigma(v) \subseteq c^\perp$, as $k = 15$. Consequently, $a \perp c$; this is a contradiction.

Therefore, $\sigma(u) \subseteq c^\perp$ for every plane u and every point c. Consequently, the point-line system of Γ is a linear space. A contradiction is now obtained as in the final part of Subsection 5.2. □

Lemma 5.16 *Given a plane u and a point $b \notin \sigma(u)$ with $b^\perp \cap \sigma(u) \neq \emptyset$, the set $b^\perp \cap \sigma(u)$ is the union of a family of planes of the projective space* Res(u).

Proof We first prove that $\sigma(u) \cap b^\perp$ contains a plane of Res(u). Let L be the set of lines m of u with $\sigma^-(m) \subseteq b^\perp$. As every point of $\sigma(u) \cap b^\perp$ belongs to three lines coplanar with b in Γ, every point of $\sigma(u) \cap b^\perp$ belongs to at least four lines of L. If L contains a triple of lines that are coplanar in the projective space Res(u), then the conclusion follows.

Suppose, on the contrary, that no three lines of L are coplanar in Res(u). Given $a \in b^\perp \cap \sigma(u)$, let l_1, l_2, l_3 be three lines of L through a, for example, the three lines of u coplanar with b in Γ. Given $a_1 \in \sigma^-(l_1)\backslash\{a\}$, let l_4, l_5 be two lines of L on a_1, different from l_1. As no three lines of L are coplanar in Res(u), neither l_4 nor l_5 meet any of l_1, l_2, l_3. Hence $| \cup_{i=1}^5 \sigma^-(l_i)| = 1 + 6 + 8 = 15$. However, the set $\cup_{i=1}^5 \sigma^-(l_i)$ is contained in $b^\perp \cap \sigma(u)$. Thus, we have reached a contradiction to Lemma 5.15.

So, $\sigma(u) \cap b^\perp$ contains a plane P of Res(u). Let a be a point in $\sigma(u) \cap b^\perp$ but not in P. Let m_1, m_2 be two of the lines of L through a. They meet P in points a_1, a_2, which belong to a line of P. As all lines of P belong to L, the plane of Res(u) spanned by a, a_1 and a_2 is also contained in $\sigma(u) \cap b^\perp$. Thus, $\sigma(u) \cap b^\perp$ is a union of planes of Res(u). □

Corollary 5.17 *Given u and b as in Lemma* 5.16, *we have $\sigma^-(l) \cap b^\perp \neq \emptyset$ for every line l of u.*

Proof This is clear, by Lemma 5.16. □

Lemma 5.18 *We have* $\sigma^-(l) \cap b^\perp \neq \varnothing$ *and* $\sigma(u) \cap b^\perp \neq \varnothing$ *for every point* b*, every plane* u *and every line* l *of* Γ.

Proof We shall only prove the claim on planes. The claim on lines will follow from this and Corollary 5.17.

Suppose to the contrary that $\sigma(u) \cap c^\perp = \varnothing$ for some plane u and a point c. We may assume to have chosen c in such a way that c has distance 2 from $\sigma(u)$ in the collinearity graph of Γ. Let $b \in c^\perp$ with $b^\perp \cap \sigma(u) \neq \varnothing$ and, given a point $a \in \sigma(u) \cap b^\perp$, let l be the line on a and b. Let v be a plane on l. As both u and v are incident to a, there is a line m through a incident with both u and v. Also, $b \in c^\perp \cap \sigma(v)$. Hence $\sigma^-(m) \cap c^\perp \neq \varnothing$, by Corollary 5.17, contrary to the assumption that $\sigma(u) \cap c^\perp = \varnothing$. □

Corollary 5.19 *The stabilizer* G_u *of a plane* u *is transitive on the set of points* $b \notin \sigma(u)$.

Proof This follows easily, by Corollary 5.14 and Lemma 5.18. □

Lemma 5.20 *We have* $k = 14$ *and* Γ *has* 75 *points.*

Proof We have $k = 7, 11, 13$ or 14 by Lemmas 5.15 and 5.16. Accordingly, for a plane u and a point $b \notin \sigma(u)$, the set $\sigma(u) \cap b^\perp$ is a plane or the union of 2, 3 or 4 planes.

If $k = 7$, then every point $a \in \sigma(u) \cap b^\perp$ belongs to exactly three lines of u contained in $b^\perp$. These are the three lines of u on a that are coplanar with b in Γ. In this case, (T) holds in Γ. However, we have assumed that (T) fails to hold in Γ. Hence $k > 7$.

Now, let N be the number of points of Γ not in $\sigma(u)$. Each of them is collinear with k points of u. On the other hand, every point of u belongs to 28 lines not in u and each of these has two points not in u. Therefore, $kN = 15{\cdot}28{\cdot}2$. Hence k divides the latter number. This rules out the cases of $k = 11$ or 13. So, $k = 14$ and $N = 60$. □

End of the proof of Proposition 5.1. Given a plane u of Γ, let b be a point of Γ not in u. As $k = 14$, $\sigma(u) \setminus b^\perp$ is a singleton. Let a be the unique point of $\sigma(u) \setminus b^\perp$. Then $G_{u,b} \leq G_{u,a} = L_3(2)$. Also, $|G_{a,u} : G_{u,b}| = 4$, as there are 15 points in u, 60 points outside u and G_u transitively permutes the latter (Corollary 5.19). However, $L_3(2)$ does not admit any subgroup of index 4 (see [7]). We have reached a final contradiction. □

6. Appendix

6.1. A lemma on linear groups

The following lemma as been exploited in the proof of Lemma 5.7. It can be obtained from the classification of the maximal subgroups of classical groups (Kleidman and Liebeck [17]), but we give an elementary proof here, not to force the reader to go through that classification.

Lemma 6.1 *For a prime power q and an integer $n \geq 2$, let L be a subgroup of $GL_n(q)$ containing $SL_n(q)$ and admitting a proper subgroup X of index a divisor d of q. Then $n = 2$, $q = 2, 3, 5$ or 11 and $d = q$.*

Proof When $n = 2$ the conclusion follows from a well known theorem of Dickson on subgroups of $L_2(q)$ (see Huppert [11]). Thus, suppose $n > 2$. We will obtain a contradiction.

As $|GL_n(q) : SL_n(q)| = q - 1$, the group $X_0 := X \cap SL_n(q)$ also has index d in $SL_n(q)$. So, we may assume without loss that $L = SL_n(q)$, namely L is quasi-simple. Replacing X with a maximal subgroup of L containing X if necessary, we may also assume that X is maximal in L. Clearly, $X \not\leq Z(L)$. Hence X is not normal in L, since L is quasi-simple. Therefore $N_L(X) = X$.

Let S_0 be a Sylow subgroup of X. Then $S_0 < S$ for some Sylow subgroup S of L and $|S : S_0| = d$. We have $|S| = q^{n(n-1)/2}$. Furthermore $N_S(X) = X_0$ as $N_L(X) = X$. Therefore the members of the conjugacy class $\mathcal{X}$ of X in L are the d conjugate of X by elements of S. The kernel of the action of L on $\mathcal{X}$ is contained in $Z(L)$, as L is quasi-simple. Clearly, $Z := Z(S) \cap S_0$ stabilizes all members of $\mathcal{X}$. Hence $Z \leq Z(L)$. However, $|Z(S)| = q$ and $|Z(L)| = (n + 1, q - 1)$. Hence $Z = 1$. This proves that $S = Z(S) \times S_0$. The latter forces S_0 to act trivially on $\mathcal{X}$. Hence $S_0 = 1$. That is, $S = Z(S)$, of order q; contradiction. □

6.2. A model of $PG(3, 2)$

It is known that a model of $PG(3, 2)$ can be built over a given set S of size 7 (see [18, page 279], for instance). As we have exploited that model twice in this paper, we recall its construction here.

A copy of $PG(2, 2)$ can be formed by taking S as the set of points and a suitable family of seven 3-subsets of S as the set of lines. There are thirty different ways of choosing such a family of 3-subsets. Accordingly, thirty copies of $PG(2, 2)$ are obtained. Let $\mathcal{S}$ be the collection of those copies of $PG(2, 2)$ and $\mathcal{L}$ the family of 3-subsets of S. The symmetric group S_7, in its natural action on S, acts transitively on $\mathcal{S}$ but the alternating group A_7 has two orbits $\mathcal{S}_1$ and $\mathcal{S}_2$ on $\mathcal{S}$, each of size 15.

We define an incidence structure Π as follows: $\mathcal{S}_1 \cup \mathcal{L} \cup \mathcal{S}_2$ is the set of elements of Π, the members of $\mathcal{S}_1$, $\mathcal{L}$ and $\mathcal{S}_2$ being called *points*, *lines* and *planes* respectively; a line $L \in \mathcal{L}$ and a point or a plane $P \in \mathcal{S}$ are said to be incident when L is a line of P, the latter being regarded as a copy of $PG(2,2)$. A point P and a plane Q are declared to be incident if and only if they, regarded as copies of $PG(2,2)$, have a line in common.

It is not difficult to see that, if $P \in \mathcal{S}_1$ and $Q \in \mathcal{S}_2$ are incident in Π, then they have a bundle of lines in common. Thus, given a point (plane) P, the function sending every plane (point) Q of Π incident to P to the centre of the bundle of lines of P incident to Q, is an isomorphism from the residue of P in Π to the copy P of $PG(2,2)$. This makes it clear that Π belongs to the Coxeter diagram A_3. Hence Π is a projective geometry [18, Corollary 7.7]. As it has uniform order 2, it is a copy of $PG(3,2)$. Clearly, the alternating group A_7 in its natural action on S is a flag-transitive automorphism group of Π. The elements of $S_7 \setminus A_7$ are correlations of Π.

Given a point or a plane P of Π, the lines of Π that are not incident to P are the conics of the copy P of $PG(2,2)$.

Two lines $L, M \in \mathcal{L}$ meet in a point of Π if and only if $|L \cap M| = 1$ in S. Thus, A_7 has two orbits on the set of lines skew to a given line L. One orbit is formed by the lines M with $L \cap M = \varnothing$ in S. The other orbit contains the lines M with $|L \cap M| = 2$.

References

[1] R.D. Baker, Partitioning the planes of $AG_{2^m}(2)$ into 2-designs, *Discrete Math.* **15** (1974), 35–40.

[2] B. Baumeister and A. Pasini, On $C_2.L$-geometries, *Geom. Dedicata* **71** (1998), 33–59 and *Geom. Dedicata* **75** (1999), 217 (Erratum).

[3] B. Baumeister, A. Del Fra, T. Meixner and A. Pasini, Flag transitive $c.Af^*$-geometries, *Contrib. Algebra Geom.* **37** (1996), 231–258.

[4] F. Buekenhout, C. Huybrechts and A. Pasini, Parallelism in diagram geometry, *Bull. Belg. Math. Soc. Simon Stevin* **1** (1994), 355–397.

[5] F. Buekenhout, A. Delandtsheer, J. Doyen, P.B. Kleidman, M.W. Liebeck and J. Saxl, Linear spaces with flag-transitive automorphism groups, *Geom. Dedicata* **36** (1990), 89–94.

[6] A. Beutelspacher, On parallelism in finite projective spaces, *Geom. Dedicata* **3** (1974), 35–40.

[7] J.H. Conway, R.T. Curtis, S.P. Norton, R.A. Parker and R.A. Wilson, *Atlas of Finite Groups*, Clarendon Press, Oxford, 1985.

[8] A. Del Fra and D. Ghinelli, Diameter bounds for locally partial geometries, *European J. Combin.* **12** (1991), 293–307.

[9] R.H.F. Denniston, Some packings of projective spaces, *Atti Accad. Naz. Lincei Rend.* **52** (1972), 36–40.

[10] D.G. Higman, Flag-transitive collineation groups of finite projective spaces, *Illinois J. Math.* **6** (1962), 434–446.

[11] B. Huppert, *Endliche Gruppen I*, Springer, Berlin 1979.

[12] C. Huybrechts, Reduction des Géométries de Type $L.L^*$, Ph.D. Thesis, Université Libre de Bruxelles, 1996.

[13] C. Huybrechts, $L.L^*$-geometries and D_n-buildings, *Discrete Math.*, to appear.

[14] C. Huybrechts and A. Pasini, A characterization of the Hall–Janko group J_2 by a $c.L^*$-geometry, in *Groups and Geometries* (eds. L. Di Martino et al.), Birkhäuser, Basel (1998), 91–106.

[15] C. Huybrechts and A. Pasini, Flag-transitive extensions of dual projective spaces, *Bull. Belg. Math. Soc. Simon Stevin* **5** (1998), 341–352.

[16] C. Huybrechts and A. Pasini, Flag-transitive extensions of dual affine spaces, *Contrib. Algebra Geom.* **40** (1999), 503–531.

[17] P.B. Kleidman and M.W. Liebeck, *Subgroup Structure of the Finite Classical Groups*, London Math. Soc. Lecture Notes Series **129**, Cambridge University Press, Cambridge, 1990.

[18] A. Pasini, *Diagram Geometries*, Oxford University Press, Oxford, 1994.

GROUP-THEORETIC CHARACTERIZATIONS OF CLASSICAL OVOIDS

A. Cossidente
Dipartimento di Matematica
Università della Basilicata
85100 Potenza
Italy
cossidente@unibas.it

O.H. King
Department of Mathematics
University of Newcastle
Newcastle Upon Tyne NE1 7RU
United Kingdom
o.h.king@ncl.ac.uk

Abstract A characterization of classical ovoids in $PG(3,q)$, for q even, is given. If an ovoid admits a point-transitive automorphism group, then the ovoid is classical.

1. Introduction

An ovoid of $PG(3,q)$, $q>2$, is a set of q^2+1 points of $PG(3,q)$, no three of which are collinear. The only known ovoids of $PG(3,q)$ are the elliptic quadrics, which exist for all q, and the Suzuki–Tits ovoids, which exist for $q=2^e$, $e \geq 3$ odd, [10, Chapter 4]. For odd q, the only ovoids are the elliptic quadrics, [10, Theorem 16.1.7]. For even q, the ovoids have been classified only for q up to and including 32, [13]. Elliptic quadrics and Suzuki–Tits ovoids are usually called "classical ovoids", [13].

There are several results characterizing ovoids in $PG(3,q)$, some involving geometry, some involving group theory.

A. Blokhuis et al. (eds.), Finite Geometries, 121–131.

Since a plane of $PG(3,q)$ meets an ovoid either in a single point or in an oval, a successful technique in studying ovoids has involved examining their plane sections.

The plane sections of an elliptic ovoid are all conics, while those of the Suzuki–Tits ovoid are all translation ovals, namely, ovals invariant under a group of elations of order q having a common axis, which are not conics. Conversely, it has been shown recently [14], that an ovoid admitting a pencil of translation ovals must be either an elliptic quadric or a Suzuki–Tits ovoid (by a pencil of an ovoid with carrier L is meant the set of ovals occurring as secant plane sections for the planes on a fixed tangent line L). The result by O'Keefe and Penttila is a refinement of a previous result by Glynn [9]. For other results in this direction see also [15], [20] and [13] for a survey on ovoids in $PG(3,q)$.

On the other hand, in 1966 Lüneburg [17], proved that if an ovoid $\mathcal{O}$ in $PG(3,q)$, q even, admits an automorphism group containing a subgroup of even order which is transitive on the points of the ovoid, then $\mathcal{O}$ is a classical ovoid.

Using linear codes, in 1987 Bagchi and Sastry [2] extended Lüneburg's result proving that if an ovoid of $PG(3,q)$, q even, admits a point-transitive automorphism group, the the ovoid must be classical.

In this paper, we give another proof of the result by Bagchi and Sastry, using only group-theoretic tools.

In particular, we use the fundamental and celebrated theorem by Aschbacher and the description of certain maximal subgroups of the 4-dimensional symplectic group $PSp(4,q)$, q even, as given in a paper by D.E. Flesner.

2. Singer cyclic groups in symplectic and Suzuki groups

In this section we give some information about Singer cyclic groups in symplectic, orthogonal and Suzuki groups.

Let q be a power of a prime p and let $V = V(4,q)$ be a vector space of dimension 4 over the Galois field $GF(q)$.

Definition 2.1 A cyclic subgroup of $GL(V)$ acting transitively on the non–zero vectors in V will be called a *Singer cyclic group* in $GL(V)$. A generator of a Singer cyclic group will be called a *Singer cycle* in $GL(V)$.

From [11], we recall the following results.

Proposition 2.2 *Let $Sp(V)$ be a symplectic group in $GL(V)$. Then there exists a Singer cyclic group S of $GL(V)$ such that $|S \cap Sp(V)| = q^2+1$.*

We call $S \cap Sp(V)$ a *Singer cyclic group* of $Sp(V)$.

Proposition 2.3 *In $PG(3,q)$, let F be the form defining an elliptic quadric and let $O^-(V)$ be the subgroup of $GL(V)$ fixing F up to a scalar multiple. Then there exists a Singer cyclic group S of $GL(V)$ such that $|S \cap O^-(V)| = q^2+1$.*

We call $S \cap O^-(V)$ a *Singer cyclic group* of $O^-(V)$.

Assume q even. In this case, the centre of $Sp(V)$ is trivial and so $PSp(V) = Sp(V)$. Moreover $O^-(V) \leq Sp(V)$, and a Singer cyclic group of $O^-(V)$ always is a Singer cyclic group of $Sp(V)$.

It turns out that $S \cap O^-(V) = S \cap Sp(V)$ preserves an elliptic quadric $\mathcal{E}_3$ of $PG(3,q)$ acting transitively on its points. We recall that an elliptic quadric of $PG(3,q)$, (any q), is an ovoid, namely a set of q^2+1 points such that no three of them are collinear [10]. In fact $S \cap O^-(V)$ has $q+1$ orbits, each of which is an elliptic quadric. At the same time $S \cap O^-(V)$ has $q+1$ orbits of totally isotropic lines of length q^2+1 and each is a regular spread (elliptic congruence) [10, 17.1]. A partial converse also holds.

Theorem 2.4 [2] *Let H be a subgroup of $Sp(4,q)$, such that the order of H is divisible by q^2+1. Suppose $q > 2$ and H is not transitive on the points of $PG(3,q)$. Then H fixes either an elliptic quadric or a Suzuki–Tits ovoid.*

On the other hand, given an ovoid $\mathcal{O}$ of $PG(3,q)$, q even, at each point of $\mathcal{O}$, there is a tangent plane containing $q+1$ tangent lines. These $(q+1)(q^2+1)$ lines form a general linear complex, Λ, say ([21]).

This way, to the ovoid $\mathcal{O}$ we can associate a symplectic polarity, say π, interchanging a point of $\mathcal{O}$ with the tangent plane at that point, and a non–tangent plane of $\mathcal{O}$ with the nucleus of the $(q+1)$–arc, which is the intersection between $\mathcal{O}$ and the plane. Also, the polarity π maps chords of $\mathcal{O}$ to external lines and vice versa. The collineation group G fixing Λ is, of course, an isomorphic copy of $Sp(4,q)$.

Assume that $q = 2^h$, $h \geq 3$ odd. Set $r = 2^{(h+1)/2}$.

Denote by $Sz(q)$ a Suzuki group of $PGL(4,q)$. Associated with $Sz(q)$ is an ovoid $\mathcal{O}$, the Suzuki–Tits ovoid, in the sense that $\mathcal{O}$ is invariant under $Sz(q)$. Using Segre's construction, the group $Sz(q)$ can always be represented as a subgroup of a 4–dimensional symplectic group $Sp(4,q)$.

We have that $|Sz(q)| = q^2(q-1)(q^2+1)$ and for $q > 2$, $Sz(q)$ admits subgroups of order q^2, $q-1$, $q+1+r$ and $q+1-r$ [12]. In particular, $Sz(q)$ has no element of order q^2+1.

The subgroups of order $q+1+r$ and $q+1-r$ are cyclic and they are irreducible subgroups of distinct Singer cyclic groups of $PGL(4,q)$ [12].

In [8], Glaubermann gave a nice descriptions of these Singer subgroups in terms of the field multiplication and the trace linear map of $GF(q^4)$ see [8, Theorem 4.2] allowing him to define $Sz(q)$ containing a Singer cyclic group of order $q+1+\epsilon r$ and a "mild perturbation" of it of order $q+1-\epsilon r$, with $\epsilon = \pm 1$.

3. Aschbacher's Theorem

In this section we identify subgroups of $Sp(4,q)$ whose order is divisible by q^2+1. Note that $Sp(4,q)$ has order $q^4(q^2-1)^2(q^2+1)$.

Aschbacher's famous theorem ([1]) essentially states that a subgroup of a classical group either lies in a member of one of eight naturally defined classes, $C_1 - C_8$, or is almost simple.

The following are the classes of maximal subgroups G of $Sp(4,q)$ with q even. This list comes, in the main, from [16].

Class C_1: G is a reducible subgroup of order $q^4(q-1)(q^2-1)$. Here we have the stabilizers of either totally isotropic or non-isotropic subspaces. A non-isotropic subspace would have dimension 2 and so be isometric to its conjugate; stabilizers of such subspaces lie inside subgroups in class C_2. This leaves totally isotropic subspaces of dimensions $1, 2$.

Class C_2: G is an imprimitive subgroup of order $2q^2(q^2-1)^2$. Here we have the stabilizer of a pair of 2-dimensional orthogonal non-isotropic subspaces .

Class C_3: $G = Sp(2,q^2).2$. Here we have the stabilizer of a field extension.

Class C_5: $G = Sp(4,q')$, where q is a power of q'. Here G is the stabilizer of a subfield structure.

Class C_8: G is the stabilizer of a form. The only possibilities here are orthogonal groups: one is the group of a hyperbolic quadratic form and has order $2q^2(q^2-1)^2$; the other is the group of an elliptic quadratic form and has order $2q^2(q^2-1)(q^2+1)$.

$\mathcal{S}$: Almost simple subgroups satisfying a number of further restrictions.

The classes C_4, C_6, C_7 do not occur here, since C_4 and C_7 would be stabilizers of tensor product structures and C_6 would be symplectic-type groups.

It is useful to have the following lemma. The proof is elementary and is thus omitted.

Lemma 3.1 *Suppose that a, b are positive integers with $c = \gcd(a,b)$. Then*

(i) $(2^a+1, 2^b+1) = 2^c+1$ *if $a/s, b/s$ are both odd, and* 1 *otherwise;*

(ii) $(2^a+1, 2^b-1) = 2^c+1$ *if* a/s *is odd and* b/s *is even, and* 1 *otherwise;*

(iii) $(2^a-1, 2^b-1) = 2^c-1$.

Theorem 3.2 *If* G *is a subgroup of* $Sp(4,q)$ *with* $q \geq 8$ *even such that* q^2+1 *divides the order of* G*, then either* G *stabilizes a regular spread of lines in* $PG(3,q)$ *or an elliptic ovoid in* $PG(3,q)$*, or* G *is almost simple.*

Proof The orders of the groups in classes C_1, C_2 and C_5, and the order of $O^+(4,q)$ are not divisible by q^2+1, so only C_3, one group in C_8, and $\mathcal{S}$ remain. □

We now turn our attention to papers by Flesner ([5],[6],[7]). Here he addresses the question of maximal subgroups of $PSp(4,2^a)$, in the main concentrating on subgroups containing central elations or non-centred skew elations. The first thing to note is that $PSp(4,2^a)$ is isomorphic to $Sp(4,2^a)$; so the theorems give us information about $Sp(4,2^a)$. The second thing is that central elations are just the images of transvections in $PSp(4,2^a)$ while non-centred skew elations are dual to elations; that is, they are the images of elations under the outer automorphism of $PSp(4,2^a)$. We need to note the first theorem, but for us it is the second which is more significant. Both theorems appear in the third paper but refer to ideas developed in the earlier papers.

Theorem 3.3 (Flesner) *The conjugacy classes of those maximal subgroups of* $PSp(4,2^a)$ *which contain central elations or non-centred skew elations are as follows:*

(a) *stabilizer of a point;*

(a*) *stabilizer of a totally isotropic line;*

(b) *maximal index orthogonal group;*

(b*) *stabilizer of a pair of hyperbolic lines;*

(c) *non-maximal index orthogonal group;*

(c*) *dual of non-maximal index orthogonal group;*

(d_r) *stabilizer of subgeometry over the maximal subfield* $GF(2^{a/r})$*, for each prime* r *dividing* a.

Theorem 3.4 (Flesner) *If* M *is a maximal subgroup of* $PSp(4,2^a)$ *that contains no central elations or non-centred skew elations, then either* $q=2$ *and* M *is isomorphic to* A_6*, or* M *contains normal subgroups* M_1

and M_2 such that $M \geq M_1 \geq M_2 \geq 1$, where M/M_1 and M_2 are of odd order, and M_1/M_2 is isomorphic to either $PSL(2,q')$ or $Sz(q')$ for some power q' of 2.

Flesner's first theorem gives subgroups in the main Aschbacher classes. It is the second theorem which addresses the question of almost simple subgroups.

Lemma 3.5 *Suppose that M is a maximal subgroup of $Sp(4,q)$, with $q \geq 8$ and even, such that M does not lie in one of the Aschbacher classes $C_1 - C_8$. Then $M_0 \leq M \leq Aut(M_0)$ for some subgroup $M_0 \cong PSL(2,q')$ or $Sz(q')$, where q' is a power of 2.*

Proof If M is such a maximal subgroup of $Sp(4,q)$, then M is almost simple and does not appear amongst the subgroups listed in Flesner's first theorem. Thus M contains normal subgroups M_1 and M_2 such that $M \geq M_1 \geq M_2 \geq 1$, where M/M_1 and M_2 are of odd order, and M_1/M_2 is isomorphic to either $PSL(2,q')$ or $Sz(q')$ for some power q' of 2. The term "almost simple" means that there is a non-abelian simple group M_0 such that $M_0 \leq M \leq Aut(M_0)$; in consequence M_0 is the unique minimal normal subgroup of M and any non-trivial normal subgroup of M contains M_0. The subgroup M_2 has odd order, so cannot contain M_0, since any non-abelian simple group has even order, and therefore $M_2 = 1$ and $M_0 \leq M_1$. Now, M_1/M_2 is simple, and so $M_1 = M_0 = PSL(2,q')$ or $Sz(q')$ for some power q' of 2 and $M \leq Aut(M_0)$.

In arguments that follow we refer to the list of subgroups of $PSL(2,p^a)$ established by Moore [19] and published in Dickson's book [4] nearly a century ago. The version stated here is for $PSL(2,q^2)$ where $q \geq 8$ is even.

Theorem 3.6 *The subgroups of $PSL(2,q^2)$ are as follows:*

(a) *cyclic subgroups of order d, where d divides $q^2 - 1$;*

(b) *dihedral subgroups of order $2d$, where d divides $q^2 - 1$;*

(c) *cyclic subgroups of order d, where d divides $q^2 + 1$;*

(d) *dihedral subgroups of order $2d$, where d divides $q^2 + 1$;*

(e) *elementary abelian subgroups of order q^2, together with all subgroups;*

(f) *subgroups of order cd, where $c = 2^m$ and $d = 2^k - 1$ for some divisor k of both m and $2e$ (here $q = 2^e$);*

(g) *subgroups isomorphic to $PSL(2,q')$ or $PGL(2,q')$, where q^2 is a power of q' and an even power in the case of $PGL(2,q')$.*

We have an immediate corollary.

Corollary 3.7 *If $q \geq 8$ is an odd power of 2 and if G is a subgroup of $PSL(2, q^2)$ divisible by $q^2 - 1$, then G satisfies one of the following:*

(i) $G = PSL(2, q^2)$;

(ii) $|G| = q^2(q^2 - 1)$;

(iii) *G is cyclic of order $q^2 - 1$;*

(iv) *G is dihedral of order $2(q^2 - 1)$.*

We find that subgroups of $PSp(4, q)$ isomorphic to $PSL(2, q^2)$ arise. It is helpful to know that these are the types in classes C_3 and C_8. The following argument establishes this using Singer cyclic subgroups and results from Flesner.

Proposition 3.8 *If $q \geq 8$ and q is an odd power of 2, then a subgroup of $PSp(4, q)$ isomorphic to $PSL(2, q^2)$ either stabilizes an elliptic quadric or a regular spread.*

Proof Let J be a subgroup of $PSp(4, q)$ isomorphic to $PSL(2, q^2)$ and let C be a cyclic subgroup of J of order $q^2 + 1$. A reducible subgroup of $PSp(4, q)$ has order dividing $q^4(q^2 - 1)(q - 1)$; so C is irreducible and it follows from [7, Lemma 3] that C is a Singer cyclic subgroup of $PSp(4, q)$. The orbits of C on the points of $PG(3, q)$ are $q + 1$ elliptic quadrics; dually, the orbits of C on the totally isotropic lines of $PG(3, q)$ are $q + 1$ regular spreads. Now let D be a cyclic subgroup of J of order $q^2 - 1$. The same theorem of Flesner implies that D must be reducible; so D stabilizes one of a point, a totally isotropic line or a non-isotropic line. Note, however, that the stabilizer of a non-isotropic line is isomorphic to $PSL(2, q) \times PSL(2, q)$ and that a cyclic subgroup of this group having order $q^2 - 1$ must fix a point on either the line or its orthogonal complement. Thus D stabilizes either a point or a totally isotropic line. Let us suppose that D stabilizes a point x and suppose that the orbit of J on $PG(3, q)$ containing J has length $a(q^2 + 1)$. Then $1 \leq a \leq q + 1$ and J_x, the stabilizer of x in J, has order $q^2(q^2 - 1)/a$. We have chosen x so that $D \leq J_x$; so a divides q^2. The corollary to Dickson's list of subgroups of $PSL(2, q^2)$ demonstrates that a must be $1, q^2/2$ or q^2, but we also have the restriction that $a \leq q + 1$. Hence $a = 1$ and J stabilizes an elliptic quadric. Since J is perfect, it must be the group $PSO^-(4, q)$ of an elliptic quadric. The second possibility for D, that it stabilizes a totally isotropic line, gives the dual conclusion:

J stabilizes a regular spread and must be the group $PSO^-(4,q)$ of a regular spread. □

Theorem 3.9 *If M is a maximal subgroup of $Sp(4,q)$, with $q \geq 8$ and even, such that q^2+1 divides the order of M, then one of the following occurs:*

(i) *$M \cong PSL(2,q^2).2$ and stabilizes a regular spread of lines in the space $PG(3,q)$;*

(ii) *$M \cong PSL(2,q^2).2$ and stabilizes an elliptic ovoid in $PG(3,q)$;*

(iii) *$M \cong Sz(q)$ and stabilizes a Suzuki–Tits ovoid.*

Proof Suppose that M does not stabilize a regular spread of lines in $PG(3,q)$ or an elliptic ovoid in $PG(3,q)$. Then, by the previous lemma, $M_0 \leq M \leq Aut(M_0)$ for some subgroup $M_0 \cong PSL(2,q')$ or $Sz(q')$, where q' is a power of 2. We write $q = 2^e$ and $q' = 2^f$.

Consider first the case where $M_0 \cong PSL(2,q')$. The order of M is $q'(q'^2-1)g$ for some divisor g of f. Let $s = (2e, 2f)$. One possibility is that $2e/s$ is even or $2f/s$ is odd, in which case $(2^{2e}+1, 2^{2f}-1) = 1$. Here q'^2-1 divides $(q^2-1)^2$ and so $q' < q^2$, but also q^2+1 divides $g < q'-1$; so $q^2 < q'$, giving a contradiction. Therefore $2e/s$ is odd and $2f/s$ is even; so $(2^{2e}+1, 2^{2f}-1) = 2^s+1$ and $(2^{2e}-1, 2^{2f}-1) = 2^s-1$. We have that q'^2-1 divides $(q^2-1)^2(q^2+1)$, so $q'^2-1 \leq (2^s-1)^2(2^s+1) < 2^{3s}-1$ and therefore $2f < 3s$. Given that $2f/s$ is even, we can only have $f = s$ and then $2e/f$ is odd. If $2e \geq 3f$, then $q^2+1 > q'^3 > (q'^2-1)g$, a contradiction. Thus we are left with just one possibility: $2e/f = 1$, that is, $q' = q^2$ and $M_0 \cong PSL(2,q^2)$. However, any subgroup of $PSp(4,q)$ isomorphic to $PSL(2,q^2)$ fixes an elliptic ovoid or a regular spread of lines (dual to an elliptic ovoid). It follows that, in this case, M fixes an elliptic ovoid or a regular spread of lines, a contradiction.

Thus we may suppose that $M_0 \cong Sz(q')$. The automorphism group of $Sz(q')$ is $Sz(q').f$; so the order of M is $q'^2(q'^2+1)(q'-1)g$ for some divisor g of f. The significant facts are that q^2+1 divides $q'^2(q'^2+1)(q'-1)f$ and $q'^2(q'^2+1)(q'-1)$ divides $q^4(q^2-1)^2(q^2+1)$; we immediately deduce that q'^2 divides q^4, so that $f \leq 2e$, and that q'^2+1 divides either q^2+1 or $(q^2-1)^2$. Let $t = (2e, f)$. Then $(2e, 2f) = t$ or $2t$. If $(2e, 2f) = t$, then $2f/t$ is even, so $(2^{2e}+1, 2^{2f}+1) = 1$ and $(2^{2e}-1, 2^{2f}+1) = 1$; but then q'^2+1 cannot divide q^2+1 or $(q^2-1)^2$, a contradiction. On the other hand, if $(2e, 2f) = 2t$, then $2e/t$ is even and therefore $(2^{2e}+1, 2^f-1) = 1$; so q^2+1 divides $(q'^2+1)g$. However, $q > e$; so $q^2+1 > f \geq g$ and therefore $(2^{2e}+1, 2^{2f}+1) \neq 1$. It follows that $(2^{2e}-1, 2^{2f}+1) = 1$

and so ${q'}^2 + 1$ divides $q^2 + 1$. Hence $f = t$ and e/f is odd. Finally $q' > g$, so $q^2 + 1 < ({q'}^2 + 1)^{3/2}$ and therefore $e/f = 1$; that is, $q' = q$. We have shown that $M_0 \cong Sz(q)$. By [18], there is only one subgroup of $PSp(4, q)$ isomorphic to $Sz(q)$ up to conjugacy, and such a subgroup is maximal. Thus in this case $M \cong Sz(q)$ and stabilizes a Suzuki–Tits ovoid. □

Corollary 3.10 *Suppose that G is a subgroup of $PSp(4, q)$, with $q \geq 8$ and even, such that $q^2 + 1$ divides the order of G. Then either G fixes an elliptic ovoid or a regular spread (dual to an elliptic ovoid) or $G \cong Sz(q)$ and fixes a unique (Suzuki–Tits) ovoid.*

Proof Let M be a maximal subgroup of $PSp(4, q)$ containing G. Then, by the previous theorem, either M fixes an elliptic ovoid or a regular spread (dual to an elliptic ovoid) or $M \cong Sz(q)$ and fixes a Suzuki–Tits ovoid. Suppose that $M \cong Sz(q)$. From the list of subgroups of $Sz(q)$ given by [12] it is clear that $G = M$. Since $Sz(q)$ has two orbits of points on $PG(3, q)$, the points on a Suzuki–Tits ovoid forming one orbit, so G fixes a unique ovoid. □

Corollary 3.11 *Suppose that a subgroup G of $PSp(4, q)$ has order divisible by $q^2 + 1$, with $q \geq 8$ and even, and stabilizes an ovoid Ω in $PG(3, q)$. Then the ovoid is classical.*

Proof By the theorems above, either $G \cong Sz(q)$ and G fixes a unique (Suzuki–Tits) ovoid, or $G \leq M \cong PSL(2, q^2).2$. We need only consider the latter case.

Let $G_1 = G \cap PSL(2, q^2)$. Then $q^2 + 1$ divides the order of G_1. Consulting the list of subgroups of $PSL(2, q^2)$, we see that the only subgroups whose order is divisible by $q^2 + 1$ are cyclic groups of order $q^2 + 1$ and dihedral groups of order $2(q^2 + 1)$. In any case G contains a cyclic subgroup of order $q^2 + 1$ and by [7] this is a subgroup of a Singer subgroup of $PSp(4, q)$. Moreover, the orbits of this cyclic subgroup are all elliptic quadrics. It follows that the orbits of G are unions of elliptic quadrics and, in particular, Ω must be an elliptic quadric. □

We are now in a position to state three theorems, all essentially proven.

Theorem 3.12 *If Ω is an ovoid in $PG(3, q)$, where $q \geq 8$ and even, such that the stabilizer of Ω in $PSp(4, q)$ acts transitively on Ω, then Ω is classical.*

Theorem 3.13 *If Ω is an ovoid in $PG(3,q)$, where $q \geq 8$ and even, admitting a group containing cyclic subgroups of orders $q+2r+1$, $q-2r+1$, where $r=\sqrt{2q}$, then Ω is classical.*

Theorem 3.14 *Suppose that G is a subgroup of $PSp(4,q)$, where $q \geq 8$ and even, containing elements of orders $q+2r+1$ and $q-2r+1$. Then one of the following occurs:*

(i) *$G \leq PSL(2,q^2).2$ and $S \leq G \leq N_{PSp(4,q)}(S)$, where S is a Singer cyclic subgroup of $PSp(4,q)$ of order q^2+1 with normalizer having order $4(q^2+1)$;*

(ii) *$G \cong PSL(2,q^2)$;*

(iii) *$G \cong PSL(2,q^2).2$;*

(iv) *$G \cong Sz(q)$;*

(v) *$G = PSp(4,q)$.*

Acknowledgements

This research was supported for the first author by the G.N.S.A.G.A of C.N.R. and the Italian Ministry of Research and Technology.

References

[1] M.G. Aschbacher, On the maximal subgroups of the finite classical groups, *Invent. Math.* **76** (1984), 469–514.

[2] B. Bagchi and N.S.N. Sastry, Even order inversive planes, generalized quadrangles and codes, *Geom. Dedicata* **22** (1987), 137–147.

[3] P. Dembowski, *Finite Geometries*, Springer, Berlin, 1968.

[4] L.E. Dickson, *Linear Groups with an Exposition of the Galois Field Theory*, Teubner, Leipzig, 1901.

[5] D.E. Flesner, Finite symplectic geometry in dimension four and characteristic two, *Illinois J. Math.* **19** (1972), 41–47.

[6] D.E. Flesner, The geometry of subgroups of $PSp_4(2^n)$, *Illinois J. Math.* **19** (1975), 48–70.

[7] D.E. Flesner, Maximal subgroups of $PSp_4(2^n)$ containing central elations or noncentered skew elations, *Illinois J. Math.* **19** (1975), 247–268.

[8] G. Glaubermann, On the Suzuki groups and the outer automorphisms of S_6, *Groups, Difference Sets, and the Monster*, Proceedings of a Special Research Quarter at the Ohio State University, Spring 1993 (eds. K.T. Arasu et al.), de Gruyter, Berlin, 1996, 57–62.

[9] D.G. Glynn, The Hering classification for inversive planes of even order, *Simon Stevin* **58** (1984), 319–353.

[10] J.W.P. Hirschfeld, *Finite Projective Spaces of Three Dimensions*, Oxford University Press, Oxford, 1985.

[11] B. Huppert, Singer-Zyklen in klassichen Gruppen, *Math. Z.* **117** (1970), 141–150.

[12] B. Huppert and N. Blackburn, *Finite Groups III*, Springer, Berlin, 1982.

[13] C.M. O'Keefe, Ovoids in $PG(3, q)$: a survey, *Discrete Math.* **151** (1996), 175–188.

[14] C.M. O'Keefe and T. Penttila, Ovoids with a pencil of translation ovals, *Geom. Dedicata* **62** (1996), 19–34.

[15] C.M. O'Keefe and T. Penttila, Ovals in translation hyperovals and ovoids, *European J. Combin.* **18** (1997), 667–683.

[16] P.B. Kleidman and M.W. Liebeck, *The Subgroup Structure of the Finite Classical Groups*, London Math. Soc. Lecture Notes Series **129**, Cambridge University Press, Cambridge, 1990.

[17] H. Lüneburg, On Möbius-planes of even order, *Math. Z.* **92** (1966), 187–193.

[18] H. Lüneburg, *Translation Planes*, Springer, Berlin, 1980.

[19] E.H. Moore, The subgroups of the generalized modular group, *Decennial Publ. Univ. Chicago* (1904), 141–190.

[20] T. Penttila and C.E. Praeger, Ovoids and translation ovals, *J. London Math. Soc.* **56** (1997), 607–624.

[21] B. Segre. On complete caps and ovaloids in three-dimensional Galois spaces of characteristic two, *Acta Arith.* **5** (1959), 315–332.

[22] D.E. Taylor, *The Geometry of Classical Groups*, Heldermann, Berlin, 1992.

[23] J. Tits, Ovoides et groups de Suzuki, *Arch. Math.* **19** (1962), 187–198.

A GENERAL FRAMEWORK FOR SUBEXPONENTIAL DISCRETE LOGARITHM ALGORITHMS IN GROUPS OF UNKNOWN ORDER

Andreas Enge
Institut für Mathematik
Universität Augsburg
86135 Augsburg
Germany
http://www.math.uni-augsburg.de/~enge
enge@math.uni-augsburg.de

Abstract We develop a generic framework for the computation of logarithms in finite class groups. The model allows us to *formulate* a probabilistic algorithm based on collecting relations in an abstract way independently of the specific type of group to which it is applied, and to *prove* a subexponential running time if a certain smoothness assumption is verified. The algorithm proceeds in two steps: first, it determines the abstract group structure as a product of cyclic groups; second, it computes an explicit isomorphism, which can be used to extract discrete logarithms.

1. Motivation

For an additively written group G, an element $\alpha \in G$ and a second element $\beta \in \langle \alpha \rangle$, the *discrete logarithm problem* consists of determining an integer l, denoted by $\log_\alpha \beta$, such that $\beta = l\alpha$. The problem has increasingly received attention since it plays an important role in cryptography. Following Diffie and Hellman's and ElGamal's constructions ([5, 6]), it is possible to base a public key cryptosystem on a finite abelian group in a way that computing logarithms in the group amounts to breaking the system. Conversely, it is widely believed that if the discrete logarithm problem in the group is hard, then the cryptosystem so constructed is indeed secure; see [3] and the literature cited there.

In a generic setting, in which no specific structural properties of the group representation can be exploited, the discrete logarithm problem is

A. Blokhuis et al. (eds.), Finite Geometries, 133–146.

of exponential complexity ([21, 24]). However, in multiplicative groups of finite fields, which were first suggested for cryptography, it can be solved by subexponential algorithms. As the name indicates, the growth of their running times is, in a sense which will be made precise in Section 3.4, between polynomial and exponential, so that relatively large keys are needed to achieve a reasonable level of security.

To avoid subexponential attacks, different finite class groups have been proposed for cryptographic use; among these groups are elliptic curves ([18, 15]), which constitute the state of the art of public key cryptography and which inherit the class group structure from the isomorphism with their Jacobians, Jacobians of hyperelliptic ([16]) and more general ([11], [1]) curves over finite fields, which are commonly seen as a possible replacement for elliptic curves in case the latter turn out to be insecure, and ideal class groups of number fields ([4]). Over the years, various algorithms have been developed which show that the discrete logarithm problem in some of these class groups can be solved in subexponential time, while others apparently resist this kind of attack. Precisely, the subexponential complexity has been proved for ideal class groups of imaginary quadratic number fields under the generalised Riemann hypothesis ([17, 12]) and unconditionally for Jacobians of large genus hyperelliptic curves over finite fields ([7]).

If nothing is known about the class group except its definition and arithmetic, then these algorithms all proceed in two stages: in the first step, the structure of the group as a direct product of cyclic groups of known orders is determined. In the second step, an explicit isomorphism is used to transfer the logarithm problem into the cyclic groups, in which it can be solved trivially by the extended Euclidean algorithm.

If the order of the base element α is known, then the first phase can be omitted, which results in a much faster algorithm. It is observed in [9] that all groups to which this kind of algorithm applies share a common structure, so that it is possible to *formulate* once and for all a generic algorithm for computing discrete logarithms in class groups of known order. Its subexponential running time can also be *proved* in general provided that the class group satisfies a certain smoothness assumption, which has to be verified separately for each type of group.

In this paper, we apply and extend the general framework of [9] to formulate a generic subexponential discrete logarithm algorithm for class groups of unknown order. It solves both computational tasks of determining the abstract group structure and of providing an explicit isomorphism of the group with its abstract model. For imaginary quadratic number fields, we recover the algorithm of [17, 12], for hyperelliptic Jacobians that of [7]. The subexponential complexity is again proved in complete generality under the same smoothness assumption; to show that logarithms in a further type of class groups can be computed in subexponential time, it is thus sufficient to verify this smoothness assumption without going over the details of the algorithm again.

2. The generic model

The setting in which the following algorithm can be applied is that of an *arithmetic formation*, introduced in [14]. The basic idea is to have a structure in which each element admits non-unique decompositions into sums of primes, which permits us to deduce a relation of logarithms

$$\sum (e_i - r_i) \log_\alpha \mathfrak{P}_i = 0$$

from a relation of elements of the form

$$\sum e_i \mathfrak{P}_i = \sum r_i \mathfrak{P}_i.$$

Strictly speaking, this only holds if G is cyclic and generated by α. In [9], further computational assumptions are introduced, and we need some more in our situation where the group order is unknown.

Let $\mathbb{M}$ be the free abelian monoid over some countable (usually infinite) set of *primes* $\mathcal{P}$, and let

$$\deg : \mathbb{M} \to \mathbb{R}^+$$

be a homomorphism of monoids, which to each element of $\mathbb{M}$ associates its *size*. We assume that all primes have a size of at least 1, so that an element of size d is decomposed into a sum of at most $\lfloor d \rfloor$ primes, counting multiplicities. Let $\sim$ be an equivalence relation on $\mathbb{M}$, compatible with addition. If $G = \mathbb{M}/\sim$ is a finite group, then it is clearly abelian, and is called an *arithmetical formation.* We assume that the discrete logarithm problem has to be solved in such a formation G, and denote by N the order of G. Since N is unknown, let N' be such that the input size of the problem is in $O(\log N')$. It is conceivable that $N \ll N'$; see [9]. As polynomial time effort may be neglected compared to subexponential one (see Section 3.4), we measure all complexities up to polynomials in $\log N'$, that is, in $O^\sim$-notation, where for a positive function f of N' we have

$$O^\sim(f) = \bigcup_{k=0}^{\infty} O\left(f(N') \log^k N'\right).$$

Thus, a (time or space) bound is polynomial if and only if it is contained in $O^\sim(1)$.

We suppose that each element of G is given by a canonical representative in $\mathbb{M}$ of polynomial bit size and denote by $\deg g$ the size of this representative. Let

$$d \geq \lfloor \max\{\deg g : g \in G\} \rfloor,$$

so that the maximum number of primes in a decomposition of a group element is bounded above by d, even counting multiplicities. Then we require that $d \in O^{\sim}(1)$ and that the arithmetic of G is realised by manipulating the canonical representatives in polynomial time.

For some *smoothness bound* S, let the *factor base* $\mathcal{P}_S = \{\mathfrak{P}_1, \ldots, \mathfrak{P}_{n_S}\}$ be the set of primes of size at most S, and call an element of the group *S-smooth* if its canonical representative is a sum of primes in $\mathcal{P}_S$. Since a distinction between smooth and non-smooth elements arises only for $S \in O^{\sim}(1)$, we assume this case. On the other hand, S must be sufficiently large so that $\mathcal{P}_S$ generates G; that is, for any $g \in G$, there are non-negative integers e_i such that

$$g \sim \sum_{i=1}^{n_S} e_i \mathfrak{P}_i.$$

In the setting of cyclic groups treated in [9], this generating property was automatically fulfilled.

Let $E \geq 5\max\{N, N'\} + d$, which implies that in fact all e_i in the relation above can be assumed to be smaller than E. On the other hand, we demand $\log E \in O^{\sim}(1)$, so that scalar multiples $e\mathfrak{P}$ for $e < E$ and $\mathfrak{P} \in \mathcal{P}_S$ can be computed in polynomial time by a double-and-add algorithm. It is also assumed that $\log n_S \in O^{\sim}(1)$, which is fulfilled for n_S subexponential in $\log N'$.

We suppose that the factor base can be constructed in $O^{\sim}(n_S^2)$ and that a smoothness test decides in $O^{\sim}(n_S)$ whether a given group element is S-smooth and if so, determines the decomposition over $\mathcal{P}_S$ in the same time bound. In a generic setting, one would expect that $\mathcal{P}_S$ can be constructed by enumerating all elements of size at most S and trial division by all smaller elements. Likewise, the smoothness test might use trial division. In practice, both phases are usually faster, but this is compensated by other parts of the algorithm.

Examples

(1) Ideal class groups of imaginary quadratic number fields.

Let $K = \mathbb{Q}(\sqrt{D})$ be a quadratic number field of discriminant $D < 0$ and $\mathcal{O} = \mathbb{Z}\left[\frac{D+\sqrt{D}}{2}\right]$ its ring of integers. The ideal class group G of

K is defined as $\mathbb{M}/\sim$, where $\mathbb{M}$ is the set of integral ideals of $\mathcal{O}$, the free abelian monoid over the set $\mathcal{P}$ of prime ideals, and $\sim$ is induced by the principal ideals. The size function is given by the dual logarithm of the norm, $\deg \mathfrak{P} = \mathrm{ld}(\mathrm{N}_{K/\mathbb{Q}}(\mathfrak{P}))$. Let $N' = |D|$; by a theorem of Siegel's, $\log N' \in (2+o(1)) \log N$ ([25]). The reduction theory of binary quadratic forms implies that $d = \mathrm{ld}\left(\sqrt{\frac{N'}{3}}\right)$ is a suitable choice, and with $E = 5N' + d$ all computational requirements are met; for details, see [9].

Assuming the validity of the Generalised Riemann Hypothesis, by Corollary 6.2 of [22] the class group is generated, for $n_S \in O(\log^2 N')$, by $\mathcal{P}_S$; that is, $S \in O(\log\log N')$, with effectively computable constants. We will assume n_S to be of subexponential instead of polynomial size, so that the generating property will be fulfilled.

(2) Jacobians of hyperelliptic curves over finite fields.

Let $K = \mathbb{F}_q$ be the finite field with q elements. We consider hyperelliptic curves over K for which the prime divisor at infinity is ramified. Denoting their genus by $g \geq 1$, these are smooth projective models of affine curves of the form

$$H = Y^2 + vY - u$$

with u, $v \in K[X]$, where u is monic of degree $2g+1$; also $v = 0$ if $\operatorname{char} K$ is odd and v is of degree at most g if $\operatorname{char} K = 2$. The K-rational part G of the Jacobian of H is isomorphic to the ideal class group of $\mathcal{O} = K[X,Y]/(H)$, that is, $G = \mathbb{M}/\sim$, where $\mathbb{M}$ is the set of integral ideals of $\mathcal{O}$ and $\sim$ is induced by the principal ideals. Interpreting the norm over $K(X)$ of an ideal of $\mathcal{O}$ as a polynomial in $K[X]$, the size of an ideal is given by the degree of its norm. We let $N' = q^g$, which by Weil's theorem is somehow close to N. By reduction theory, $d = g$, and a theorem of Artin's ([2], § 24, Formula (8)) implies that $E = 5(2g+1)q^g + g$ is a suitable choice so that all computational assumptions are fulfilled; for details, see [9]. Choosing a factor base with $S \geq \lceil 2\log_q(4g-2) \rceil$, the generating property also holds, see [7].

Clearly, Example 2 can be extended to Jacobians of more general curves over finite fields. For instance, the generating property for $S \geq \lceil 2\log_q(4g-2) \rceil$ has been proved for arbitrary curves in [8], Theorem 6.2. Since, to date, the arithmetic has been worked out for only few further classes of curves ([11, 1]) and the smoothness result needed for proving the subexponential running time has been verified only for hyperelliptic curves, we omit the details.

3. The algorithm

As explained in Section 1, the algorithm proceeds in two stages. The first phase determines the group structure as an abstract product of cyclic groups of known order with, as a byproduct, the group order. The second phase is in fact an algorithm for computing the image of any given group element under the isomorphism. Both phases are probabilistic. If a bound Θ on the group order N such that $\Theta/2 < N \leq \Theta$ is known, then they always output the correct answer. If only the input data α, $\beta \in G$ of the discrete logarithm problem are known, then the first phase may terminate with a wrong group structure. In case this structure is compatible with the discrete logarithm problem nevertheless, the second phase halts with the correct logarithm; otherwise, it reports failure, and the algorithm has to be run again. So, for computing discrete logarithms, the algorithm never returns a wrong result; for computing the group order and structure it can only be relied upon if the additional information Θ is available. Notice that Θ can be computed in polynomial time for imaginary quadratic number fields by evaluating a truncated Euler product of the analytic class number formula ([17], p. 468; [22], Theorem 6.3). A similar approach should also be successful for hyperelliptic Jacobians.

The algorithm is based on the observation that the group homomorphism

$$\mathbb{Z}^{n_S} \to G, \quad (e_1, \ldots, e_{n_S}) \mapsto \sum_{i=1}^{n_S} e_i \mathfrak{P}_i$$

is surjective because $\mathcal{P}_S$ is a generating set. Thus, its kernel Γ is a full-dimensional lattice of determinant N, and $G \simeq \mathbb{Z}^{n_S}/\Gamma$. We call the elements of Γ *relations*.

3.1. The group structure

The first phase of the algorithm consists of finding a basis for Γ and bringing it into a special form.

Algorithm 1

(1) Choose a smoothness bound S and construct the factor base, with $n = n_S$,

$$\mathcal{P}_S = \{\mathfrak{P}_1, \ldots, \mathfrak{P}_n\}.$$

(2) Find $14n\left(1 + \operatorname{ld} E + \frac{1}{2} \operatorname{ld} n\right)$ relations by repeatedly selecting random vectors $e = (e_1, \ldots, e_n)^T \in \{0, \ldots, E-1\}^n$ until the canonical representative of $\sum_{i=1}^{n} e_i \mathfrak{P}_i$ is S-smooth and decomposes as

$\sum_{i=1}^{n} r_i \mathfrak{P}_i$, so that $r - e \in \Gamma$. Write the relation vectors $r - e$ as columns into a matrix A.

(3) Compute the rank of A. If it is less than n, return to Step (2).

(4) There is a matrix $\Delta = \operatorname{diag}(d_1, \ldots, d_n)$ with $d_i \in \mathbb{N}$, $1 \leq d_1 | \cdots | d_n$, and unimodular integral square matrices T and U of the appropriate dimensions such that $TAU = (\Delta | 0)$; Δ is called the *Smith normal form* or *elementary divisor form* of A. Compute Δ and T.

(5) If an approximation $\Theta/2 < N \leq \Theta$ is known and $d_1 \cdots d_n > \Theta$, then return to Step (2). If $d_1 \cdots d_n \leq \Theta$, then G is proven to be of the type $\mathbb{Z}/d_1\mathbb{Z} \times \cdots \times \mathbb{Z}/d_n\mathbb{Z}$. Otherwise, this is very likely to be the correct group structure; see Section 3.3.

The correctness of Step (5) follows from the observation that $\operatorname{Im} A \subseteq \Gamma$, whence $\det \Gamma = N$ divides $d_1 \cdots d_n = |\det A|$ and $\operatorname{Im} A = \Gamma$ if and only if $|\det A| = N$.

3.2. The isomorphism and discrete logarithms

Assume that Algorithm 1 has determined the correct group structure. Then the following algorithm computes $\varphi(\gamma)$ for $\gamma \in G$, where

$$\varphi : G \to \mathbb{Z}/d_1\mathbb{Z} \times \cdots \times \mathbb{Z}/d_n\mathbb{Z}$$

is an isomorphism. It expresses γ over the factor base by an approach similar to the relation collection phase of Algorithm 1 and exploits the knowledge of the matrix T.

Algorithm 2

(1) Select random vectors $e = (e_1, \ldots, e_n)^T \in \{0, \ldots, E-1\}^n$ until the canonical representative of $\gamma + \sum_{i=1}^{n} e_i \mathfrak{P}_i$ is S-smooth and decomposes as $\sum_{i=1}^{n} r_i \mathfrak{P}_i$. Let $c = r - e$; then $\gamma = \sum_{i=1}^{n} c_i \mathfrak{P}_i$ in G.

(2) Let $\varphi(\gamma) = Tc \bmod \Delta$, where by $(v_1, \ldots, v_n)^T \bmod \Delta$ we understand $(v_1 + d_1\mathbb{Z}, \ldots, v_n + d_n\mathbb{Z})^T$.

Notice that φ is well-defined: if $\gamma = \sum c_i \mathfrak{P}_i = \sum c_i' \mathfrak{P}_i$, then $c - c' \in \Gamma = \operatorname{Im} A$, so that $T(c - c') \in \operatorname{Im} \Delta$; that is, $Tc - Tc' \equiv 0 \pmod{\Delta}$.

If $\beta = l\alpha$ is the element whose logarithm is sought, then we compute $\varphi(\alpha)$ and $\varphi(\beta)$ and try to solve

$$\varphi(\beta) \equiv l\varphi(\alpha) \pmod{\Delta}. \tag{3.1}$$

If this is possible, then we have found the correct discrete logarithm; otherwise, Algorithm 1 did not determine the correct group structure, and we have to restart the algorithm. Beware that, if no approximation Θ is known, then even in the first case the group structure determined by Algorithm 1 may be wrong. Notice that Algorithm 1 has to be carried out only once and for all and can thus be seen as a precomputation step. Once it has been accomplished, the group structure is known and additional discrete logarithms can be computed by Algorithm 2 alone, which is considerably faster.

3.3. Success probability and running time analysis

The analysis of the algorithms is very similar to the special case treated in [7]; we therefore do not give all the details.

Let N_S denote the number of S-smooth elements of G. Heuristically, one would expect that the probability of finding a relation in Step (2) of Algorithm 1 is about $\frac{N_S}{N}$, and since E is chosen sufficiently large, the deviation from this estimate can be shown to be small. Observe that the vectors $-e$ lie in the cube $\{1-E, \ldots, 0\}^n$ and that the coefficient vectors r of smooth elements lie in the cube $\{0, \ldots, d\}^n$. Since each obtained relation is of the form $c = r - e$, only relations inside the Minkowski sum $W^+ = \{1 - E, \ldots, d\}^n$ of these two cubes may be reached; moreover, the number of pairs $(e, r) = (r - c, r)$ yielding a fixed relation c is at most N_S. Hence, the probability of obtaining a relation is bounded above by $\frac{|\Gamma \cap W^+| N_S}{E^n}$, which can be shown to be no larger than $\frac{6 \sqrt[40]{e} N_S}{5N} \in \frac{N_S}{N}\Theta(1)$ by estimating the cardinality of the intersection of a cube and a lattice. Similarly, there are exactly N_S pairs (e, r) yielding a fixed relation $c \in W^- = \{d+1-E, \ldots, 0\}^n$, so that the probability of finding a relation is bounded below by $\frac{|\Gamma \cap W^-| N_S}{E^n} \geq \frac{4 N_S}{5 \sqrt[40]{e} N} \in \frac{N_S}{N}\Theta(1)$. The same analysis applies to Step (1) of Algorithm 2, which amounts to finding one additional relation since the polynomial time effort needed to solve (3.1) is negligible.

Our next aim is to estimate the probability with which one run of Steps (1) to (4) of Algorithm 1 determines the correct group structure, that is, that Step (2) determines the full lattice Γ. Assume that after finding i relations some sublattice $\Gamma_i \subseteq \Gamma$ is reached. The $(i+1)$-th relation is called *useful* if it increases the dimension of Γ_i by 1 or, after the maximal dimension is reached, it decreases its determinant. In the first case, let Γ_i' be a full-dimensional lattice such that $\Gamma_i \subsetneq \Gamma_i' \subsetneq \Gamma$, in the second case, let $\Gamma_i' = \Gamma_i$. Then all relations in $\Gamma \backslash \Gamma_i'$ are useful, and the probability of finding a useless relation is bounded above by

$\frac{|\Gamma_i' \cap W^+| N_S}{E^n} \leq \frac{7\sqrt[40]{e} N_S}{10N}$. Hence the conditional probability that a found relation is useless is at most $\frac{7\sqrt[40]{e}N_S}{N} / \frac{4N_S}{5\sqrt[40]{e}N} < \frac{12}{13}$.

To reach a full-dimensional lattice, exactly n useful relations are needed. After that, Hadamard's bound shows that the determinant of the obtained lattice is at most $(\sqrt{n}E)^n$ since the coefficients of the relations are bounded by E in absolute value. As each further useful relation reduces the determinant by a factor of at least 2, at most $n\left(\operatorname{ld} E + \frac{1}{2}\operatorname{ld} n\right)$ additional useful relations are needed. Modelling the usefulness of relations by a random variable X which is binomially distributed with parameters $\left(14n\left(1 + \operatorname{ld} E + \frac{1}{2}\operatorname{ld} n\right), 1/13\right)$, we see that one run of Algorithm 1 is successful with probability at least

$$\begin{aligned} P\left(X \geq n\left(1 + \operatorname{ld} E + \frac{1}{2}\operatorname{ld} n\right)\right) &\geq 1 - P\left(|X - E(X)| \geq \frac{1}{13}E(X)\right) \\ &\geq 1 - \frac{\operatorname{Var}(X)}{\frac{1}{169}E(X)^2} \\ &\qquad \text{by Tchebychev's inequality} \\ &= 1 - \frac{168}{n\left(1 + \operatorname{ld} E + \frac{1}{2}\operatorname{ld} n\right)} \to 1 \\ &\qquad \text{for } n \to \infty. \end{aligned}$$

Since n will be chosen such that $n \to \infty$ for $N' \to \infty$, the success probability is in $1 - o(1)$, and only $1 + o(1)$ executions of the algorithm will be needed on average.

Taking into account that one linear combination of n primes can be computed in $O(n \log E)$, the analysis above shows that the expected running time for Step (2) of Algorithm 1 is in

$$O\left(n \log E \cdot n(\log E + \log n) \cdot \frac{N}{N_S}\right) \subseteq O^{\sim}\left(n^2 \frac{N}{N_S}\right)$$

and that the expected running time of Algorithm 2 is in $O^{\sim}\left(n\frac{N}{N_S}\right)$. By assumption, Step (1) of Algorithm 1 can be accomplished in $O^{\sim}(n^2)$, which is dominated by Step (2).

Following [13], it can be shown that the linear algebra problems in Steps (3) and (4) of Algorithm 1 can be solved slightly faster than assumed in [7]. Let $m = 14n\left(1 + \operatorname{ld} E + \frac{1}{2}\operatorname{ld} n\right)$, $f(k) = k \log^2 k \log\log k$, and let $|A|$ denote the maximal absolute value among the coefficients of A. Using fast multiplication techniques, the rank of A and its Hermite normal form B, which is an intermediate step for computing the elementary divisor form Δ, are determined in

$$O\left(mn^2 f(n \log(n|A|))\right) \subseteq O^{\sim}\left(n^4\right),$$

and Δ together with the companion matrix S are computed in

$$O\left(n^3 \log |B| f(\log |B|)\right) \subseteq O^{\sim}\left(n^3\right),$$

since $|B| \leq N$ and $\log N \in O^{\sim}(1)$. Thus, the total expected running time of Algorithm 1 is

$$O^{\sim}\left(n^4 + n^2 \frac{N}{N_S}\right),$$

which dominates the expected running time of

$$O^{\sim}\left(n \frac{N}{N_S}\right)$$

of Algorithm 2.

3.4. Subexponentiality

Clearly, $n = n_S$ grows with S, while $\frac{N}{N_S}$ decreases with S, so that the optimal algorithmic complexity is obtained for some intermediate value of S. It turns out that the best choice for n is subexponential. Recall the definition

$$L_{N'}(\rho) = e^{\rho\sqrt{\log N' \log\log N'}}$$

of the subexponential function with respect to the input size $\log N'$ and a parameter $\rho > 0$, which interpolates between the polynomial $e^{\rho\sqrt{(\log\log N')^2}} = (\log N')^{\rho}$ and the exponential function $e^{\rho\sqrt{(\log N')^2}} = (N')^{\rho}$. Notice that the inclusion $O^{\sim}(1) \subseteq L_{N'}(o(1))$ holds.

Theorem 3.1 *Assume that a* smoothness result *of the following form is available: S can be chosen such that*

$$n \in O\big(L_{N'}(\rho + o(1))\big),$$
$$\frac{N}{N_S} \in O\big(L_{N'}(\sigma + o(1))\big),$$

for some constants ρ, $\sigma > 0$. Then the expected running time of Algorithm 1 becomes

$$O\big(L_{N'}(\max\{4\rho, 2\rho + \sigma\} + o(1))\big)$$

and that of Algorithm 2 becomes

$$O\big(L_{N'}(\rho + \sigma + o(1))\big).$$

To keep the presentation concise, only the complexity of a complete run of Algorithms 1 and 2 is considered in the following, corresponding to the first discrete logarithm problem solved in the group. Further logarithms can be computed faster by Algorithm 2 alone, and Theorem 3.1 shows how to determine the corresponding complexity.

Examples

(1) Ideal class groups of imaginary quadratic number fields of discriminant D.

Proposition 4.4 of [23] shows that under the Generalised Riemann Hypothesis, the choice $S = \lceil \log L_N(\rho) \rceil$ yields $n \in O(L_N(\rho))$ and $\sigma = \frac{1}{2\rho}$. Letting $\rho = \frac{1}{2}$, which turns out to be optimal, we recover the running time of

$$O\big(L_N(2 + o(1))\big) = O\big(L_{|D|}(\sqrt{2} + o(1))\big)$$

of [17, 12].

(2) Jacobians of hyperelliptic curves over finite fields

Theorem 6 of [10] shows that the choice $S = \lceil \log_q L_{q^g}(\rho) \rceil$ unconditionally yields $\frac{N}{N_S} \in O\big(L_{q^g}(\sigma + o(1))\big)$ for $\sigma = \frac{1}{2\rho}$. Unfortunately, the need for rounding up S implies that no tighter bound than $n \in O(qL_{q^g}(\rho))$ can be shown. Thus, q must be subexponential in $g \log q$, which is only possible for g large compared to q. More precisely, $n \in O\left(L_{q^g}\left(\rho + \frac{1}{\sqrt{\vartheta}}\right)\right)$ if all instances under consideration satisfy $g \geq \vartheta \log q$, [7, Theorem 8]. Letting $\rho = \frac{1}{2}\left(\sqrt{1 + \frac{1}{\vartheta}} - \sqrt{\frac{1}{\vartheta}}\right)$, we have proved the following result.

Theorem 3.2 *Algorithms 1 and 2 compute discrete logarithms in Jacobians of hyperelliptic curves of genus g over finite fields $\mathbb{F}_q$ satisfying $g \geq \vartheta \log q$ for some positive constant ϑ in expected time in*

$$O\left(L_{q^g}\left(2\left(\sqrt{1 + \frac{1}{\vartheta}} + \sqrt{\frac{1}{\vartheta}}\right) + o(1)\right)\right).$$

If $g/\log q$ tends to infinity for the instances under consideration, for example, if q is constant and $g \to \infty$, then the complexity is

$$O\big(L_{q^g}(2 + o(1))\big).$$

Especially for small values of ϑ this is better than the complexity proved in [7] of

$$O\left(L_{q^g}\left(\frac{5}{\sqrt{6}}\left(\sqrt{1+\frac{3}{\vartheta}}+\sqrt{\frac{3}{\vartheta}}\right)+o(1)\right)\right)$$

since we have based our analysis on faster algorithms for the linear algebra. For $\vartheta = 1$, for instance, the constant in the subexponential function drops from 7.61... to 4.82...

4. Outlook and remarks

Algorithms 1 and 2 generalise the previous subexponential algorithms for computing the group structure and discrete logarithms in imaginary quadratic ideal class groups and hyperelliptic Jacobians to arbitrary arithmetical formations satisfying a few reasonable computational assumptions. This general framework is intended to simplify the development and the analysis of discrete logarithm algorithms for further class groups of unknown order. Specially interesting examples might be provided by more general Jacobians over finite fields. To apply the generic algorithm to a new class of arithmetical formations, it suffices to provide a description of their arithmetic meeting the requirements of Section 2. The proof of the subexponential running time then boils down to verifying the smoothness assumption of Theorem 3.1.

We remark that applying the results of Section 3.3 on the complexity of the linear algebra steps one may show that the discrete logarithm problem described in [19], which is set in some structure different from, but close to a group, can be solved slightly faster. Repeating the analysis of [19], which is in fact only valid for $g/\log q \to \infty$, and extending it for arbitrary values of ϑ, we obtain the following result.

Theorem 4.1 *There is a subexponential algorithm computing the regulator and discrete logarithms in the infrastructure of real quadratic hyperelliptic curves (that is, curves for which the infinite prime divisor splits) of genus g over finite fields $\mathbb{F}_q$ of odd characteristic satisfying $g \geq \vartheta \log q$ for some positive constant ϑ. Its expected running time is in*

$$O\left(L_{q^g}\left(2\left(\sqrt{1+\frac{1}{\vartheta}}+\sqrt{\frac{1}{\vartheta}}\right)+o(1)\right)\right) =$$

$$O\left(L_{q^{2g+2}}\left(\sqrt{2}\left(\sqrt{1+\frac{1}{\vartheta}}+\sqrt{\frac{1}{\vartheta}}\right)+o(1)\right)\right).$$

If $g/\log q$ tends to infinity for the instances under consideration, then the complexity is

$$O\big(L_{q^g}(2+o(1))\big) = O\big(L_{q^{2g+2}}(\sqrt{2}+o(1))\big).$$

Using the theory of real quadratic hyperelliptic function fields of characteristic 2 developed in [20], it should be possible to extend the theorem to curves over arbitrary finite fields.

References

[1] S. Arita, Algorithms for computations in Jacobian group of C_{ab} curve and their application to discrete-log-based public key cryptosystems, *Conference on The Mathematics of Public Key Cryptography* (eds. A. Odlyzko, G. Walsh, and H. Williams), Toronto, 1999.

[2] E. Artin, Quadratische Körper im Gebiete der höheren Kongruenzen II, *Math. Z.* **19** (1924), 207–246.

[3] D. Boneh, The decision Diffie–Hellman problem, *Algorithmic Number Theory ANTS-III* (ed. J.P. Buhler), Lecture Notes in Computer Science **1423**, Springer, Berlin, 1998, 48–63.

[4] J. Buchmann and H.C. Williams. A key-exchange system based on imaginary quadratic fields, *J. Cryptology* **1** (1988), 107–118.

[5] W. Diffie and M.E. Hellman, New directions in cryptography, *IEEE Trans. Inform. Theory* **22** (1976), 644–655.

[6] T. ElGamal, A public key cryptosystem and a signature scheme based on discrete logarithms, *IEEE Trans. Inform. Theory* **31** (1985), 469–472.

[7] A. Enge, Computing discrete logarithms in high-genus hyperelliptic Jacobians in provably subexponential time, 1999, *Math. Comp.*, to appear.

[8] A. Enge, Hyperelliptic Cryptosystems: Efficiency and Subexponential Attacks, PhD thesis, Universität Augsburg, 2000.

[9] A. Enge and P. Gaudry, A general framework for subexponential discrete logarithm algorithms, Research Report LIX/RR/00/04, Laboratoire d'Informatique de l'École Polytechnique, Palaiseau, June 2000, submitted.

[10] A. Enge and A. Stein, Smooth ideals in hyperelliptic function fields, 2000, *Math. Comp.*, to appear.

[11] S.D. Galbraith, S. Paulus, and N.P. Smart, Arithmetic on superelliptic curves, 2000, *Math. Comp.*, to appear.

[12] J.L. Hafner and K.S. McCurley, A rigorous subexponential algorithm for computation of class groups, *J. Amer. Math. Soc.* **2** (1989), 837–850.

[13] J.L. Hafner and K.S. McCurley, Asymptotically fast triangularization of matrices over rings, *SIAM J. Comput.* **20** (1991), 1068–1083.

[14] J. Knopfmacher, *Abstract Analytic Number Theory*, North-Holland Mathematical Library **12**, North-Holland, Amsterdam, 1975.

[15] N. Koblitz, Elliptic curve cryptosystems, *Math. Comp.* **48** (1987), 203–209.

[16] N. Koblitz, Hyperelliptic cryptosystems, *J. Cryptology* **1** (1989), 139–150.

[17] K.S. McCurley. Cryptographic key distribution and computation in class groups, *Number Theory and Applications* (ed. R.A. Mollin), NATO ASI Series C: Mathematical and Physical Sciences **265**, Kluwer, Dordrecht, 1989, 459–479.

[18] V.S. Miller. Use of elliptic curves in cryptography, *Advances in Cryptology — CRYPTO '85*, (ed. H.C. Williams), Lecture Notes in Computer Science, **218**, Springer, Berlin, 1986, 417–426.

[19] V. Müller, A. Stein and C.Thiel, Computing discrete logarithms in real quadratic congruence function fields of large genus, *Math. Comp.* **68** (1999), 807–822.

[20] V. Müller, S.A. Vanstone, and R. Zuccherato, Discrete logarithm based cryptosystems in quadratic function fields of characteristic 2, *Des. Codes Cryptogr.* **14** (1998), 159–178.

[21] V.I. Nechaev, Complexity of a determinate algorithm for the discrete logarithm, *Math. Notes* **55** (1994), 165–172.

[22] R.J. Schoof, Quadratic fields and factorization, *Computational Methods in Number Theory, Part II* (eds. H.W. Lenstra and R. Tijdeman), Mathematisch Centrum, Amsterdam, 1982, 235–286.

[23] M. Seysen, A probabilistic factorization algorithm with quadratic forms of negative discriminant, *Math. Comp.* **48** (1987), 757–780.

[24] V. Shoup, Lower bounds for discrete logarithms and related problems, *Advances in Cryptology — EUROCRYPT '97* (ed. W. Fumy), Lecture Notes in Computer Science **1233**, Springer, Berlin, 1997, 256–266.

[25] C.L. Siegel, Über die Classenzahl quadratischer Zahlkörper, *Acta Arith.* **1** (1936), 83–86.

GENERALIZED QUADRANGLES AND PENCILS OF QUADRICS

Dina Ghinelli
Dipartimento di Matematica
Università degli studi di
Roma "La Sapienza"
Piazzale Aldo Moro 2
I 00185 Roma
Italy
dina@mat.uniroma1.it

Stefan Löwe
T-Nova Deutsche Telekom
Innovationsgesellschaft mbH
Technologiezentrum
Goslarer Ufer 35
10589 Berlin
Germany
Stefan.Loewe@telekom.de

Abstract Starting with a pencil of quadrics not all singular in $PG(2,q)$, q odd, such that the pencil has reducible base, a generalized quadrangle of order (q^4, q^2) having property (G) is constructed. The purpose of the present paper is to show that the only kind of pencil that will work is a pencil of elliptic quadrics.

1. Introduction

Generalized quadrangles, as a special case of generalized polygons, were introduced by Tits in 1959 [22]. There he gave what are considered the classical examples. The first non-classical examples were also found by Tits in the mid-1960s and first appeared in P. Dembowski [3]. Other non-classical examples were constructed in [1, 7, 11, 12].

A. Blokhuis et al. (eds.), Finite Geometries, 147–158.

In the 1980s starting with a construction of Kantor [8], several new families with parameters (q^2, q) have been discovered, often with restrictions on the prime power q. As Payne pointed out in the survey [13], "the stories of these discoveries blend in an interesting way, connecting the study of translation planes, spreads of the 3-dimensional projective space over a finite field, flocks of a quadratic cone, and even generalized hexagons!".

The reader is referred to [14] and [15]–[21] for definitions, results, and references to the vast literature on finite generalized quadrangles.

A unifying method for the construction of all known GQs with a regular point was given in [10]. Applications of this method, which is based on covering of nets, can be found in [6] and [5]. Certain central extensions of abelian groups, called in [10] (see also [9]) *fourgonal extensions*, yield further applications of the method. There it is noted that one can construct a generalized quadrangle from the following data:

(i) a field $F = \mathrm{GF}(q)$, q odd;

(ii) n symmetric $2n \times 2n$ matrices $S_1, \ldots, S_n$ over F;

(iii) q^n $2n \times 2n$ matrices $\{M_a \mid a \in F^n\}$ over F, such that

   (a) $M_0 = 0$ and $M_1 = 1$;

   (b) $M_a - M_b$ is regular for different $a, b \in F^n$;

   (c) $M_a S_j = S_j M_a^{\mathrm{tr}}$ for all $j = 1, \ldots, n$ and $a \in F^n$;

   (d) for pairwise different a, b, $c \in F^n$, each of the two sets

$$\{(M_a - M_b)S_j \mid j = 1, \ldots n\},$$
$$\{(M_a - M_b)(M_b - M_c)^{-1}(M_c - M_a)S_j \mid j = 1, \ldots n\},$$

   of quadratic forms has only the null vector as a common zero.

It will be seen in Section 4 how to construct a generalized quadrangle from these data.

Remark 1.1 (A) By (iii)(c), all matrices $(M_a - M_b)S_j$ and $(M_a - M_b)(M_b - M_c)^{-1}(M_c - M_a)S_j$ are symmetric.

(B) If every condition except (iii)(a) is fulfilled, we may replace M_a by $M_a^* = (M_a - M_0)(M_1 - M_0)^{-1}$ and S_j by $S_j^* = (M_1 - M_0)S_j$. The starred system then fulfills (i)–(iii).

(C) For a regular matrix $(r_{i,j})$ define $S_i^* = \sum_j r_{i,j} * S_j$. Then (i)–(iii) hold for $\{S_i^*\}$ if and only if they hold for $\{S_i\}$.

(D) For a regular matrix N, put $S_i^* = NS_iN^{\mathrm{tr}}$ and $M_a^* = NM_aN^{-1}$. Then (i)–(iii) hold for the starred system if and only if they hold for the unstarred.

In this article we shall be concerned with $n = 2$. Here, classical terminology is appropriate. If S and T are two symmetric 4×4 matrices, we call their linear span the pencil $\mathsf{P}(S,T)$. The set $\{x \mid xSx^{\mathrm{tr}} = xTx^{\mathrm{tr}} = 0\}$ is called the base of P and is the set of common zeros of the quadratic forms contained in it. Since we think projectively, we call a base empty if it only contains the zero vector. Thus (iii)(d) means that certain pencils must have an empty base.

2. Pencils with empty base

In this section we use the results of Bruen and Hirschfeld in [2]. The base of a pencil is a quartic curve. In order to be empty, it must split over some extension field into more than one component; so it is reducible. If a pencil consists only of singular quadratic forms, its base is not empty. To construct generalized quadrangles, we therefore have to deal with non-singular, reducible pencils in a projective 3-space. In general, there are six classes of quadrics in the projective 3-space: a plane, a plane pair, a line, a cone, a ruled quadric, and an ovoid. The numbers of quadrics of the various types contained in a pencil P will be denoted by n_1, n_2^+, n_2^-, n_3, n_4^+, n_4^-, respectively. The indices indicate the rank of the quadric. From the classification of non-singular, reducible pencils in [2] we obtain that there are only four types of pencils with empty base. With the same enumeration as in [2]), these are as follows:

(1d)	$n_2^- = 2,\ n_4^+ = q-1;$
(1e)	$n_4^- = q+1;$
(3d)(i)	$n_2^- = 1,\ n_3 = 2,\ n_4^+ = \frac{1}{2}(q-3),\ n_4^- = \frac{1}{2}(q-1);$
(3d)(iii)	$n_2^- = 1,\ n_4^+ = \frac{1}{2}(q-1),\ n_4^- = \frac{1}{2}(q+1).$

Note that a pencil with empty base is uniquely characterized by the number of singular quadrics it contains. Observe also that the number of ruled quadrics never equals the number of ovoids.

It is useful to recall the next result.

Lemma 2.1 *Let S be a non-singular, symmetric 4×4 matrix over F. Then $\{x \mid xSx^{\mathrm{tr}} = 0\}$ is an ovoid if and only if* $\det S$ *is a non-square in F.*

Putting everything together we arrive at the theorem.

Theorem 2.2 *Let $\mathsf{P}(S,T)$ a reducible pencil with empty base, let M be a non-singular matrix such that MS and MT are symmetric, and let* *the pencil* $M\mathsf{P}(S,t)$ *have empty base. Then* $\det M$ *is a square, and there exists a matrix* N *such that*

$$M\mathsf{P}(S,T) = N\mathsf{P}(S,T)N^{\mathrm{tr}}.$$

Proof Since M is non-singular, $\mathsf{P}(S,T)$ and $M\mathsf{P}(S,T)$ have the same number of singular quadrics. As both have empty base, they must be of the same type. If $\det M$ was a square, an ovoid in $\mathsf{P}(S,T)$ would, by Lemma 2.1, become a ruled quadric in $M\mathsf{P}(S,T)$. This is, however, impossible. □

3. Admissible pencils

We call a pencil $\mathsf{P}(S,T)$ *admissible for the* q^2 *matrices* $\{M_a \mid a \in F^2\}$, with $M_0 = 0, M_1 = 1$, if

(i) M_aS and M_aT are symmetric for all $a \in F^2$;

(ii) $M_a - M_b$ is invertible for all $a \neq b$;

(iii) the pencils $(M_a - M_b)\mathsf{P}(S,T)$ have empty base for all $a \neq b$.

We call a pencil P *admissible* if and only if there exists a set of matrices such that P is admissible for it. If $S = S_1$ and $T = S_2$ have the properties (i)–(iii) from Section 1, $\mathsf{P}(S,T)$ must be an admissible pencil. The purpose of this section is to show that an admissible pencil is of type (1e). We assume S and T as given in the standard form according to [2]. Let $x^2+2\beta x+1$ and $x^2+x+\mu$ be irreducible polynomials over F, where μ is a non-square. Let ν and $1+\gamma^2$ be non-squares and $\delta^2 = \nu(1+\gamma^2)$. Define

$$e = \begin{pmatrix} \nu & 0 \\ 0 & 1 \end{pmatrix}, \quad f = \begin{pmatrix} 1 & \beta \\ \beta & 1 \end{pmatrix}, \quad g = \begin{pmatrix} 0 & \frac{1}{2} \\ \frac{1}{2} & 0 \end{pmatrix}, \quad h = \begin{pmatrix} 1 & \frac{1}{2} \\ \frac{1}{2} & \mu \end{pmatrix},$$

Note that all the matrices are invertible. The standard form of the three pencils may now be read off from Table 1.

We should add that (1e) is not in the standard form given by [2]; since, in that case, $\det T = \nu(\nu\gamma)^2$ and $\det(S+xT) = \nu^3\gamma^2(\nu\gamma^2+2\delta x+x^2)^2$, the pencil contains solely elliptic quadrics.

Lemma 3.1 *Let* k *be a regular* 2×2 *matrix over* F *and let*

$$S = \begin{pmatrix} k & \mathbf{0} \\ \mathbf{0} & \mathbf{0} \end{pmatrix}.$$

Table 1. Admissable pencils

Type	S	T
(1d)	$\begin{pmatrix} f & 0 \\ 0 & 0 \end{pmatrix}$	$\begin{pmatrix} 0 & 0 \\ 0 & f \end{pmatrix}$
(1e)	$\begin{pmatrix} 2\delta\nu e - 2\nu^2(2+\gamma^2)g & 0 \\ 0 & e+2\delta g \end{pmatrix}$	$\begin{pmatrix} \nu e - 2\nu\delta g & 0 \\ 0 & 2g \end{pmatrix}$
(3d)(i)	$\begin{pmatrix} f & 0 \\ 0 & 0 \end{pmatrix}$	$\begin{pmatrix} g & 0 \\ 0 & f \end{pmatrix}$
(3d)(iii)	$\begin{pmatrix} h & 0 \\ 0 & 0 \end{pmatrix}$	$\begin{pmatrix} g & 0 \\ 0 & h \end{pmatrix}$

Let M be a regular matrix such that $MS = SM^{\mathrm{tr}}$. Then M is a block upper-triangular matrix

$$M = \begin{pmatrix} W & X \\ \mathbf{0} & Z \end{pmatrix},$$

and $Wk = kW^{\mathrm{tr}}$.

Proof This is an elementary calculation. □

For this type of matrices we consider the impact of T.

Lemma 3.2 *Let k and ℓ be two 2×2-matrices where ℓ is regular. Let*

$$T = \begin{pmatrix} k & \mathbf{0} \\ \mathbf{0} & \ell \end{pmatrix}, \quad M = \begin{pmatrix} W & X \\ \mathbf{0} & Z \end{pmatrix}$$

such that $MT = TM^{\mathrm{tr}}$. Then $X = \mathbf{0}$, $Wk = kW^{\mathrm{tr}}$, $Z\ell = \ell Z^{\mathrm{tr}}$.

Proof This is also an elementary calculation. □

Lemmas 3.1 and 3.2 apply to all three types of matrices S being considered; hence the matrices $\{M_a\}$ of an admissible pencil not of type (1e) are of the form

$$M_a = \begin{pmatrix} W_a & \mathbf{0} \\ \mathbf{0} & Z_a \end{pmatrix}.$$

For $a \neq b$, both $W_a - W_b$ and $Z_a - Z_b$ are regular; also, $W_0 = Z_0 = \mathbf{0}$ and $W_1 = Z_1 = \mathbf{1}$. It follows that the matrices $\{W_a\}$ are pairwise different and regular for $a \neq 0$; the same holds for the matrices $\{Z_a\}$. Let us exploit these properties further.

Lemma 3.3 *A pencil* $\mathsf{P}(S,T)$ *of type* (3d)(i) *is not admissible.*

Proof Suppose that P is admissible for the q^2 matrices

$$M_a = \begin{pmatrix} W_a & \mathbf{0} \\ \mathbf{0} & Z_a \end{pmatrix},$$

and let

$$M = \begin{pmatrix} W & \mathbf{0} \\ \mathbf{0} & Z \end{pmatrix}$$

be one of those matrices. Then $Wf = fW^{\mathrm{tr}}$ and $Wg = gW^{\mathrm{tr}}$. Since $f - 2\beta g = \mathbf{1}$, the matrix W is symmetric with $Wg = gW^{\mathrm{tr}}$. It follows that

$$W = \begin{pmatrix} x & y \\ y & x \end{pmatrix}$$

for x, $y \in F$; hence the set $\{W_a\}$ of q^2 matrices is uniquely determined. However, we would need that all except one of these matrices are regular, which clearly does not hold for the matrices W we have found. □

Lemma 3.4 *A pencil* $\mathsf{P}(S,T)$ *of type* (3d)(iii) *is not admissible.*

Proof Suppose that P is admissible for the q^2 matrices

$$M_a = \begin{pmatrix} W_a & \mathbf{0} \\ \mathbf{0} & Z_a \end{pmatrix},$$

and let

$$M = \begin{pmatrix} W & \mathbf{0} \\ \mathbf{0} & Z \end{pmatrix}$$

be one of those matrices. Then $Wg = gW^{\mathrm{tr}}$ and $Wh = hW^{\mathrm{tr}}$. Since

$$h - g = \begin{pmatrix} 1 & 0 \\ 0 & \mu \end{pmatrix}$$

we must determine all W such that $Wg = gW^{\mathrm{tr}}$ and

$$W \begin{pmatrix} 1 & 0 \\ 0 & \mu \end{pmatrix} = \begin{pmatrix} 1 & 0 \\ 0 & \mu \end{pmatrix} W^{\mathrm{tr}}.$$

It follows that

$$W = \begin{pmatrix} x & y \\ \mu y & x \end{pmatrix};$$

so the set $\{W_a\}$ is again uniquely determined. Note that this time all but one of these matrices are regular. By Theorem 2.2, there exists a matrix N such that $M\mathsf{P} = N\mathsf{P}N^{\mathrm{tr}}$. Since S is the only singular quadric in P, we must have $MS = NSN^{\mathrm{tr}}$. Writing N in block form

$$N = \begin{pmatrix} N_{11} & N_{12} \\ N_{21} & N_{22} \end{pmatrix},$$

it follows that $Wh = N_{11}hN_{11}^{\mathrm{tr}}$. Since h is regular, $\det W = x^2 - \mu y^2$ is a square. By [4], the latter expression can assume values which are non-squares. □

Lemma 3.5 *A pencil* $\mathsf{P}(S,T)$ *of type* (1d) *is not admissible.*

Proof We start as in the proofs of the previous two lemmas. For $a \neq b$, let $M = M_a - M_b$ and write

$$M = \begin{pmatrix} W & \mathbf{0} \\ \mathbf{0} & Z \end{pmatrix}.$$

Again, there exists a regular matrix N such that $M\mathsf{P} = N\mathsf{P}N^{\mathrm{tr}}$. Now, in P, only the quadrics S and T are singular; so either $MS = NSN^{\mathrm{tr}}$ or $MS = NTN^{\mathrm{tr}}$. In both cases, $Wf = N_{11}fN_{11}^{\mathrm{tr}}$. Since f is regular, $\det W$ is a square.

Let us now determine all possible $W = W_a - W_b$ such that $Wf = fW^{\mathrm{tr}}$. It is advantageous to transform the problem using the symmetric matrix

$$B = \begin{pmatrix} 1 & 1 \\ 1 & -1 \end{pmatrix}.$$

We have $B^{-1} = \frac{1}{2}B$, and

$$f^* = BfB^{-1} = \begin{pmatrix} 1+\beta & 0 \\ 0 & 1+\beta \end{pmatrix}.$$

Put $V = BWB^{-1}$. Then $V^{\mathrm{tr}} = BW^{\mathrm{tr}}B^{-1}$, and hence $Vf^* = f^*V^{\mathrm{tr}}$. It follows that the possible V are among

$$\begin{pmatrix} x & y \\ \frac{1-\beta}{1+\beta}y & z \end{pmatrix}$$

with $x, y, z \in F$. Since $\det W$ is a square, $\det V = \det W$ is a square as well. Recall that a and b are arbitrary and that $M_0 = \mathbf{0}$. Suppose that, for certain a and b, we have that

$$V_a = BW_aB^{-1} = \begin{pmatrix} x & y_a \\ \frac{1-\beta}{1+\beta}y_a & z_a \end{pmatrix}, \quad V_b = BW_bB^{-1} = \begin{pmatrix} x & y_b \\ \frac{1-\beta}{1+\beta}y_b & z_b \end{pmatrix}.$$

Then $\det V = \det(V_a - V_b) = \frac{\beta-1}{\beta+1}(y_a - y_b)^2$. But this is not a square; for, $x^2 + 2\beta x + 1$ is irreducible, hence $\beta^2 - 1$ is a non-square, whereas $\frac{\beta-1}{\beta+1}(\beta^2 - 1)$ is a square. This means that the upper left-hand entries of the matrices $\{V_a\}$ must be pairwise different; so we can find at most q of these matrices. □

The result of the previous three lemmas is the following.

Theorem 3.6 *An admissible pencil is of type* (1e).

It is now timely to show that there exist matrices $\{M_a\}$ for which a pencil of type (1e) is admissible. For example, after choosing an F-basis $1, i$ for $\mathrm{GF}(q^2)$ such that $\nu = i^2$ is a primitive element of K, we may identify $\mathrm{GF}(q^2)$ with F^2 and put

$$A_{x,y} = \begin{pmatrix} x & \nu y \\ y & x \end{pmatrix}, \quad M_{x,y} = \mathbf{1}_2 \otimes A_{x,y} ,$$

where $\mathbf{1}_2$ is the 2×2 unit matrix. It is easy to check properties (i)–(iii) of Section 1.

4. Property (G)

From now on we shall consider generalized quadrangles arising from the totally elliptic pencil P of type(1e), given in standard form. We consider briefly a connection between the algebraic and geometric structure of such a GQ and then proceed to show that it must have property (G).

Property (G) *Let K_1, L_1 be distinct concurrent lines of a generalized quadrangle $\mathcal{Q}$. The pair (K_1, L_1) is said to have Property* (G) *if, whenever four distinct lines (K_1, K_2, K_3, K_4) are pairwise non-concurrent and (L_1, L_2, L_3, L_4) are four distinct lines for which K_i meets L_j for $2 \le i + j \le 7$, then also K_4 meets L_4. A generalized quadrangle $\mathcal{Q}$ is said to have Property* (G) *at a point p provided each pair (K, L) of distinct lines through p has Property* (G).

Let $\mathsf{P}(S, T)$ be a pencil of type (1e). By Remark 1.1 of Section 1, we may assume for our purposes that S and T are given in the standard form as shown in Table 1. Let $\{M_a \mid a \in F^2\}$ be a set of matrices such that (i)–(iii) of Section 1 hold with respect to S and T. Let $E = \begin{pmatrix} 0 & -1 \\ 1 & 0 \end{pmatrix}$.

On the set $G = F^2 \times F^8$ we define a group structure with the 2-cocycle $\Sigma = (E \otimes S, E \otimes T)$. For every matrix M_a let $U_a = \{(x, xM_a) \in F^8 \mid x \in F^4\}$; let $U_\infty = \{(0, x) \in F^8 \mid x \in F^4\}$. Every U_b defines the subgroups

$A_b = \{[0, y] \mid y \in U_b\}$ and $A_b^* = \{[\alpha, y] \mid y \in U_b, \alpha \in F^2\}$. Define a point-line incidence structure $\mathcal{Q} = (\mathcal{P}, \mathcal{L}, \mathcal{I})$ as follows.

Points are of three types:

(p1) elements of G;

(p2) right cosets $A_b^* g$ ($b \in F^2 \cup \{\infty\}$, $g \in G$);

(p3) one symbol ∞.

Lines are of two types:

(L1) right cosets $A_b g$ ($b \in F^2 \cup \{\infty\}$, $g \in G$);

(L2) symbols $[A_b]$ ($b \in F^2 \cup \{\infty\}$).

Incidence is defined as follows:

(I1) a point g of type (p1) is incident with thes lines $A_b g$ ($b \in F^2 \cup \{\infty\}$) of type (L1);

(I2) a point $A_b^* g$ of type (p2) is incident with the line $[A_b]$ of type (L2) and with each line $A_b h g$ contained in $A_b^* g$, for $b \in F^2 \cup \{\infty\}$, $g \in G$, $h \in A_b^*$;

(I3) the point ∞ is incident with each line $[A_b]$ of type (L2);

(I4) there are no further incidences.

Then $\mathcal{Q}$ is a generalized quadrangle with parameters (q^4, q^2) and G is a group of automorphisms of $\mathcal{Q}$ acting regularly on the points of type (p1).

Lemma 4.1 *Let $a \neq b$ be two elements in $F \cup \{\infty\}$. If $x_a = [0, u_a] \in A_a$ and $x_b = [0, u_b] \in A_b$, then $A_a x_b$ and $A_b x_a$ are concurrent in $\mathcal{Q}$ if and only if $\Sigma(u_a, u_b) = 0$.*

Proof Let $[\alpha, u] \in A_a x_b \cap A_b x_a$. Then there are elements $[0, v_a] \in A_a$ and $[0, v_b] \in A_b$ such that

$$[\alpha, u] = [0, v_a][0, u_b] = [0, v_b][0, u_a].$$

Thus

$$[\Sigma(v_a, u_b), v_a + u_b] = [\Sigma(v_b, u_a), v_b + u_a],$$

which implies that $v_a + u_b = v_b + u_a$ and $\Sigma(v_a, u_b) = \Sigma(v_b, u_a)$. From the first identity we get $v_a - u_a = v_b - u_b \in U_a \cap U_b = \{0\}$; hence

$v_a = u_a$ and $v_b = u_b$. Substituting this in $\Sigma(v_a, u_b) = \Sigma(v_b, u_a)$ we obtain $\Sigma(u_a, u_b) = \Sigma(u_b, u_a)$. Since Σ is antisymmetric, $\Sigma(u_a, u_b) = -\Sigma(u_b, u_a)$, thus, as $p \neq 2$ we have $\Sigma(u_a, u_b) = 0$.

Conversely, if $\Sigma(u_a, u_b) = 0$ then $\Sigma(u_b, u_a) = -\Sigma(u_a, u_b) = 0$ as well. Therefore

$$x_a x_b = (u_a, 0)(u_b, 0) = (u_a + u_b, 0) = (u_b, 0)(u_a, 0) = x_b x_a,$$

which yields $x_a x_b \in A_a x_b \cap A_b x_a$ and so the two lines of $\mathcal{Q}$ meet. □

The following observation will be crucial.

Lemma 4.2 *Let S and T be of type* (1e) *as given in Table 1. Then the minimal polynomial of ST^{-1} is $\left(X^2 - 2\delta X + (\delta^2 - \nu)\right)^2$.*

Proof We have

$$S = \begin{pmatrix} 2\delta\nu^2 & -\nu^2(2+2\gamma^2) & 0 & 0 \\ -\nu^2(2+2\gamma^2) & 2\delta\nu & 0 & 0 \\ 0 & 0 & \nu & \delta \\ 0 & 0 & \delta & 1 \end{pmatrix},$$

$$T^{-1} = \begin{pmatrix} -\frac{1}{\nu^2\gamma^2} & -\frac{\delta}{\nu^2\gamma^2} & 0 & 0 \\ -\frac{\delta}{\nu^2\gamma^2} & -\frac{1}{\nu\gamma^2} & 0 & 0 \\ 0 & 0 & 0 & 1 \\ 0 & 0 & 1 & 0 \end{pmatrix}.$$

Therefore,

$$ST^{-1} = \begin{pmatrix} \delta & -\nu & 0 & 0 \\ -1 & \delta & 0 & 0 \\ 0 & 0 & \delta & \nu \\ 0 & 0 & 1 & \delta \end{pmatrix},$$

which proves the assertion. □

This lemma has the following consequence.

Lemma 4.3 *Let x, y, x', y' be nonzero vectors in F^4, equipped with the canonical scalar product. Suppose that $\langle x, y\rangle S \perp x'$, $\langle x, y\rangle T \perp x'$, and $\langle xS, xT\rangle \perp y'$. Then $\langle yS, yT\rangle \perp y'$.*

Proof This is clear if x and y are linearly dependent. So let us assume they are not. Let $V = \langle x, y\rangle S$ and $W = \langle x, y\rangle T$. We shall show that $\langle xS, xT\rangle = \langle yS, yT\rangle$.

Because $x' \neq \mathbf{0}$ is perpendicular to V and W, they cannot span the whole of F^4. Suppose that $\dim(V \cap W) = 1$. Then there exist v,

$w \in \langle x, y \rangle$ such that $vS = wT$. By Lemma 4.2, ST^{-1} has no eigenvector over F; hence v and w must be linearly independent. We may therefore assume that $xS = yT$. Since $\dim(V + W) = 3$, the set $\{xS, yS, xT\}$ is a basis of $V + W$. Substitute xST^{-1} for y and multiply by T^{-1} to obtain the set $\{x, xST^{-1}, x(ST^{-1})^2\}$ of linearly independent vectors. This contradicts Lemma 4.2. Therefore, $V = W$. This implies that the subspaces $\langle xS, xT \rangle$ and $\langle yS, yT \rangle$ are contained in V. But ST^{-1} has no eigenvector over F; so both are in fact equal to V. □

We are now ready to show that $\mathcal{Q}$ has Property (G); so the notation in its statement is used. Let $K_1 = [A_a]$ and $L_1 = [A_b]$. Since G acts regularly on the points not collinear with ∞, we may assume that K_2 and L_2 intersect in $[0, 0]$. It follows that $K_2 = A_b$ and $L_2 = A_a$. Let $K_3 \cap L_2 = [0, u_a]$, $K_4 \cap L_2 = [0, v_a]$, $K_2 \cap L_3 = [0, u_b]$, and $K_2 \cap L_4 = [0, v_b]$. Since L_i must meet K_1 we have $L_3 = A_a[0, u_b]$ and $L_4 = A_a[0, v_b]$. Similarly, $K_3 = A_b[0, u_a]$ and $K_4 = A_b[0, v_a]$. From Lemma 4.1 we obtain that $\Sigma(u_a, u_b) = \Sigma(u_a, v_b) = \Sigma(v_a, u_b) = 0$, and we must show that $\Sigma(v_a, v_b) = 0$ as well. By definition,

$$\Sigma((x, xM_a), (y, yM_b)) = (x(M_a - M_b)Sy^{\mathrm{tr}}, x(M_a - M_b)Ty^{\mathrm{tr}}).$$

Let $u_i = (x_i, x_iM_i)$ and $v_i = (y_i, y_iM_i)$; $i = a, b$. Applying Lemma 4.3 with $x = x_a(M_a - M_b)$, $y = y_a(M_a - M_b)$, $x' = x_b$, and $y' = y_b$ we deduce the final result.

Theorem 4.4 *A generalized quadrangle constructed from a pencil of quadrics has Property* (G).

Acknowledgements

This research was supported by the G.N.S.A.G.A of C.N.R. and the Italian Ministry of Research and Technology (project: *Strutture geometriche, combinatoria e loro applicazioni*).

References

[1] R.W. Ahrens and G. Szekeres, On a combinatorial generalization of 27 lines associated with a cubic surface, *J. Austral. Math. Soc.* **10** (1969), 485–492.

[2] A.A. Bruen and J.W.P. Hirschfeld, Intersections in projective space II: Pencils of quadrics *European J. Combin.* **9** (1988), 255–270.

[3] P. Dembowski, *Finite Geometries*, Springer, Berlin, 1968.

[4] L.E. Dickson, *Linear Groups with an Exposition of the Galois Field Theory*, Teubner, Leipzig, 1901 (Dover Publications, New York, 1958).

[5] D. Ghinelli and S. Löwe, Generalized quadrangles with a regular point and association schemes, *Linear Algebra Appl.* **226–228** (1995), 87–104.

[6] D. Ghinelli and U. Ott, Nets and generalized quadrangles, *Geom. Dedicata* **52** (1994), 1–14.

[7] M. Hall, Affine generalized quadrilaterals, *Studies in Pure Mathematics* (ed. L. Mirsky), Academic Press, 1971, 113 –116.

[8] W.M. Kantor, Generalized quadrangles associated with $G_2(q)$, *J. Combin. Theory Ser. A* **29** (1980), 212–219.

[9] S. Löwe, Fourgonal extensions, *Geom. Dedicata* **69** (1998), 67–81.

[10] S. Löwe, Über die Konstruktion von verallgemeinerten Vierecken mit einem regulären Punkt, Habilitationsschrift, Braunschweig, 1992.

[11] S.E. Payne, Nonisomorphic generalized quadrangles, *J. Algebra* **18** (1971), 201–212.

[12] S.E. Payne, Quadrangles of order $(s-1, s+1)$, *J. Algebra* **22** (1972), 97–119.

[13] S.E. Payne, A census of generalized quadrangles, *Finite geometries, Buildings and Related Topics* (eds. W.M. Kantor et al.), Oxford University Press, Oxford, 1990, 29–36.

[14] S.E. Payne and J.A. Thas, *Finite Generalized Quadrangles*, Research Notes in Mathematics **110**, Pitman, London, 1984.

[15] J.A. Thas, Recent developments in the theory of finite generalized quadrangles, *Med. Konink. Acad. Wetensch. België* **56** (1994), 99–113.

[16] J.A. Thas, Generalized quadrangles of order (s, s^2). I. *J. Combin. Theory Ser. A* **67** (1994), no. 2, 140–160.

[17] J.A. Thas, Generalized polygons, *Handbook of Incidence Geometry*, North-Holland, Amsterdam, 1995, 383–431.

[18] J.A. Thas, Generalized quadrangles of order (s, s^2): recent results, *Discrete Math.* **208/209** (1999), 577–587.

[19] J.A. Thas, Generalized quadrangles of order (s, s^2). II. *J. Combin. Theory Ser. A* **79** (1997), 223–254.

[20] J.A. Thas, 3-regularity in generalized quadrangles: a survey, recent results and the solution of a longstanding conjecture, *Rend. Circ. Mat. Palermo Suppl.* **53**, (1998), 199–218.

[21] J.A. Thas, Generalized quadrangles of order (s, s^2). III. *J. Combin. Theory Ser. A* **87** (1999), 247–272.

[22] J. Tits, Sur la trialité et certaines groupes qui s'en déduisent, *Inst. Hautes Etudes Sci. Publ. Math.* **2** (1959), 14–60.

THE DESARGUESIAN PLANE OF ORDER THIRTEEN

M. Giulietti
Dipartimento di Matematica
Università di Perugia
Perugia
Italy
giuliet@dipmat.unipg.it

J.W.P. Hirschfeld
School of Mathematical Sciences
University of Sussex
Brighton BN1 9QH
United Kingdom
jwph@sussex.ac.uk

G. Korchmáros
Dipartimento di Matematica
Università della Basilicata
85100 Potenza
Italy
korchmaros@unibas.it

Abstract

The algebraic curve associated to an arc in $PG(2, q)$, with q odd, is examined using both properties of the curve itself as well as properties of the arc. The key case of $(q-1)$-arcs means that the behaviour of the associated sextic curves needs to be studied. The case of $PG(2, 13)$ is examined in detail. There is a geometric bijection between 12-arcs and their duals. The latter lead to optimal sextic curves; the former lead to sextics whose set of rational points make them 'look like' quartics.

A. Blokhuis et al. (eds.), Finite Geometries, 159–170.

1. Introduction

The Hasse–Weil and Stöhr–Voloch theorems both give upper bounds on the number of rational points on an algebraic curve over $\mathbf{F}_q$. Particular cases of both results can be stated as follows.

Theorem 1.1 *The number N of rational points on a non-singular plane curve $\mathcal{C}$ of genus g and degree d satisfies the following:*
(i) (Hasse–Weil) $N \leq q + 1 + 2g\sqrt{q}$;
(ii) (Stöhr–Voloch) $N \leq \frac{1}{2}\{2g - 2 + (q+2)d\}$.

The key idea of Stöhr and Voloch is to consider the set of rational points P of a curve $\mathcal{C}$ such that the Frobenius image of P lies on its tangent ℓ_P:

$$\mathcal{T}(\mathcal{C}) = \{P \in \mathcal{C}(\overline{\mathbf{F}}_q) \mid P^q \in \ell_P\} \supset \mathcal{C}(\mathbf{F}_q).$$

Here, the curve associated to $\mathcal{T}(\mathcal{C})$ has interesting properties.

A *k-arc* in $PG(2,q)$ is a set of k points no 3 collinear. Equivalent concepts are a set of k vectors in ${\mathbf{F}_q}^3$ no 3 linearly dependent and a $[k, 3, k-2]_q$ code. A k-arc is *complete* if it is not contained in a $(k+1)$-arc. Let $m(2,q)$, $m'(2,q)$ denote the largest and second largest k for which a complete k-arc exists in $PG(2,q)$. Then $m(2,q)$ is $q+1$ or $q+2$ according as q is odd or even. For $q \leq 29$, the values of $m'(2,q)$ are given in Table 1.

Table 1. The second largest arc

q	7	8	9	11	13	16	17	19	23	25	27	29
$m'(2,q)$	6	6	8	10	12	13	14	14	17	21	22	24

From Table 1, $m'(2,q) = q-1$ for $q = 7, 9, 11, 13$ and $m'(2,q) < q-1$ for $16 \leq q \leq 29$. Also, $m'(2,q) < q-1$ for $q > 83$; see [3, Chapter 10]. So, it is conjectured that $m'(2,q) < q-1$ for $q > 13$. So, the case of $q = 13$ warrants further investigation.

Segre's classical theorem is used to connect arcs and curves.

Theorem 1.2 *Let $\mathcal{K}$ be a k-arc in $PG(2,q)$, and let $\mathcal{K}'$ be its dual.*

(i) *The $kt = k(q+2-k)$ tangents through the points of $\mathcal{K}$ lie on an algebraic envelope Γ' whose dual curve Γ is of degree t or $2t$ according as q is even or odd.*

(ii) *The envelope* Γ' *contains no bisecant of* $\mathcal{K}$ *and so no pencil with vertex* P *in* $\mathcal{K}$.

(iii) *For* q *odd, the* t *tangents to* $\mathcal{K}$ *through a point* P *of* $\mathcal{K}$ *each count twice in the intersection of* Γ' *with the pencil* $\mathcal{L}_P$ *of lines through* P. *Dually, each line* l *of* $\mathcal{K}'$ *is a tangent at* t *distinct points of* Γ.

(iv) *For* q *odd,* Γ' *may contain components of multiplicity two, but does not consist entirely of double components.*

(v) *If* $k > 2t$, *that is,* $k > \frac{2}{3}(q+2)$, *then*

  (a) Γ' *is unique;*

  (b) $\mathcal{K}$ *is incomplete if and only if* Γ' *has a rational linear component.*

Denote the plane curve determined by a polynomial F by $\mathbf{v}(F)$; more precisely,

$$\mathbf{v}(F) = (\mathbf{V}(F), (F)),$$

where $\mathbf{V}(F)$ is the set of rational points of the curve and (F) is the ideal generated by F in $\mathbf{F}_q[X_0, X_1, X_2]$.

Is it possible to extract further general information from the connection between arcs and curves?

Consider $\mathcal{K}'$ as in Theorem 1.2 and apply the idea of Stöhr and Voloch. There are three curves to consider:

$$\begin{aligned} \Gamma &= \mathbf{v}(F(X_0, X_1, X_2)); & (1.1) \\ \Delta &= \mathbf{v}(G) = \mathbf{v}(X_0{}^q \partial F/\partial X_0 + X_1{}^q \partial F/\partial X_1 + X_2{}^q \partial F/\partial X_2); & (1.2) \\ \Lambda &= \mathbf{v}(L) = \mathbf{v}(l_1 \cdot l_2 \cdot \ldots \cdot l_k), & (1.3) \end{aligned}$$

where $\mathcal{K}' = \{l_1, l_2, \ldots, l_k\}$.

Since Γ meets each l_i twice at precisely t rational points P, that is, $I(P, \Gamma \cap l_i) = 2$ and since $I(P, \Gamma \cap \Delta) \geq 2$ for any such point P, so Noether's theorem holds; that is, since, for $P \in \Gamma \cap \Lambda$, the relation $I(P, \Delta \cap \Gamma) \geq I(P, \Lambda \cap \Gamma)$ holds, there exist polynomials A and B such that

$$G = AL + BF. \qquad (1.4)$$

Note that A and B are not unique, since $A' = A - \lambda F,\ B' = B + \lambda L$ is also a solution.

2. A particular sextic in $PG(2,13)$

For the unique complete 12-arc in $PG(2,13)$, the associated curve can be put in the form $\Gamma = \mathbf{v}(F)$ with

$$F = X^6 + Y^6 + Z^6 + 3X^2Y^2Z^2;$$

see [2]. Then $\Delta = \mathbf{v}(G)$ with

$$G = X^{18} + Y^{18} + Z^{18} + X^2Y^2Z^2(X^{12} + Y^{12} + Z^{12}).$$

The Hessian $\Phi = \mathbf{v}(H)$ of Γ is given by

$$H = X^6Y^6 + X^6Z^6 + Y^6Z^6 + 2X^2Y^2Z^2(X^6 + Y^6 + Z^6).$$

The curve Γ is non-singular and has 54 rational points: 18 on the sides of the triangle of reference and 36 other points. The 18 form

$$\mathcal{J}_{18} = \{(0,t,1),(t,0,1),(t,1,0) \mid t^6 = -1\};$$

that is, $t = \pm 2, \pm 5, \pm 6$. The 36 are

$$\begin{array}{llll} (\pm 1, \pm 2, 1), & (\pm 2, \pm 1, 1), & (\pm 2, \pm 5, 1), & (\pm 5, \pm 2, 1), \\ (\pm 3, \pm 5, 1), & (\pm 5, \pm 3, 1), & (\pm 4, \pm 6, 1), & (\pm 6, \pm 4, 1), \\ (\pm 6, \pm 6, 1). & & & \end{array}$$

It should be noted that Γ is optimal in that it attains the upper bound in the Stöhr–Voloch theorem, Theorem 1.1. Here, $g = 10$, $q = 13$, $d = 6$, $N = 54$. So this result may be stated as follows.

Theorem 2.1 *The largest number of rational points on a plane curve of genus* 10 *over* $\mathbf{F}_{13}$ *is* 54.

By definition, Δ has the same 54 rational points as Γ. However, Φ has only the three rational points $(1,0,0), (0,1,0), (0,0,1)$.

The 36 points of Γ lie by threes on 12 lines, each of which is a tangent to Γ at all three points. The tangents and the points on them are given in Table 2.

Now, with $\Lambda = \mathbf{v}(L)$,

$$\begin{aligned} L = \Pi L_i &= \{Z^2 - (X+Y)^2\}\{Z^2 - (X-Y)^2\} \\ &\quad \times\{Z^2 - (3X+4Y)^2\}\{Z^2 - (3X-4Y)^2\} \\ &\quad \times\{Z^2 - (4X+3Y)^2\}\{Z^2 - (4X-3Y)^2\} \\ &= X^{12} + Y^{12} + Z^{12} - 2(X^6Y^6 + X^6Z^6 + Y^6Z^6) \\ &\quad -5X^2Y^2Z^2(X^6 + Y^6 + Z^6) + 5X^4Y^4Z^4. \end{aligned}$$

Table 2. Triple tangents to Γ and points of contact

$L_1 = X + Y + Z$	$(1, -2, 1)$,	$(-2, 1, 1)$,	$(6, 6, 1)$
$L_2 = X - Y + Z$	$(1, 2, 1)$,	$(-2, -1, 1)$,	$(6, -6, 1)$
$L_3 = -X + Y + Z$	$(-1, -2, 1)$,	$(2, 1, 1)$,	$(-6, 6, 1)$
$L_4 = -X - Y + Z$	$(-1, 2, 1)$,	$(2, -1, 1)$,	$(-6, -6, 1)$
$L_5 = 3X + 4Y + Z$	$(2, -5, 1)$,	$(-4, 6, 1)$,	$(-5, -3, 1)$
$L_6 = 3X - 4Y + Z$	$(2, 5, 1)$,	$(-4, -6, 1)$,	$(-5, 3, 1)$
$L_7 = -3X + 4Y + Z$	$(-2, -5, 1)$,	$(4, 6, 1)$,	$(5, -3, 1)$
$L_8 = -3X - 4Y + Z$	$(-2, 5, 1)$,	$(4, -6, 1)$,	$(5, 3, 1)$
$L_9 = 4X + 3Y + Z$	$(-5, 2, 1)$,	$(6, -4, 1)$,	$(-3, -5, 1)$
$L_{10} = 4X - 3Y + Z$	$(-5, -2, 1)$,	$(6, 4, 1)$,	$(-3, 5, 1)$
$L_{11} = -4X + 3Y + Z$	$(5, 2, 1)$,	$(-6, -4, 1)$,	$(3, -5, 1)$
$L_{12} = -4X - 3Y + Z$	$(5, -2, 1)$,	$(-6, 4, 1)$,	$(3, 5, 1)$

Write

$$\begin{aligned} S_i &= X^i + Y^i + Z^i, \\ D &= X^6Y^6 + X^6Z^6 + Y^6Z^6, \\ T &= X^2Y^2Z^2. \end{aligned}$$

Note the identities:

$$S_6{}^2 = S_{12} + 2D, \quad S_6S_{12} = S_{18} + DS_6 - 3T^3.$$

Then

$$\begin{aligned} F &= S_6 + 3T, \\ G &= S_{18} + TS_{12}, \\ H &= D + 2TS_6, \\ L &= S_{12} - 2D - 5TS_6 + 5T^2. \end{aligned} \tag{2.1}$$

Theorem 2.2 *In* $PG(2, 13)$, *with* F, L, G *as in (2.1),*

$$G = A_\lambda L + B_\lambda F$$

with

$$\begin{aligned} A_\lambda &= \lambda S_6 + (3\lambda - 5)T = \lambda F - 5T, \\ B_\lambda &= (1 - \lambda)S_{12} + (-1 + 2\lambda)D + (3 + 5\lambda)TS_6 + (5 - 5\lambda)T^2 \\ &= S_{12} - D + 3TS_6 + 5T^2 - \lambda L. \end{aligned}$$

Proof For $P \in \Gamma \cap \Lambda$,

$$I(P, \Lambda \cap \Gamma) = 2 \leq I(P, \Delta \cap \Gamma).$$

Therefore, by Noether's Theorem, both A_λ and B_λ exist. Choose

$$\begin{aligned} A_\lambda &= a_1 S_6 + b_1 T, \\ B_\lambda &= a_2 S_{12} + b_2 D + c_2 T S_6 + d_2 T^2. \end{aligned}$$

Then the coefficients are as in the statement of the theorem. □

Corollary 2.3 *For $P \in \Delta \cap \Gamma$, the intersection number $I(P, \Delta \cap \Gamma) = 2$.*

Proof By Bézout's theorem, $\sum(P, \Delta \cap \Gamma) = 6 \times 18 = 108 = 2 \times 54$.□

Because $\deg H = 2 \deg A_\lambda$, it may be conjectured that there is a relation between them, but the following result follows by manipulation.

Lemma 2.4 *There do not exist λ and C such that ${A_\lambda}^2 = CF + cH$.*

Note that

$$\begin{aligned} A_3 &= 3(X^2+Y^2+Z^2)(X^2+3Y^2-4Z^2)(X^2-4Y^2+3Z^2), \\ A_{-2} &= -6(-4X^2+Y^2+Z^2)(X^2-4Y^2+Z^2)(X^2+Y^2-4Z^2), \\ A_{-5} &= -6(3X^2+Y^2+Z^2)(X^2+3Y^2+Z^2)(X^2+Y^2+3Z^2). \end{aligned} \tag{2.2}$$

The three conics in such an $\mathcal{A}_\lambda = \mathbf{v}(A_\lambda)$ are mutually apolar in pairs; hence the 12 points formed by the three sets of intersections of these pairs of conic form a 12-arc projectively equivalent to $\mathcal{K}$; see [1]. In particular, the 12-arc $\mathcal{K}_1$ from $\mathcal{A}_3$ consists of the poles of L_i, $i = 1, \ldots, 12$ with respect to $\mathbf{v}(X^2+Y^2+Z^2)$ or, equally well, with respect to the conics $\mathbf{v}(X^2+3Y^2-4Z^2)$ or $\mathbf{v}(X^2-4Y^2+3Z^2)$. Let $\mathcal{K}_1' = \{L_1, \ldots, L_{12}\}$. The arc and its dual are given in Table 3. Each line of $\mathcal{K}_1'$ is tangent to two of these three conics, but is tangent to none of the conics in A_{-2} or A_{-5}.

Lemma 2.5 *The dual 12-arc $\mathcal{K}_1'$ consists of four triangles inscribed to the 12-arc $\mathcal{K}_1$; namely, 16β, 25γ, 389, 47α.*

Given a complete 12-arc $\mathcal{K}$, it is the set of intersections of three conics $\mathcal{C}, \mathcal{C}', \mathcal{C}''$. The tangent to one of the conics, say $\mathcal{C}$, at an intersection point P of, say $\mathcal{C}$ and $\mathcal{C}'$, is also a tangent to $\mathcal{C}''$ at an intersection point of $\mathcal{C}''$ and $\mathcal{C}$. Hence, at the 12 points of intersection of pairs of the three conics, there are 12 tangents, not 24, and they form a dual 12-arc $\mathcal{K}'$. The six pairs of triangles of $\mathcal{K}'$ are each in perspective from one of three points that form the diagonal triangle for each of the three quadrangles that make up $\mathcal{K}$. The group $G(\mathcal{K})$ of projectivities fixing $\mathcal{K}$ has order 72 and is an extension of $\mathbf{S}_4$ by $\mathbf{Z}_3$. The three subgroups $\mathbf{S}_4$ may be

Table 3. The 12-arc $\mathcal{K}_1$ and its dual $\mathcal{K}_1'$

	$\mathcal{K}_1$	$\mathcal{K}_1'$	
1	$(1,1,1)$	$X+Y+Z$	6β
2	$(1,-1,1)$	$X-Y+Z$	5γ
3	$(-1,1,1)$	$-X+Y+Z$	89
4	$(-1,-1,1)$	$-X-Y+Z$	7α
5	$(3,4,1)$	$3X+4Y+Z$	25
6	$(3,-4,1)$	$3X-4Y+Z$	16
7	$(-3,4,1)$	$-3X+4Y+Z$	47
8	$(-3,-4,1)$	$-3X-4Y+Z$	38
9	$(4,3,1)$	$4X+3Y+Z$	39
α	$(4,-3,1)$	$4X-3Y+Z$	4α
β	$(-4,3,1)$	$-4X+3Y+Z$	1β
γ	$(-4,-3,1)$	$-4X-3Y+Z$	2γ

viewed as fixing the three quadrangles that are intersections of pairs of the conics, whereas $\mathbf{Z}_3$ fixes each of the 4 triangles of $\mathcal{K}'$.

The other 12-arcs are

$$\begin{aligned}\mathcal{K}_2 &= \{(\pm1,\pm4,1),(\pm4,\pm1,1),(\pm3,\pm3,1)\};\\ \mathcal{K}_3 &= \{(\pm1,\pm3,1),(\pm3,\pm1,1),(\pm4,\pm4,1)\}.\end{aligned}$$

3. Hurwitz's theorem

The picture is completed by understanding the pencil $\mathcal{P}$ of sextics given by $\mathcal{A}_\lambda$, and four associated pencils of cubics. Let

$$F_\mu = X^6+Y^6+Z^6+\mu X^2Y^2Z^2, \tag{3.1}$$

$$C_\rho = X^3+Y^3+Z^3+\rho XYZ, \tag{3.2}$$
$$C_\rho^{(0)} = -X^3+Y^3+Z^3+\rho XYZ, \tag{3.3}$$
$$C_\rho^{(1)} = X^3-Y^3+Z^3+\rho XYZ, \tag{3.4}$$
$$C_\rho^{(2)} = X^3+Y^3-Z^3+\rho XYZ. \tag{3.5}$$

Here, $F_\infty = X^2Y^2Z^2$ and similarly for $\rho=\infty$ in the cubics. Write $\mathcal{F}_\mu = \mathbf{v}(F_\mu)$, $\mathcal{C}\rho = \mathbf{v}(C_\rho)$ and $\mathcal{C}_\rho^{(i)} = \mathbf{v}(C_\rho^{(i)})$.

The non-singular sextic $\mathcal{F}_3$ has canonical series g_9^{18}; so this series is cut out by all cubics in the plane.

As in (2.2), each of $\mathcal{F}_{-3},\mathcal{F}_4,\mathcal{F}_{-1}$ is a set of three conics; more precisely, $A_\lambda = F_\mu$ with $\mu = 3-5\lambda^{-1}$. The intersections of pairs of the three conics in any one set forms a complete 12-arc; they are the singular points of the corresponding sextic. The correspondence between 12-arcs and dual 12-arcs is given in Table 4.

Table 4. Correspondence between dual 12-arcs and 12-arcs

Curve whose triple tangents form a dual 12-arc	Curve whose singular points form a 12-arc
$\mathcal{F}_3$	$\mathcal{F}_{-3}$
$\mathcal{F}_{-4}$	$\mathcal{F}_4$
$\mathcal{F}_1$	$\mathcal{F}_{-1}$

Each pencil of cubics has four inflexional triangles. One is common to all four pencils, namely, the triangle of reference $\mathbf{v}(XYZ)$. This leaves three triangles for each of the four pencils. The $4 \times 9 = 36$ lines forming these triangles are the $3 \times 12 = 36$ lines of the dual 12-arcs associated to the three sextics $\mathcal{F}_3, \mathcal{F}_{-4}, \mathcal{F}_1$. Each dual 12-arc takes one triangle from each pencil of cubics. The three triangles for each cubic are given in Table 5. Then the three connected dual 12-arcs are

$$\mathcal{K}_1' = \{t_1, t_2, t_3, t_4\},\ \mathcal{K}_2' = \{t_1', t_2', t_3', t_4'\},\ \mathcal{K}_3' = \{t_1'', t_2'', t_3'', t_4''\}.$$

The cubic $\mathcal{C}_\rho$ is a triangle for $\rho = \infty, -3, 4, -1$, equianharmonic for $\rho = 0, 6, 5, 2$ and harmonic for $\rho = 1, 3, -6, -5, -4, -2$.

Table 5. Inflexional triangles

$\mathcal{C}_\rho$	$(X+Y+Z)(3X-4Y+Z)(-4X+3Y+Z)$	t_1
	$(3X+Y+Z)(-4X-4Y+Z)(X+3Y+Z)$	t_1'
	$(-4X+Y+Z)(X-4Y+Z)(3X+3Y+Z)$	t_1''
$\mathcal{C}_\rho^{(0)}$	$(-X+Y+Z)(-3X-4Y+Z)(4X+3Y+Z)$	t_2
	$(-3X+Y+Z)(4X-4Y+Z)(-X+3Y+Z)$	t_2'
	$(4X+Y+Z)(-X-4Y+Z)(-3X+3Y+Z)$	t_2''
$\mathcal{C}_\rho^{(1)}$	$(X-Y+Z)(3X+4Y+Z)(-4X-3Y+Z)$	t_3
	$(3X-Y+Z)(-4X+4Y+Z)(X-3Y+Z)$	t_3'
	$(-4X-Y+Z)(X+4Y+Z)(3X-3Y+Z)$	t_3''
$\mathcal{C}_\rho^{(2)}$	$(-X-Y+Z)(-3X+4Y+Z)(4X-3Y+Z)$	t_4
	$(-3X-Y+Z)(4X+4Y+Z)(-X-3Y+Z)$	t_4'
	$(4X-Y+Z)(-X+4Y+Z)(-3X-3Y+Z)$	t_4''

The sextic $\mathcal{F}_3$ is a 4-cover for $\mathcal{C}_3$ via the map $\phi : (x, y, z) \mapsto (x^2, y^2, z^2)$; that is, each of the 9 inflexions of $\mathcal{C}_3$ is covered by two of the 18 points of $\mathcal{J}_{18}$, whereas each of the other 9 points of $\mathcal{C}_3$ is covered by four of the other 36 points of $\mathcal{F}_3$; see Table 6. Similarly, $\mathcal{F}_3$ covers $\mathcal{C}_{-3}^{(0)}$ by the map $\phi_0 : (x,y,z) \mapsto (-x^2, y^2, z^2)$; $\mathcal{F}_3$ covers $\mathcal{C}_{-3}^{(1)}$ by the map $\phi_1 : (x,y,z) \mapsto (x^2, -y^2, z^2)$; $\mathcal{F}_3$ covers $\mathcal{C}_{-3}^{(2)}$ by the map $\phi_2 : (x,y,z) \mapsto (x^2, y^2, -z^2)$.

These covers give an illustration of Hurwitz's theorem, which can be expressed as follows. For curves $\mathcal{C}$, $\mathcal{C}'$ defined over $\mathbf{F}_q$, let $\theta : \mathcal{C} \to \mathcal{C}'$ be a rational map, with $\mathbf{F}_q(\mathcal{C})$ and $\mathbf{F}_q(\mathcal{C}')$ the corresponding function fields. Let G be the group of automorphisms τ of $\mathbf{F}_q(\mathcal{C})$ such that $\tau\theta = \theta$. Then G acts on the places of $\mathcal{C}$ and $n = |G| = [\mathbf{F}_q(\mathcal{C}) : \mathbf{F}_q(\mathcal{C}')]$. Let $m_1, \ldots, m_s$ be the lengths of the *short* orbits, that is, those of length less than n. Then, with g the genus of $\mathcal{C}$ and g' the genus of $\mathcal{C}'$,

$$2g - 2 = n(2g' - 2) + \sum_1^s (n - m_i). \tag{3.6}$$

In the case that $\mathcal{C} = \mathcal{F}_3$, $\mathcal{C}' = \mathcal{C}_3$, the group G consists of the elements $(x,y,z) \mapsto (\pm x, \pm y, z)$. Then $n = 4$, $g' = 1$, $g = 10$, $s = 9$, $m_i = 2$ for all i; so (3.6) is satisfied for these values. The short orbits consist of the pairs of points of $\mathcal{F}_3$ mapping to an inflexion of $\mathcal{C}_3$; see Table 6.

Table 6. $\mathcal{F}_3$ covers $\mathcal{C}_3$

$(\pm1,\pm2,1)$	$\mapsto$	$(1,4,1)$;	$(\pm5,\pm2,1)$	$\mapsto$	$(-1,4,1)$;
$(\pm2,\pm1,1)$	$\mapsto$	$(4,1,1)$;	$(\pm3,\pm5,1)$	$\mapsto$	$(-4,-1,1)$;
$(\pm2,\pm5,1)$	$\mapsto$	$(4,-1,1)$;	$(\pm5,\pm3,1)$	$\mapsto$	$(-1,-4,1)$;
$(\pm4,\pm6,1)$	$\mapsto$	$(3,-3,1)=(-1,1,4)$;			
$(\pm6,\pm4,1)$	$\mapsto$	$(-3,3,1)=(1,-1,4)$;			
$(\pm6,\pm6,1)$	$\mapsto$	$(-3,-3,1)=(1,1,4)$.			
$(0,\pm2,1)$	$\mapsto$	$(0,4,1)$;	$(0,\pm5,1)$	$\mapsto$	$(0,-1,1)$;
$(\pm2,0,1)$	$\mapsto$	$(4,0,1)$;	$(\pm5,0,1)$	$\mapsto$	$(-1,0,1)$;
$(\pm2,1,0)$	$\mapsto$	$(4,1,0)$;	$(\pm5,1,0)$	$\mapsto$	$(-1,1,0)$;
$(0,\pm6,1)$	$\mapsto$	$(0,-3,1)=(0,1,4)$;			
$(\pm6,0,1)$	$\mapsto$	$(-3,0,1)=(1,0,4)$;			
$(\pm6,1,0)$	$\mapsto$	$(-3,1,0)=(1,4,0)$.			

The properties of the complete 12-arc and its dual can be summarised in the following theorem.

Theorem 3.1 *In* $PG(2,13)$,

(i) *a complete* 12*-arc* $\mathcal{K}_1$ *uniquely defines a dual* 12*-arc* $\mathcal{K}_1'$, *and conversely;*

(ii) *a complete* 12*-arc* $\mathcal{K}_1$ *uniquely defines two others* $\mathcal{K}_2, \mathcal{K}_3$; *these are orbits of the group fixing* $\mathcal{K}$.

(iii) *a complete dual* 12*-arc* $\mathcal{K}'_1$ *uniquely defines two others* $\mathcal{K}'_2, \mathcal{K}'_3$*; these are the tritangents to the two other curves in the pencil of sextics defined by* $\mathcal{K}'_1$ *consisting of* 36 *points;*

(iv) *the* 14 *members of* $\mathcal{P}$ *are given in Table 7 with the number* N_0 *of their rational points other than those of* $\mathcal{J}_{18}$*;*

(v) *the* 183 *points of the plane are partitioned as* $18+21+3\times36+3\times12$.

Table 7. The pencil of sextic curves

Curve	Description	N_0
$\mathcal{F}_\infty$	triangle of reference	21
$\mathcal{F}_3, \mathcal{F}_{-4}, \mathcal{F}_1$	non-singular, optimal	36
$\mathcal{F}_{-3}, \mathcal{F}_4, \mathcal{F}_{-1}$	three conics	12
$\mathcal{F}_0$	non-singular, 18 inflexions with 6-point contact	0
$\mathcal{F}_2, \mathcal{F}_{-2}, \mathcal{F}_5, \mathcal{F}_{-5}, \mathcal{F}_6, \mathcal{F}_{-6}$	non-singular	0

4. A sextic that looks like a quartic

Consider the pencil of sextics $\mathcal{R}_\lambda = \mathbf{v}(R_\lambda)$ given by

$$R_\lambda = F_{-3} + \lambda(X^2 - Y^2)(3Z^2 - X^2)(Y^2 - 3Z^2).$$

The base of this pencil, as an intersection divisor W of any two members of the pencil, has degree 36, and is

$$W = \mathcal{K}_0 + 2\mathcal{K}_1,$$

where $\mathcal{K}_0$ lies in $PG(2, 13^2)\backslash PG(2, 13)$, with $|\mathcal{K}_0| = 12$; so the rational support of W is precisely $\mathcal{K}_1$. Geometrically, this is because each of the six lines in $\mathcal{R}_\infty$ meets two of the conics in $\mathcal{R}_0$ in the same pair of points and meets the third conic in two non-rational points.

Since $\mathcal{R}_0 = \mathbf{v}(F_{-3})$ is three conics meeting pairwise in four points, it contains 30 rational points. Since $\mathcal{R}_\infty$ is a set of six lines forming the sides of a quadrangle, it contains 73 points. The sextics $\mathcal{R}_\lambda$ for $\lambda = \pm1, \pm2, \pm5, \pm6$ each contain precisely 12 rational points, namely those of $\mathcal{K}_1$. For $\lambda = \pm3$, the sextic $\mathcal{R}_\lambda$ is singular with four nodes; it contains 24 rational points apart from the four nodes. For $\lambda = \pm4$, the sextic $\mathcal{R}_\lambda$ is non-singular with 42 rational points. Hence the plane

$PG(2,13)$ is partitioned as follows:

$$\begin{aligned}183 &= 12+1\times(30-12)+1\times(73-12)+8\times(12-12)\\ &\quad +2\times(28-12)+2\times(42-12)\\ &= 12+18+61+0+32+60.\end{aligned}$$

Each of the eight $\mathcal{R}_\lambda$ with 12 rational points have a remarkable property, illustrated for $\mathcal{R}_1$, in the following result.

Theorem 4.1 *The non-singular sextic curve over* $\mathbf{F}_{13}$ *given by*

$$\begin{aligned}R_1 &= (X^2+Y^2+Z^2)(X^2+3Y^2-4Z^2)(X^2-4Y^2+3Z^2)\\ &\quad +(X^2-Y^2)(3Z^2-X^2)(Y^2-3Z^2).\end{aligned}$$

has the property that, for every line ℓ *in* $PG(2,13)$*, the sum* S *of the intersection numbers at rational points is*

$$S=\sum_{P\in\ell\cap PG(2,13)} I(P,\ell\cap\mathcal{R}_1)\leq 4.$$

Proof For any point P of $\mathcal{K}_1$ whose tangent to $\mathcal{R}_1$, say, is ℓ_P, the intersection divisor

$$\ell_P\cdot\mathcal{R}_1=2P+2P'+Q+Q',$$

where P' is another point of $\mathcal{K}_1$ and Q,Q' are non-rational points of $\mathcal{R}_1$; more precisely, Q,Q' are $\mathbf{F}_{13^2}$-rational points conjugate over $\mathbf{F}_{13}$. So, $S=4$ for such a line ℓ_P; for any other line of $PG(2,13)$, the sum $S=2,1,0$. The lines for which $S=4$ are precisely the six lines of $\mathcal{R}_\infty$; each is a tangent to $\mathcal{R}_1$. □

Remark 4.2 Note that $(q+1)$-arcs in $PG(2,q)$, of particular interest for q even, have the form $\mathcal{D}(f)=\{(f(t),t,1)\mid t\in\mathbf{F}_q\cup\{\infty\}\}$ and consist of the rational points of the curve $\mathcal{C}$ with affine equation $X=f(Y)$. Each line of $PG(2,q)$ contains at most two rational points of $\mathcal{D}(f)$. However, a line ℓ through $P_0=(1,0,0)$ has the property $\sum_{P\in\ell\cap PG(2,q)} I(P,\ell\cap\mathcal{C})=\deg\mathcal{C}$.

Remark 4.3 J.F. Voloch observed that the curve $\mathbf{v}(2X^6+4Y^6+Z^6)$ has no rational points over $\mathbf{F}_{13}$; this follows immediately since $x^6=0,1,-1$ in $\mathbf{F}_{13}$.

Acknowledgements

This research was supported, for the second author, by a NATO grant and, for the first and third authors, by the G.N.S.A.G.A of C.N.R. and the Italian Ministry of Research and Technology.

References

[1] A.H. Ali, J.W.P. Hirschfeld and H. Kaneta, The automorphism group of a complete $(q-1)$-arc in $PG(2,q)$, *J. Combin. Des.* **2** (1994), 131–145.

[2] A. Bichara and G. Migliori, Alcuni osservazioni sui $(q-1)$-archi completi di $S_{2,q}$, *Boll. Un. Mat. Ital.* **2-D** (1983), 21–27.

[3] J. W. P. Hirschfeld, *Projective Geometries over Finite Fields*, Second edition, Oxford University Press, Oxford, 1998.

[4] J. W. P. Hirschfeld and G. Korchmáros, On the embedding of an arc into a conic in a finite plane, *Finite Fields Appl.* **2** (1996), 274-292.

[5] J. W. P. Hirschfeld and G. Korchmáros, On the number of rational points on an algebraic curve over a finite field, *Bull. Belg. Math. Soc. Simon Stevin* **5** (1998), 313–340.

[6] K. O. Stöhr and J. F. Voloch, Weierstrass points and curves over finite fields, *Proc. London Math. Soc.* **52** (1986), 1–19.

TWO CHARACTERIZATIONS OF THE HERMITIAN SPREAD IN THE SPLIT CAYLEY HEXAGON

Eline Govaert
Department of Pure Mathematics and Computer Algebra
Ghent University
Galglaan 2
9000 Gent
Belgium
egovaert@cage.rug.ac.be

Hendrik Van Maldeghem
Department of Pure Mathematics and Computer Algebra
Ghent University
Galglaan 2
9000 Gent
Belgium
hvm@cage.rug.ac.be

Abstract

From the Hermitian spread in the generalized hexagon $H(q)$, we construct a certain geometry Γ_S, which is a generalized quadrangle. The fact that Γ_S is a generalized quadrangle turns out to characterize the Hermitian spread as a spread of $H(q)$. Furthermore, we give a characterization of this spread using the group of projectivities induced by the spread lines.

1. Introduction

A *weak generalized* n*-gon* Γ is a point-line incidence geometry whose incidence graph has girth $2n$ and diameter n, for some natural number n, $n \geq 2$. A weak generalized n-gon is called a *generalized* n*-gon* if it is thick, that is, if every vertex in the incidence graph has valency greater than 2. Generalized polygons were introduced by Tits [4]. For an extensive survey including most proofs, we refer the reader to [5].

A. Blokhuis et al. (eds.), Finite Geometries, 171–176.

We say that Γ has *order* (s,t) if every line contains $s+1$ points, and every point is incident with exactly $t+1$ lines. Distances are measured in the incidence graph; the distance function is denoted by δ. Elements at maximal distance are called *opposite.* A *spread* of a generalized $2m$-gon is a set of mutually opposite lines, such that every element of Γ is at distance at most m from at least one line of the spread. For any element x, we denote by $\Gamma_i(x)$ the set of elements at distance i from x, by $x^\perp$ the set of elements at distance 2 from x and by $x^{\perp\!\!\perp}$ the set of elements not opposite x. If two elements u, v of Γ are at distance 4 in an n-gon, $n > 4$, then the unique element of $\Gamma_2(u) \cap \Gamma_2(v)$ will be denoted by $u \bowtie v$. If two elements u and v are not opposite, then there is a unique element incident with u and nearest to v; we denote this element by $\mathrm{proj}_u v$ and call it the *projection of v onto u.* For two opposite lines L and M, this projection map defines a bijection, denoted by $[L;M]$, from the point set of L to the point set of M. For lines $L_0, \ldots, L_k$ with L_i opposite L_{i+1} for $0 \leq i < k$, the composition $[L_0;L_1]\ldots[L_{k-1};L_k]$ is called a *projectivity* from L_0 to L_k. If $L_0 = L_k$, then we obtain a permutation of the point set of L_0. The set of all such permutations on the point set of L_0 is a group, called the group of projectivities of L_0. If L_0 is a line of a spread $\mathcal{S}$, then, in the above definition of projectivity, one can require that all the lines L_i belong to $\mathcal{S}$, in this way we obtain the *group of projectivities of L_0 with respect to $\mathcal{S}$*, denoted by $\Pi_{\mathcal{S}}(L_0)$.

Let Γ be a generalized hexagon. For two opposite points x and y of Γ, we denote by x^y the set $\Gamma_2(x) \cap \Gamma_4(y)$, and call this set a *trace.* If every trace is determined by any two of its points, the hexagon is called *point-distance-2-regular.* For two opposite points (lines) x and y of Γ, we denote by $\langle x,y \rangle$ the set $\Gamma_3(x) \cap \Gamma_3(y)$, and call this set a *line (point) regulus.* If every regulus is determined by any two of its elements, the hexagon is called *distance-3-regular.* In a distance-3-regular hexagon, we define $R(L,M)$ to be the unique line regulus containing the opposite lines L and M.

In this paper, we will only meet the quadrangle $Q(5,q)$, of order (q,q^2), and the split Cayley hexagon $H(q)$, of order (q,q). The points and lines of $Q(5,q)$ are respectively the points and lines of the nonsingular elliptic quadric $Q^-(5,q)$. The hexagon $H(q)$ has a representation on the nonsingular quadric $Q = Q(6,q)$ in $\mathrm{PG}(6,q)$ in the following way. Let $Q(6,q)$ be defined by the equation $X_3^2 = X_0X_4 + X_1X_5 + X_2X_6$. The points of $H(q)$ are all the points of $Q(6,q)$, the lines are those lines of $Q(6,q)$ whose Grassmann coordinates satisfy

$$p_{34} = p_{12}, \quad p_{35} = p_{20}, \quad p_{36} = p_{01},$$
$$p_{03} = p_{56}, \quad p_{13} = p_{64}, \quad p_{23} = p_{45}.$$

Two points are collinear on the quadric if and only if they are not opposite in the hexagon. The hexagon $H(q)$ is both 3-regular and point-distance-2-regular. The split Cayley hexagon $H(q)$ has a spread for all values of q. This spread, which is called the *Hermitian spread*, is constructed by Thas [3] as follows. Let γ be a hyperplane of $\mathrm{PG}(6,q)$ intersecting Q in an elliptic quadric $Q^-(5,q)$. Then the lines of $H(q)$ lying in γ constitute a spread of both $H(q)$ and $Q(5,q)$.

Note that a line regulus $R(L,M)$ defines a 3-space $\mathcal{L}$ intersecting Q in a hyperbolic quadric $Q^+(3,q)$. Let α be the polarity associated with the quadric Q. Then the plane $\mathcal{L}^\alpha$ intersects Q in a nondegenerate conic C which is the point regulus $\langle L,M\rangle$. Let N be a line of $H(q)$ opposite every line of $R(L,M)$ and let γ' be the 5-space spanned by L, M and N. Then γ' intersects Q either in a hyperbolic quadric $Q^+(5,q)$ (in this case, it is easy to see that the lines of $H(q)$ in γ' are the lines of a unique subhexagon of order $(1,q)$), or in a cone $pQ(4,q)$ (containing the set $p^{\perp\!\!\perp}$) or in an elliptic quadric $Q^-(5,q)$ (containing a Hermitian spread of the hexagon). Note that in the last case, γ' intersects $\mathcal{L}^\alpha$ in a line not intersecting the conic C, so there are exactly $(q^2-q)/2$ Hermitian spreads containing the regulus $R(L,M)$.

2. Characterizations

Let $\mathcal{S}$ be a spread of the generalized hexagon $\Gamma = H(q)$, and define the following geometry Γ'. The points of Γ' are the points of Γ on lines of the spread. For a point p not on any line of $\mathcal{S}$, we denote by $V_p^{\mathcal{S}}$ the set of $q+1$ points of Γ' collinear with p. Now the lines of Γ' are the lines of the spread together with the sets $V_p^{\mathcal{S}}$, $p \in \Gamma\setminus\Gamma'$. Incidence is containment. It is easy to check that Γ' is a generalized quadrangle of order (q,q^2) if and only if the spread $\mathcal{S}$ satisfies the following property.

($\diamond$) Let L_1, L_2 and L_3 be three different lines of the spread $\mathcal{S}$, and x_1 a point on L_1. Put $x_2 = \mathrm{proj}_{L_2}x_1$ and $x_3 = \mathrm{proj}_{L_3}x_2$. If $\delta(x_1,x_3)=4$, then necessarily $x_1 \bowtie x_2 = x_1 \bowtie x_3 = x_2 \bowtie x_3$.

Property ($\diamond$) says that a configuration as in Figure 1 (where the bold lines are spread lines) is forbidden.

Theorem 2.1 *A spread $\mathcal{S}$ of the finite generalized hexagon $H(q)$ is Hermitian if and only if the geometry Γ' is a generalized quadrangle, which occurs if and only if $\mathcal{S}$ satisfies property ($\diamond$).*

Proof If $\mathcal{S}$ is the Hermitian spread of $H(q)$, then Γ' is indeed the generalized quadrangle $Q(5,q)$. So assume we have a spread $\mathcal{S}$ of $H(q)$ satisfying property ($\diamond$). It is enough to prove that, for any two lines of $\mathcal{S}$, the regulus defined by these lines is contained in $\mathcal{S}$; see [1].

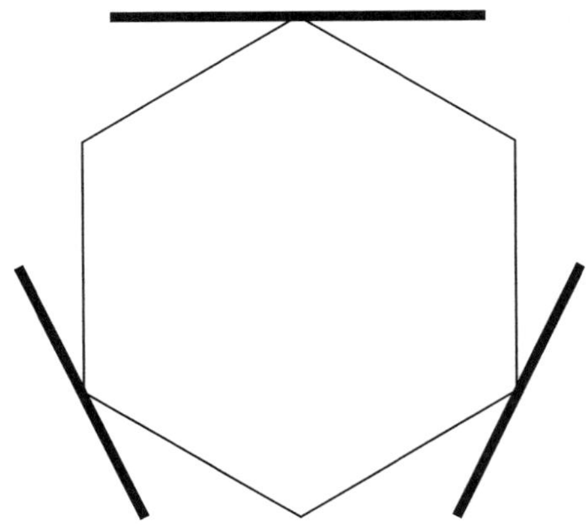

Figure 1. A forbidden configuration in $H(q)$

Let p be a point of $\Gamma = H(q)$ not on a line of the spread. We claim that $V_p^{\mathcal{S}}$ is in fact a distance-2-trace in Γ. Let a and a' be two different points of $V_p^{\mathcal{S}}$ and suppose by way of contradiction that the trace defined by a and a' contains a point b, $b \notin V_p^{\mathcal{S}}$. Let L be the line of $\mathcal{S}$ through a, and let L' be an arbitrary line of Γ through b, different from bp. Let finally L'' be the unique spreadline that is concurrent with L' and put $y = \mathrm{proj}_{L''}a$. Note that $\mathrm{proj}_a y \neq L$. Because of the distance-2-regularity, the trace defined by a and b is equal to p^y, so $\delta(y, a') = 4$. If we denote by N the spreadline through a', then $\mathrm{proj}_{a'}y \neq N$. But now we obtain a configuration forbidden by ($\Diamond$), by considering the spread lines L, N and L'', together with the ordinary hexagon through a, p, a' and y, a contradiction. This shows our assertion.

Let L_0 and L_1 be two different lines of $\mathcal{S}$, and M an arbitrary line of the regulus defined by these two lines, $L_0 \neq M \neq L_1$. We show that $M \in \mathcal{S}$. Let p and p' be two different points at distance 3 from L_0, L_1 and M. By the previous paragraph, we know that $V_p^{\mathcal{S}} = p^{p'}$ and $V_{p'}^{\mathcal{S}} = p'^{p}$. But this implies that both $\mathrm{proj}_M p$ and $\mathrm{proj}_M p'$ have to lie on lines of $\mathcal{S}$. Because spread lines are opposite, this is only possible if $M \in \mathcal{S}$, which shows that $R(L_0, L_1)$ is contained in $\mathcal{S}$. The theorem is proved. □

Note that, for a spread $\mathcal{S}$ of the hexagon $H(q)$, the groups $\Pi_{\mathcal{S}}(L)$, $L \in \mathcal{S}$ are all isomorphic. So we can define $\Pi_{\mathcal{S}} = \Pi_{\mathcal{S}}(L)$, for L an arbitrary line of the spread.

Theorem 2.2 *A spread $\mathcal{S}$ of the finite generalized hexagon $H(q)$ is Hermitian if and only if $\Pi_{\mathcal{S}}$ is a Singer group.*

Proof Let $\mathcal{S}$ be the Hermitian spread. Since $\mathcal{S}$ is also a spread of the quadrangle $Q(5,q)$, the result follows from [2]. To prove the converse, we first show that, for two opposite lines L_0 and L_1 of $H(q)$, there is a bijective correspondence between the set of Hermitian spreads containing the regulus $R(L_0,L_1)$ and the set of Singer groups in $\mathrm{PGL}_2(q)$.

Put $\mathrm{G} = \mathrm{G}_2(q)$. Let M be a line of $H(q)$ opposite L_0 and L_1, with $M \notin R(L_0,L_1)$, such that L_0, L_1 and M are contained in a Hermitian spread $\mathcal{S}_1$, and put $H_1 = \Pi_{\mathcal{S}_1}$. Let H_2 be an arbitrary Singer group acting on L_0. Since every two Singer groups in $\mathrm{PGL}_2(q)$ are conjugate, there exists an element $\sigma' \in \mathrm{PGL}_2(q)$ for which $H_1^{\sigma'} = H_2$. Now choose σ in G_{L_0} such that $\sigma/\Gamma_1(L_0) = \sigma'$. Since the pointwise stabilizer of a line in $H(q)$ acts transitively on the lines opposite this line, which follows from the Moufang condition, as may be seen in [5, Lemma 5.2.4 (ii)], we can choose an element β fixing L_0 pointwise such that $L_1^{\sigma\beta} = L_1$. Now $\sigma\beta$ maps $\mathcal{S}_1$ to a Hermitian spread $\mathcal{S}_2$ containing $R(L_0,L_1)$ for which $\Pi_{\mathcal{S}_2} = H_2$. This shows that every Singer group is the group belonging to a Hermitian spread containing $R(L_0,L_1)$. Furthermore, this spread is unique, since there are as many Hermitian spreads through a certain regulus as there are different Singer groups in $\mathrm{PGL}_2(q)$, namely $(q^2-q)/2$.

Now, let $\mathcal{S} = \{L_0, L_1, \ldots, L_{q^3}\}$ be a spread of $H(q)$ such that $G = \Pi_{\mathcal{S}}$ is a Singer group. By the previous paragraph, we can define $\mathcal{S}_H$ to be the unique Hermitian spread containing $R(L_0,L_1)$ such that $G = \Pi_{\mathcal{S}_H}$. Let A be the set of lines M of the hexagon opposite L_0 and L_1 such that the projectivity $\beta_M := [L_0;M][M;L_1][L_1;L_0]$ belongs to $G\setminus\{e\}$. Let N be an arbitrary line of the hexagon opposite L_0 and L_1. If $N \in R(L_0,L_1)$, then $\beta_N = e$, so $N \notin A$. Suppose $N \notin R(L_0,L_1)$. Let γ be a 5-space containing L_0, L_1 and N. Then γ intersects the quadric $Q(6,q)$ either in an elliptic quadric (case 1), a hyperbolic quadric (case 3) or in a cone $pQ(4,q)$, with $p \in \langle L_0, L_1\rangle$ (case 2).

(1) Note that all the lines of $\mathcal{S}_H$ not belonging to $R(L_0,L_1)$ are contained in A. In this way, we obtain q^3-q elements of A. If N does not belong to $\mathcal{S}_H$, then L_0, L_1 and N define a Hermitian spread corresponding to a group $G' \neq G$, as in the first paragraph of the proof; so $\beta_N \notin G\setminus\{e\}$ and $N \notin A$.

(2) Clearly, N lies at distance 3 from p. Let p' be the projection of p onto N. Then p' is a fixed point of β_N. Suppose there is a fixed point w different from p' on N. Put $w_0 = \mathrm{proj}_{L_0} w$ and $w_1 = \mathrm{proj}_{L_1} w$. Because w is a fixed point, we have $\delta(w_0,w_1) = 4$. Put $w' = w_0 \bowtie w_1$. Note that $\delta(w,w') = 6$. Because of the 2-regularity, $(w')^w = (w')^p$. This implies that w lies at distance 4 from the point of $(w')^p$ on the unique line L of $R(L_0,L_1)$ not opposite N, which is only possible if $p' \in L$, in which case

every point on N is fixed. So, if p' lies on a line of $R(L_0, L_1)$, $\beta_N = e$; if not, then β_N has exactly one fixed point.

(3) Let Γ' be the subhexagon of order $(1,q)$ defined by the intersection of γ and Q, and let p, p' be the unique two points belonging to $\langle L_0, L_1\rangle \cap \Gamma'$. Clearly, $\delta(p, N) \neq 1$. If $\delta(p, N) = 3$, then L_0, L_1 and N are contained in a hyperplane corresponding to case (2). Hence we may suppose that $\delta(p, N) = 5$. Put $w = \mathrm{proj}_N p$ and $v = p \bowtie w$. Again, v is a point of $p^{p'}$ not on the lines L_0 or L_1. If w lies on a line of $R(L_0, L_1)$, then $w \in (p')^p$ and again $\beta_N = e$. So suppose vw is not a line of $R(L_0, L_1)$. Put $u_i = \mathrm{proj}_{L_i} p$ and $u_i' = \mathrm{proj}_{L_i} p'$, $i = 0, 1$. Then $\delta(w, u_0') = \delta(w, u_1') = 4$; so w is a fixed point of β_N. Also $w' = \mathrm{proj}_N u_0 = \mathrm{proj}_N u_1$, so w' is a second fixed point of β_N. Suppose β_N has a fixed point f different from w and w'. Put $f_i = \mathrm{proj}_{L_i} f$, $i = 0, 1$, and $f' = f_0 \bowtie f_1$; here, $\delta(f_0, f_1) = 4$ since f is a fixed point. Note that $\delta(f, f') = 6$. Because of the 2-regularity, we have $(f')^p = (f')^f$. But this is a contradiction, since the point of $(f')^p$ on the line of $R(L_0, L_1)$ through v lies opposite f. So, in this case, β_N has exactly two fixed points.

It follows that A contains exactly $q^3 - q$ lines. Also, β_N is the identity if and only if N is a line of $R(L_0, L_1)$ ($N \neq L_0, L_1$) or N is concurrent with a line of $R(L_0, L_1)$. Hence, since lines of a spread are mutually opposite, at most $(q-1)$ lines of $\mathcal{S} \setminus \{L_0, L_1\}$ can give the identity. Consequently, at least (and hence exactly) $q^3 - q$ lines of $\mathcal{S}$ belong to A. Hence the spreads $\mathcal{S}$ and $\mathcal{S}_H$ can only differ in lines of the regulus $R(L_0, L_1)$. But now, applying the same argument to a regulus $R(L_0, M)$, with M a line of A, shows that $\mathcal{S} = \mathcal{S}_H$. □

Acknowledgements

The first author is a Research Assistant and the second author is a Research Director, both of the National Fund for Scientific Research – Flanders (Belgium).

References

[1] I. Bloemen, J.A. Thas and H. Van Maldeghem, Translation ovoids of generalized quadrangles and hexagons, *Geom. Dedicata* **72** (1998), 19–62.

[2] B. De Bruyn, Generalized quadrangles with a spread of symmetry, *European J.Combin.* **20** (1999), 759–771.

[3] J.A. Thas, Polar spaces, generalized hexagons and perfect codes, *J. Combin. Theory Ser. A* **29** (1980), 87–93.

[4] J. Tits, Sur la trialité et certains groupes qui s'en déduisent, *Inst. Hautes Études Sci. Publ. Math.* **2** (1959), 14–60.

[5] H. Van Maldeghem, *Generalized Polygons*, Monographs in Math. **93**, Birkhäuser Verlag, Basel, 1998.

EPIMORPHISMS OF GENERALIZED POLYGONS, PART 2: SOME EXISTENCE AND NON-EXISTENCE RESULTS

Ralf Gramlich
Vakgroep Discrete Wiskunde
TU Eindhoven
Den Dolech 2
5600 MB Eindhoven
The Netherlands
squid@win.tue.nl

Hendrik Van Maldeghem
Vakgroep Zuivere Wiskunde en Computeralgebra
Universiteit Gent
Galglaan 2
9000 Gent
Belgium
hvm@cage.rug.ac.be

1. Introduction and preliminaries

In Part 1 of this paper (see [7]), we studied the general theory of epimorphisms of generalized polygons, with emphasis on epimorphisms that do not preserve the diameter. Let us briefly recall the situation and relevant definitions.

A *(weak) generalized* n*-gon*, or *polygon* if we do not want to specify n, with $n \geq 2$, is a point-line geometry with an incidence graph of diameter n and girth $2n$; we sometimes say *gonality* n, where the gonality is in general half of the girth. Usually, as in the present article, one restricts oneself to the thick case; that is, every point and every line is incident with at least three elements. In this setting, a generalized n-gon is a geometry with the properties that it does not contain any ordinary k-gons with $k < n$, that any two elements are contained in an ordinary n-gon, and that there exists at least one ordinary $(n + 1)$-gon in the

A. Blokhuis et al. (eds.), Finite Geometries, 177–200.

geometry. The motivation for restricting oneself to the thick case is given by the fact that the non-thick case can be reduced to the thick case, except in some trivial cases; see (1.6.2) of [13]. Also, in the thick case, for a given generalized polygon $\mathfrak{P}$ there exist cardinals s and t such that every line is incident with precisely $s+1$ points and every point is incident with precisely $t+1$ lines. We call the pair (s,t) the *order* of $\mathfrak{P}$. If both s and t are infinite but countable, then we say that $\mathfrak{P}$ has countable infinite order. Generalized polygons were introduced by Tits [11]. We refer to [13] for a survey on the subject.

An *epimorphism* from a generalized m-gon $\mathfrak{P}$ to a generalized n-gon $\mathfrak{Q}$ consists of two maps: a surjective map from the point set of $\mathfrak{P}$ onto the point set of $\mathfrak{Q}$ and another surjective map from the line set of $\mathfrak{P}$ onto the line set of $\mathfrak{Q}$ such that these maps preserve incidence. We also assume that the induced map on the flags of $\mathfrak{P}$ is surjective, where a *flag* is an incident point-line pair.

Closely related to epimorphisms, as was proved in [7], is the concept of (partial) distance-n-ovoid in a generalized m-gon $\mathfrak{P}$, $2 < 2n \leq m$. A *partial distance-n-ovoid* is just a set of points at mutual distance $\geq 2n$; the distances are always measured in the incidence graph and we denote the corresponding function usually by δ, except if we explicitly mention otherwise. If moreover every element of $\mathfrak{P}$ is at distance $\leq n$ from at least one element of the partial distance-n-ovoid, then we call it a *distance-n-ovoid.* We omit the prefix "distance-n" if n is clear, obvious or not important. Dually, one defines a *(partial) distance-n-spread.* A *distance-n-ovoid-spread pairing* is a pair (O,S) consisting of a distance-n-ovoid O of $\mathfrak{P}$ and a distance-n-spread S of $\mathfrak{P}$ such that every element of O is incident with an element of S and every element of S is incident with an element of O. An example is given by the absolute elements of a polarity in a generalized $2n$-gon, see [13] (7.2.6). A *partition into distance-n-ovoid-spread pairings* is a partition of the point set of $\mathfrak{P}$ into distance-n-ovoids, together with a partition of the line set into distance-n-spreads such that every ovoid and every spread is contained in a distance-n-ovoid-spread pairing together with some spread and some ovoid, respectively, of the two partitions.

Given an epimorphism θ from some m-gon $\mathfrak{P}$ to some n-gon $\mathfrak{Q}$, we assign to θ a certain type $vwxyz$, where $v,w,x,y,z \in \{b,s,i\}$, as follows. If θ is bijective or surjective but not bijective onto the point set of $\mathfrak{Q}$, then we set $v=b$ or $v=s$. Similarly we define w with the line set, and x with the flag set. If the restriction of θ to every point row, that is, the set of points incident with a certain line l, is bijective onto the corresponding point row (defined by the image of the line l), then we put $y=b$; if all such restrictions are surjective or injective, and at least one is not bijective, then we put $y = s$ or $y = i$. Dually, we define z. In [7], Theorem 2.1, it is shown that, in general, if $vwx \neq sss$, then only the types *bbbbb*, *bsbib* (and dually *sbbbi*), *ssbii*, *bssis* (and dually *sbssi*) are possible. The special type *sssbb* is called a *local isomorphism.* Moreover, if there is a *bsbib*, *sbbbi*, *ssbii* or *sssbb* epimorphism from a generalized m-gon onto a generalized n-gon, then $2n \leq m$.

Our main result reads as follows.

Main Result *Given any generalized m-gon $\mathfrak{P}$ of countable infinite order, $m \geq 4$, and a natural number $2 \leq n \leq \frac{m}{2}$, there exists a generalized n-gon $\mathfrak{Q}$ and a bsbib, sbbbi, ssbii and sssbb epimorphism from $\mathfrak{P}$ onto $\mathfrak{Q}$.*

We present a detailed proof of the hardest case, namely the case of an *sssbb* epimorphism. The other cases are easier and we give the main ideas by putting $n = 3$. The case of *bssis*/*sbssi* epimorphisms is not treated, but only stated in a remark, because the proofs run along the same lines.

The proof of our Main Result can be found in Section 2. In Section 3 we prove a geometric analogue of the group-theoretic fact that every group is the epimorphic image of a free group and, as a converse of our main result, that for any generalized n-gon of countable infinite order, $n \geq 2$, there exists a generalized $2n$-gon admitting a local isomorphism onto this n-gon; however, we prove it only for the case $n = 3$. In Section 4 we take a closer look at the finite case and prove many nonexistence results, but we also give a few examples (proving existence), mainly related to generalized digons. For precise statements, we refer the reader to Section 4.

2. Constructing epimorphisms given the pre-image

In this section we prove, in some cases only state, the existence of *bsbib*, *ssbii*, *sbbbi*, *bssis*, *sbssi*, *sssbb* epimorphisms: see Section 1. We do that by some induction process in generalized polygons with countable infinite order. Such m-gons exist for every gonality m, see [12].

We start with the case *sssbb*. From [7], we recall the following characterization; see Theorem 2.3 of [7].

Fact 2.1 *There is a bijective correspondence between the class of local isomorphisms from a given generalized m-gon $\mathfrak{P}$ onto some generalized n-gon $\mathfrak{Q}$ and the class of partitions of $\mathfrak{P}$ in distance-n-ovoid-spread pairings.*

Since local isomorphisms are related to ovoid-spread pairings, we first prove a useful general lemma. Recall that the *double* of a geometry $\mathfrak{G}$ is the geometry where points are the points and lines of $\mathfrak{G}$, where the lines are the flags of $\mathfrak{G}$ and where incidence is containment.

Lemma 2.2 *Let $\mathfrak{P} = (\mathcal{P}, \mathcal{L}, \mathcal{F})$ be a generalized m-gon and let $\mathfrak{Q} = (\mathcal{F}, \mathcal{P} \cup \mathcal{L}, \mathcal{G})$ be the dual of the double of $\mathfrak{P}$, where $\mathcal{G} = \{\{f, x\} : x \in f \in \mathcal{F}\}\}$. Let $O_\mathcal{F}$ be a set of flags of $\mathfrak{P}$, and let $O_\mathcal{P}$ and $O_\mathcal{L}$ be the sets of points and lines contained in an element of $O_\mathcal{F}$. Then the pair $(O_\mathcal{P}, O_\mathcal{L})$ is a distance-n-ovoid-spread pairing in $\mathfrak{P}$ if and only if $O_\mathcal{F}$, as a set of points of $\mathfrak{Q}$, is a distance-$2n$-ovoid in $\mathfrak{Q}$.*

Also, if P is a partition of distance-n-ovoid-spread pairings of $\mathfrak{P}$, then the union of the set of flags corresponding with the pairings as above is, as a set of points of $\mathfrak{Q}$, a distance-2-ovoid of $\mathfrak{Q}$.

Proof Let the distance function measured in the incidence graphs of $\mathfrak{P}$ and $\mathfrak{Q}$ be denoted by $\delta_\mathfrak{P}$ and $\delta_\mathfrak{Q}$.

First suppose that $(O_\mathcal{P}, O_\mathcal{L})$ is a distance-n-ovoid-spread pairing in $\mathfrak{P}$. Let $f = \{x, a\}$ and $g = \{y, b\}$ be two arbitrary elements in $O_\mathcal{F}$, with $x, y \in O_\mathcal{P}$ and $a, b \in O_\mathcal{L}$. Assume, by way of contradiction, that $2k := \delta_\mathfrak{Q}(f, g) < 4n$. Then one of x, a is at distance $k - 1 < 2n - 1$ from one of y, b in $\mathfrak{P}$. Since $x, y \in O_\mathcal{P}$, they are at distance at least $2n$ from each other; similarly for a and b. If $\delta_\mathfrak{P}(x, b) = k - 1$, then $\delta_\mathfrak{P}(x, y) = k < 2n$, a contradiction. Similarly if $\delta_\mathfrak{P}(y, a) = k - 1$.

Now let c be any line of $\mathfrak{Q}$. Then there exist $z \in O_\mathcal{P}$ and $d \in O_\mathcal{L}$ such that $\delta_\mathfrak{P}(c, z) \leq n$ and $\delta_\mathfrak{P}(c, d) \leq n$. Since distances between elements of the same type have different parity from distances between elements of different type, we may assume without loss of generality that $l := \delta_\mathfrak{P}(c, z) < \delta_\mathfrak{P}(c, d) \leq n$. If h is the unique element of $O_\mathcal{F}$ containing z, then $\delta_\mathfrak{Q}(c, h) = 2\ell + 1 < 2n + 1$, and hence $\delta_\mathfrak{Q}(c, h) \leq 2n - 1$, since this distance must always be odd. If h' is any point of $\mathfrak{Q}$, then consider any line c' incident with h'. By the foregoing, there is an element h'' of $O_\mathcal{F}$ with $\delta_\mathfrak{Q}(c', h'') \leq 2n - 1$, hence $\delta_\mathfrak{Q}(h', h'') \leq 2n$. We have shown that $O_\mathcal{F}$ is a distance-$2n$-ovoid of $\mathfrak{Q}$.

Now suppose that $O_\mathcal{F}$ is a distance-$2n$-ovoid of $\mathfrak{Q}$. We show that $O_\mathcal{P}$ is a distance-n-ovoid of $\mathfrak{P}$. Let x, y be two arbitrary elements of $O_\mathcal{P}$ and let f, g be the two elements of $O_\mathcal{F}$ containing x, y, respectively (they are uniquely determined). Suppose, again by way of contradiction, that $2k := \delta_\mathfrak{P}(x, y) \leq 2n - 2$. Then $\delta_\mathfrak{Q}(f, g) \leq 4k + 2 \leq 4n - 2$, a contradiction. Now let a be any point or line of $\mathfrak{P}$. Then there is an $h = \{u, c\} \in O_\mathcal{F}$ with $2l - 1 := \delta_\mathfrak{Q}(a, h) \leq 2n - 1$ (and $u \in O_\mathcal{P}$, $c \in O_\mathcal{L}$). Consequently $\{l - 1, l\} \ni \delta_\mathfrak{P}(a, u) \leq n$. Hence $O_\mathcal{P}$ is a distance-n-ovoid of $\mathfrak{P}$. Similarly, $O_\mathcal{L}$ is a distance-n-spread of $\mathfrak{P}$. The first part of the lemma is proved.

The second statement of the lemma is now easy. □

Theorem 2.3 *Let $\mathfrak{P}$ be a generalized m-gon with countable infinite order. Then there is a local isomorphism onto a generalized n-gon for any integer $2 \leq n \leq \frac{m}{2}$.*

Proof Before embarking on the proof, we state another lemma. Let us denote by $\mathfrak{Q}$ the generalized $2m$-gon obtained by dualizing the double of $\mathfrak{P}$, as in the previous lemma. We also use the same notation as above for the distance functions. First we shall show the following.

(*) *Let c be an arbitrary line of $\mathfrak{Q}$. Let $\{x_1, x_2, \dots, x_k\}$ be a finite collection of points with $\delta_{\mathfrak{Q}}(c, x_i) \geq 2n+1$, for all $i \in \{1, 2, \dots, k\}$. Then there are infinitely many lines d with the properties:*

(i) $\delta_{\mathfrak{Q}}(c, d) \leq 2n - 2$;

(ii) *there are infinitely many points z incident with d satisfying the inequality $\delta_{\mathfrak{Q}}(z, x_i) \geq 4n$, for all $i \in \{1, 2, \dots, k\}$.*

Indeed, suppose that there are only a finite number of lines d with the given properties. Let a be any line at distance $2n - 4$ from c. Let a^* be a line concurrent with a and suppose that there are infinitely many points z on a^* with $\delta_{\mathfrak{Q}}(z, x_i) < 4n$, for some $i \in \{1, 2, \dots, k\}$ depending on z. Then there exists $i \in \{1, 2, \dots, k\}$ with x_i at distance $< 4n$ from infinitely many points of a^*. Hence x_i is at distance $< 4n$ from all points of a^*. Since there are infinitely many such lines a^* concurrent with a, we conclude that there exists $j \in \{1, 2, \dots, k\}$ with the property that x_j is at distance $< 4n$ from all points of infinitely many lines concurrent with a. Hence x_j is at distance $< 4n$ from all points at distance ≤ 3 from a. If $n = 2$, then $a = c$. If $\delta_{\mathfrak{Q}}(x_j, c) = l$, then $l + 3 < 4n = 8$, hence $l < 5$, contradicting $\delta_{\mathfrak{Q}}(x_j, l) \geq 2n + 1$.

If $n > 2$, then we consider the line a_1 concurrent with a at distance $2n - 6$ from c. From the previous paragraph we know that for every line a_1^* concurrent with a_1 and at distance $2n - 4$ from c there is a number $j^* \in \{1, 2, \dots, k\}$ such that all points z at distance ≤ 3 from a_1^* satisfy $\delta_{\mathfrak{Q}}(x_{j^*}, z) < 4n$. Since there are infinitely many such lines a_1^*, there is an integer $j_1 \in \{1, 2, \dots, k\}$ such that $\delta_{\mathfrak{Q}}(x_{j_1}, z) < 4n$ for all points z at distance ≤ 5 from a_1. Continuing like that, we conclude that there must be $j' \in \{1, 2, \dots, k\}$ such that $\delta_{\mathfrak{Q}}(x_{j'}, z) < 4n$ for all points z at distance $\leq 2n - 1$ from c. But then it easily follows (as in the case $n = 2$ above) that $\delta_{\mathfrak{Q}}(x_{j'}, c) \leq 2n - 1$, a contradiction. Our lemma (*) is proved.

Now we will construct a partition of distance-n-ovoid-spread pairings in $\mathfrak{P}$. Using Lemma 2.2 above, it suffices to construct a family of distance-$2n$-ovoids in $\mathfrak{Q}$ such that the union of all these ovoids is a distance-2-ovoid in $\mathfrak{Q}$. Let us number the lines of $\mathfrak{Q}$. The construction goes by induction.

The case $l = 0$ corresponds to the initial step of the induction. Therefore, suppose we have already constructed l distance-$2n$-ovoids $O_1, O_2, \ldots, O_l$, then we construct, also by induction, the ovoid O_{l+1}. Consider the first line b of $\mathfrak{Q}$ (remember that we numbered the lines!) that does not yet contain a point of $O := O_1 \cup O_2 \cup \ldots \cup O_l$. For each $i \in \{1, 2, \ldots, l\}$, there is at most one point of O_i collinear with some point of b: since $n \geq 2$, so the points of O_i have mutual distance ≥ 8. Hence there are only a finite number of points of b collinear with points of O. As first point x_1 of O_{l+1}, we can take a point on b not collinear with any point of O. Suppose that we have already chosen k points $x_1, x_2, \ldots, x_k$ of O_{l+1}. Then we consider the first line a of $\mathfrak{Q}$ with the property that $\delta_{\mathfrak{Q}}(a, x_i) \geq 2n+1$, for all $i \in \{1, 2, \ldots, k\}$. Such a line a exists, otherwise we already have an ovoid; however, it is shown below in Remark 2.4 that every ovoid has an infinite number of points. By (*), there are infinitely many lines b with $\delta_{\mathfrak{Q}}(a, b) \leq 2n-2$ incident with infinitely many points at distance $\geq 4n$ from all members of $\{x_1, x_2, \ldots, x_k\}$. Only finitely many amongst them are incident with points of O, since, for each $i \in \{1, 2, \ldots, l\}$, there is at most one point at distance $\leq 2n-1$ from a, and each such point is incident with at most two lines at distance $\leq 2n-2$ from a. So there remain infinitely many lines on which we can make a valid choice for x_{k+1}. Continuing in this way, it is clear that we obtain a distance-$2n$-ovoid O_{l+1} of $\mathfrak{Q}$. Also, it is clear that the union of all these ovoids is a distance-2-ovoid of $\mathfrak{Q}$. The theorem now follows from Fact 2.1. □

Remark 2.4 We show that every distance-$2n$-ovoid O in $\mathfrak{Q}$, with $\mathfrak{Q}$ as above, has infinitely many points. Indeed, it certainly must contain at least one point x. Let a be a line at distance $2n-1$ from x, and let b be any line concurrent with a and at distance $2n+1$ from x (there are infinitely many such lines). Then there is a point x_b of O at distance $\leq 2n-1$ from b. Clearly $x \neq x_b$. Since the distance between x and x_b must at least be $4n$, the distance between b and x_b must be equal to $2n-1$. This implies easily that for a different choice of b we have a different point x_b and hence O must contain infinitely many points.

Similarly one shows that every ovoid or spread of any infinite generalized polygon must be infinite itself.

In view of [7, Corollary 2.6] that states the restriction $2n \leq m$ for an *sssbb* epimorphism from a generalized m-gon onto a generalized n-gon, we cannot expect more than the preceding result. The following two theorems only involve projective planes as images, but see Remark 2.7.

Theorem 2.5 *Let $\mathfrak{P} = (\mathcal{P}, \mathcal{L}, \mathcal{F})$ be a generalized m-gon, $m \geq 6$, with countable infinite order. Then there exist bsbib or sbbbi homomorphisms onto projective planes.*

Proof We will construct a partition of the line set $\mathcal{L}$ into subsets S_α, where α lies in some index set which we can choose to be the natural numbers, satisfying the incidence conditions of Theorem 2.4 of [7] for a *bsbib* homomorphism. Let us repeat these conditions for convenience. Each S_α is a set of lines at mutual distance ≥ 6. We say that a set of points P is incident with a set of lines L, and this is denoted by $P\mathtt{I}L$, if some member of P is incident with some member of L. The incidence conditions are as follows:

(i) for any two points p, q, there is a unique set S_α with $\{p\}\mathtt{I}S_\alpha\mathtt{I}\{q\}$;

(ii) for any two sets S_{α_0} and S_{α_1}, there is a unique point p such that $S_{\alpha_0}\mathtt{I}\{p\}\mathtt{I}S_{\alpha_1}$.

Now number the points and lines of $\mathfrak{P}$ (separately). Define $S_1 := \{l_1\}$. Now let n_1 and n_2 be the biggest numbers such that the incidence condition holds for the first n_1 points and n_2 lines. Let $n := \min\{n_1, n_2\}$. Moreover, suppose that there have already finitely many sets S_α (of lines at mutual distance ≥ 6) been constructed and that each S_α contains a finite number of lines. Furthermore, assume that no pair of points or lines violates the uniqueness part of the incidence condition, that is, for any two points p, q either $\{p\}\mathtt{I}S_\alpha\mathtt{I}\{q\}$ holds for a unique α or no such α exists; dually for lines. Let S be the union of all S_α already constructed. This is a finite set of lines.

If $n = n_1$, then consider the point p_{n+1}. Take any point p_i, $i \leq n$, such that p_i and p_{n+1} do not satisfy the incidence condition (i). If these two points are collinear in $\mathfrak{P}$, then just generate a new set S_α consisting of the joining line. This line might intersect lines contained in other S_α, but it cannot intersect two distinct lines of the same S_α, since otherwise these lines would be at distance 4. If p_i and p_{n+1} are not collinear, then, by finiteness of S, we find lines a and b at distance ≥ 6 through p_i and p_{n+1}, not intersecting any line of S, except maybe in the points p_i and p_{n+1}. Generate the next S_α defined by $\{a, b\}$. Here, also a and b might intersect lines of other S_α, but one line cannot intersect two distinct lines of the same S_α. But it is also impossible to have lines $a', b' \in S_\alpha$ with a intersecting a' and b intersecting b'. Indeed, by construction, we would have $a'\mathtt{I}p_i$ and $b'\mathtt{I}p_{n+1}$, which is impossible, since we assumed that p_i and p_{n+1} do not satisfy the incidence condition. Repeat this for all points p_i, $i \leq n$. Obviously, we still have a finite number of sets S_α each with finite cardinality. Moreover, the incidence condition is now satisfied by the first $n+1$ points and still no pair of points or lines violates the uniqueness part of the incidence condition.

If $n = n_2$, then we consider the line l_{n+1}. We can assume that l_{n+1} is already contained in some S_{α_0}; otherwise we construct a new one consisting precisely of l_{n+1}, which is possible as above. Take any line l_i, $i \leq n$, such that l_i and l_{n+1} do not satisfy the incidence condition. In particular, the lines l_i and l_{n+1} do not intersect. Hence we find a line a intersecting l_i, at distance ≥ 6 from each element of S_{α_0}, and such that a does not intersect any other line already contained in some S_α, by the finiteness of S. As l_i and l_{n+1} did not satisfy the incidence condition, no line in the S_α containing l_i intersects any line of S_{α_0}. Add this line a to the set S_{α_0}. Repeating the procedure eventually results in the first $n+1$ lines satisfying the incidence condition. Still, we only have a finite number of sets S_α each of finite cardinality. Moreover, the incidence condition is now satisfied by the first $n+1$ lines and no pair of points or lines violates the uniqueness part of the incidence condition.

Now we have increased n by at least one. Continuing the above process covers the whole polygon $\mathfrak{P}$, and we are done. Dually, we get an *sbbbi* homomorphism. □

Theorem 2.6 *Let $\mathfrak{P} = (\mathcal{P}, \mathcal{L}, \mathcal{F})$ be a generalized m-gon, $m \geq 6$, with countable infinite order. Then there exist ssbii homomorphisms onto projective planes.*

Proof We use the following method. We start the construction similar to the construction of a *bsbib* homomorphism and consider $\mathfrak{P}^{\text{dual}}$ at some point. Continuing then eventually leads to an *ssbii* homomorphism. More precisely, suppose we have already constructed n_0 sets S_α, and that there exists at least one S_α, $1 \leq \alpha \leq n_0$, with $|S_\alpha| > 1$. Then $\mathfrak{P}^{\text{dual}}$, together with the induced sets O_α (we change notation!) satisfy the dual condition. The only thing we have to show is that we can construct sets S_β in $\mathfrak{P}^{\text{dual}}$, with at least one S_α of cardinality > 1, so that the conditions of Theorem 2.4 of [7] are satisfied. These conditions say precisely that the "quotient geometry", which is the geometry obtained from $\mathfrak{P}$ by taking as points the sets O_α and as lines the sets S_β, with the incidence relation as defined in the previous proof, is a projective plane. They are similar to the conditions in the previous proof; in particular, we refer to the *incidence condition for two points* p, q as the condition that there exists precisely one set S_β incident with the O_α's containing p and q, respectively; dually for the incidence condition for two lines. The result will be an *ssbii* epimorphism, since it is neither bijective on the points nor on the lines.

We use the same notation as in the previous proof regarding incidence of two sets. Moreover, if P is a set of points and L a set of lines, then we denote by $P\mathrm{I}_!L$ (or $L\mathrm{I}_!P$) that a unique point of P is incident with a unique line of L and that no other incidences exist between points of P and lines of L.

Let n_1 and n_2 be the biggest numbers such that the incidence condition holds for the first n_1 points and n_2 lines (remember that points and lines have been numbered). Let $n := \min\{n_1, n_2\}$. Moreover, suppose that

(i) only finitely many sets S_β have already been constructed and that each S_β contains a finite number of lines (the sets O_α satisfy this condition and will not be changed any more; any point not yet contained in some O_α is regarded as some O_α of cardinality 1);

(ii) no pair of points or lines violates the uniqueness part of the incidence condition, that is, for any two points p, q either $p \in O_{\alpha_1}\mathrm{I}_!S_\beta\mathrm{I}_!O_{\alpha_2} \ni q$ holds for unique α_1, α_2, β or no such indices exists, and dually for lines;

(iii) there do not exist a line l and a set $S_\beta \not\ni l$ with $O_{\alpha_1}\mathrm{I}S_\beta\mathrm{I}O_{\alpha_2}$ and $O_{\alpha_1}\mathrm{I}l\mathrm{I}O_{\alpha_2}$.

The rest of the proof is similar to the proof of the previous theorem, except that we have to take care of condition (iii) at each step. Let us sketch the proof; then we also see where condition (iii) is needed.

If $n = n_1$, then we consider the point p_{n+1}. Take any point p_i, $i \leq n$, such that $O_{\alpha_2} \ni p_i$ and $O_{\alpha_1} \ni p_{n+1}$ do not satisfy the incidence condition. If there are two points in O_{α_1} and O_{α_2} that are collinear in $\mathfrak{P}$, then generate a new set S_β consisting of the joining line l (note that such a joining line is unique if it exists, by the construction of the O_α). As in the previous proof, this line cannot intersect two distinct lines of the same S_β. Moreover, this line is incident with points from some O_α, but again by the construction of the O_α it cannot destroy condition (ii). Indeed, this would imply the existence of points q_1, q_2 of the same O_α such that $q_1\mathrm{I}l$ and $q_2 \in O_\alpha\mathrm{I}_!S_\beta\mathrm{I}_!O_{\alpha_1} \ni p_{n+1}$ or $q_2 \in O_\alpha\mathrm{I}_!S_\beta\mathrm{I}_!O_{\alpha_2} \ni p_i$ which contradicts condition (iii). If O_{α_1} and O_{α_2} do not contain collinear points we find lines a and b at distance ≥ 6 through points of O_{α_1} and O_{α_2}, respectively, such that, firstly, a and b do not intersect any line already contained in some S_β except in the points of O_{α_1} and O_{α_2}, and such that, secondly, neither a nor b are incident with some $O_\alpha \neq O_{\alpha_1}, O_{\alpha_2}$ with $|O_\alpha| > 1$: this is possible since there are only a finite number of α's with $|O_\alpha| > 1$. Generate the next S_β consisting of a and b. As in the previous proof, we repeat this for all points p_i, $i \leq n$, we still have a finite number of sets S_α each with finite cardinality, and the incidence condition is now satisfied by the first $n+1$ points and still no pair of points or lines violates the uniqueness part of the incidence condition (ii). Furthermore, condition (iii) is still satisfied since (iii) could only get violated if $\{l, m\}$ were incident with O_α with $|O_\alpha| > 1$. If $n = n_2$, then consider the line l_{n+1}. Take any line l_i, $i \leq n$, such that l_i and l_{n+1} do not satisfy the incidence condition. As in the previous proof, we may assume that l_{n+1} is already contained in some S_{β_0}. Also, we find a line a intersecting l_i, at distance ≥ 6 from each element of S_{β_0}, such that a does not intersect any other line already contained in some S_β, and such that a is not incident with a point belonging to an O_α with $|O_\alpha| > 1$: this must ensure that (iii) remains true. We add this line a to the set S_β containing l_{n+1}. Repeating this procedure for all $i \leq n$, we eventually see that the first $n+1$ lines satisfy the incidence condition and no pair of points or lines violates the uniqueness part of the incidence condition. Still, we only have a finite number of sets S_β each of finite cardinality. Furthermore, condition (iii) is still satisfied. Now we have increased n by at least one. Continuing as above finally will cover the whole polygon. $\square$

Remark 2.7 (i) The constructions given above also work for other infinite orders, by transfinite induction. The differences in the proofs are only of technical nature.

(ii) Constructing a *bsbib* homomorphism onto an n-gon, $n \geq 4$, can be done in a similar fashion and is, in fact, easier. Sloppily, for n even we have to ensure that there is a unique path from any point to any line of length at most $n-1$ and, for n odd, that there are unique paths between any pair of points and any lines of lengths at most $n-1$. But this is easily accomplished by induction. Let n be even. Indeed, for a point p and a line l at too big a distance there exists a line l' with $\delta(l, l') \geq 2n$ and $d(p, l') \leq n-1$; identify l and l'. Continuing this gives the wanted result. The case n odd works analogously. Similarly, *sbbbi* and *ssbii* onto n-gons are easily constructed for $n \geq 4$. We do not give the full explicit construction.

(iii) There also exist slight modifications of the construction used to show Theorem 2.5 that prove the existence of *bssis* and *sbssi* homomorphisms from m-gons onto n-gons. In that case, the only restriction is $m > n$; see [7].

(iv) Aart Blokhuis [1] has constructed partitions of the point set of any semifinite quadrangle (that is, an infinite generalized quadrangle with a finite number of points on a line; no examples are known yet; in fact, it is a conjecture that these do not exist) into ovoids and dually, partitions of the line set of such a quadrangle into spreads. These give rise to ovoid-spread pairings and local isomorphisms of semifinite quadrangles onto (semifinite) generalized digons.

Note that no semifinite generalized polygon admits a partition into distance-n-ovoid-spread pairings for odd n, since a generalized n-gon with n odd has the same number of points per line as it has lines per point. The existence of such a partition in a semifinite generalized polygon would contradict that by Theorem 2.3 of [7]. In fact, this might be a possible method to disprove existence of certain semifinite generalized polygons. A similar remark can be made about semifinite octagons of order (s, ∞), with $s \in \{2, 3, 4\}$, and ∞ some infinite cardinal number, because there are no semifinite quadrangles of order (s, ∞), s as above, by results of Cameron [4], Kantor (unpublished, see also Brouwer [3]) and Cherlin [5].

Unfortunately, all our efforts to disprove the existence of some semifinite generalized polygons with this method have been fruitless.

3. Constructing epimorphisms given the image

The following result is inspired by group theory. It is well-known that any group is a homomorphic image of some free group. We now prove an analogue for generalized polygons. Therefore, we should have a precise definition of what a free polygon is. Let us define a *free polygon* as a generalized n-gon, $n \geq 2$, constructed by the process described by Tits in [12] (4.4) (see also [13] (1.3.13)) by starting with a (proper) *partial n-gon*, that is, a connected geometry of diameter $> n$ that does not have ordinary j-gons as subgeometries for any $j < n$.

Theorem 3.1 *Let $\mathfrak{P}$ be any generalized n-gon. Then there exists a free generalized n-gon $\mathfrak{P}'$ having $\mathfrak{P}$ as a homomorphic image.*

Proof Let $\mathfrak{P} = (\mathcal{P}, \mathcal{L}, \mathcal{F})$. We will obtain $\mathfrak{P}'$ by a free construction, hence it will certainly be a free polygon. To start a free construction we must give ourselves a (proper) partial n-gon. We may do this in various ways by adding points and lines to $\mathfrak{P}$. We must choose the image in $\mathfrak{P}$ for each element we add to $\mathfrak{P}'$ and make sure that incident elements in the partial n-gon have incident images. The easiest way to accomplish this is to add just one point x on a line L of $\mathfrak{P}$ and declare the image of x equal to some arbitrarily chosen point of L (in $\mathfrak{P}$). Now we carry out the free construction as given in [12]. It is enough to give the image of every element we add at each step of the construction, and to prove that the image of every new flag is a flag in $\mathfrak{P}$. The construction is such that, at step k, we add a chain of $n-2$ new elements between two old elements a, b which are at distance $n+1$ at step $k-1$. Let A, B be the images of a, b. If $A \neq B$, let $(A, X_1, X_2, \ldots, X_m, B)$ be the unique path of length $m+1 < n$ in $\mathfrak{P}$ joining A and B. Otherwise, let X_1 be any element of $\mathfrak{P}$ incident with $A = B$ and put $m = 1$. Note that $m \equiv n \bmod 2$. Let $(a, x_1, x_2, \ldots, x_{n-2}, b)$ be the chain we have added in step k. Then we select X_i as the image of x_i, for $i \in \{1, 2, \ldots, m\}$. For $i \in \{m+1, m+3, \ldots, n-3\}$, we choose the image of x_i to be equal to B, and for $i \in \{m+2, m+4, \ldots, n-2\}$, we choose the image of x_i to be equal to X_m. The theorem is now clear. □

The following is a converse of Theorem 2.3. For the sake of simplicity, we restrict ourselves to $m = 6$ and $n = 3$, but see Remark 3.3.

Theorem 3.2 *Let $\mathfrak{P} = (\mathcal{P}, \mathcal{L}, \mathcal{F})$ be a projective plane of countable infinite order. Then there exists a generalized hexagon $\mathfrak{H}$ admitting a local isomorphism onto $\mathfrak{P}$.*

Proof Fix a point a of $\mathfrak{P}$ and define $\mathfrak{A} := (\mathcal{P}, [a], \mathcal{F}_{|\mathcal{P}\times[a]})$ where $[a]$ is the line pencil of a. Let $\mathfrak{H}_0$ be the disjoint union of countably infinitely many copies of $\mathfrak{A}$. We have a map π from $\mathfrak{H}_0$ to $\mathfrak{P}$ mapping a point of $\mathfrak{H}_0$ onto the corresponding point of $\mathfrak{P}$ (given by the embedding of the copy of $\mathfrak{A}$ containing this point into $\mathfrak{P}$). The map on the lines is given by the map on the point rows.

Now we will construct incidence structures $\mathfrak{H}_k$, $k \in \mathbb{N}$, to obtain a generalized hexagon $\cup_{k\in\mathbb{N}_0}\mathfrak{H}_k$ (the symbol "$\cup$" must be interpreted as follows: the point sets of all $\mathfrak{H}_k$ will be identical, but we will define lines as subsets of points and sometimes add points to a line; for each such line, we take the union of all points we add in each step) in such a way that the map described above becomes a local isomorphism. We identify lines with their point rows, which is possible since we are working with partial linear geometries. Incidence then is given by containment.

Let us proceed by induction on k. To this end number the points and lines of $\mathfrak{H}_0$ (together as one set; denote the elements with x_i, $i \in \mathbb{N}$). Moreover, for each line of $\mathcal{L}\backslash[a]$, we number the points on its point row and, likewise, for each point of $\mathcal{P}\backslash\{a\}$, we number the lines of its line pencil. Finally, also number the copies of $\mathfrak{A}$ in $\mathfrak{H}_0$. For all $k \in \mathbb{N}_0$ for which $\mathfrak{H}_k$ already has been defined, let n_k be the biggest number such that all points and lines of $\mathfrak{H}_k$ with numbers lower that n_k are mutually at distance ≤ 6. For a line l of $\mathcal{L}\backslash[a]$ and $k \in \mathbb{N}_0$ for which $\mathfrak{H}_k$ exists, let $n_{l,k}$ be the maximal number such that, for every point p_i, $i \leq n_{l,k}$, of l, and for every pre-image l^* of l, there is some pre-image of p_i in some copy of $\mathfrak{A}$ in $\mathfrak{H}_k$ with number smaller than or equal to k contained in l^* (where the copies of $\mathfrak{A}$ in $\mathfrak{H}_k$ are numbered in the same way as in $\mathfrak{H}_0$). If l does not have a pre-image yet, set $n_{l,k} := 0$. Dually, define $n_{p,k}$ for all points p of $\mathcal{P}\backslash\{a\}$. Note that all points do have pre-images.

Let us call a copy of $\mathfrak{A}$ in $\mathfrak{H}_k$ *untouched* if every point except the pre-image of a in that copy is incident with a unique line (which is necessarily also incident with the inverse image of a in that copy).

Suppose that there do not exist circuits of length smaller than 12 in $\mathfrak{H}_k$, that objects having the same image are at mutual distance ≥ 6 (in particular, no point row contains two points with the same image, and no line pencil contains two lines with the same image) and that there are infinitely many untouched copies in $\mathfrak{H}_k$. The geometry $\mathfrak{H}_0$ serves as initial step of the induction ($k = 0$); moreover, in the initial step of the induction set $\mathfrak{H}_k := \mathfrak{H}_0$ for all $k \in \mathbb{N}_0$ with $k \leq \min\left(\{n_0\} \cup \{n_{\pi(x_i),0} : x_i \in \mathfrak{H}_0 \text{ and } 1 \leq i \leq k\}\right)$.

Let k be the maximal natural number such that

$$k \leq \min\left(\{n_k\} \cup \{n_{\pi(x_i),k} : x_i \in \mathfrak{H}_k \text{ and } 1 \leq i \leq k\}\right).$$

Define $k_0 := k$. Set $\mathfrak{H}_{k_0+1} := \mathfrak{H}_{k_0}$. In the following we will add lines to $\mathfrak{H}_{k_0+1}$. Note that at the moment $n_{k_0+1} = n_{k_0}$ and $n_{\pi(x_i),k_0} = n_{\pi(x_i),k_0+1}$ for all $x_i \in \mathfrak{H}_{k_0+1} = \mathfrak{H}_{k_0}$.

If $k_0 = n_{k_0+1}$, consider the geometrical object x_{k_0+1} of $\mathfrak{H}_{k_0+1}$. Take the first x_i, $i \leq k_0$, with $\delta(x_i, x_{k_0+1}) > 6$. If x_i and x_{k_0+1} are of the same type, we will put them at distance 6, otherwise at distance 5. Suppose they are both of the same type, lines say. We claim that we can arrange it so that these two lines contain points p and q, respectively, with $\delta(p,q) \geq 8$ and $\pi(p) \neq \pi(q)$ and that the cardinalities of the line pencils of p and q are both ≤ 2. Indeed, consider any of these lines. If it has an infinite point row, this is obvious; we can even find points lying on just one line. If it is finite, however, we can just add an admissible point, "admissible" in the sense that images of lines have to be lines, from one of the untouched copies of $\mathfrak{A}$ in $\mathfrak{H}_{k_0+1}$. The claim is proved. So, we can find a point r (lying on only one line) in an untouched copy of $\mathfrak{A}$ such that neither $\pi(p)$, $\pi(q)$, $\pi(r)$ nor $\pi(p)$, $\pi(r)$, a nor $\pi(q)$, $\pi(r)$, a are lying on one line in $\mathfrak{P}$ and such that there is no line having $\pi(p)\pi(r)$ or $\pi(q)\pi(r)$ as image at distance ≤ 5 from p or r, respectively from q or r. The choice of r is possible, since by assumption there are infinitely many untouched copies available. Call $\{p,r\}$ and $\{q,r\}$ lines and include them in the numbering with numbers higher than $k + 2^{k_0}$ and as multiples of $2^{k_0} + 1$. The two lines x_i and x_{k_0+1} are now at distance 6 and all conditions mentioned in the above paragraph are still satisfied. Similar constructions handle the cases where x_i and x_{k_0+1} are both points or are objects of different types. Repeat until $n_{k_0+1} > k_0$.

Suppose $k_0 = n_{\pi(x_i),k_0+1}$ for some $i \leq k_0$. If $\pi(x_i)$ is a point, then for each suitable pre-image c of $\pi(x_i)$ (suitable here means a pre-image of $\pi(x_i)$ contained in a copy of $\mathfrak{A}$ with number smaller than or equal to k_0 in $\mathfrak{H}_{k_0+1}$ that is not yet incident with a pre-image of line number $n_{\pi(x_i),k_0+1}+1$ in $[\pi(x_i)]$) we can add a suitable line b to $\mathfrak{H}_{k_0+1}$ such that $c\mathrm{I}b$ and the number of $\pi(b)$ in $[\pi(x_i)]$ is equal to $n_{\pi(x_i),k_0+1}+1$. Moreover, we keep an infinite number of untouched copies, since we add only finitely many lines. We include these new lines in the numbering, again with numbers higher than $k+2^{k_0}$ and as multiples of $2^{k_0}+1$. If $\pi(x_i)$ is a line, then for each pre-image of $\pi(x_i)$ we find pre-images of the point d with number $n_{\pi(x_i)}+1$ in $[\pi(x_i)]$ in untouched copies of $\mathfrak{A}$ and we can add them to the pre-images of $\pi(x_i)$ (again we can do that in such a way that infinitely many untouched copies remain). Finally, for all points x_i of $\mathfrak{H}_{k_0+1}$ with $i \leq k_0+1$, make sure that the pre-image of $\pi(x_i)$ contained in the (k_0+1)st copy of $\mathfrak{A}$ in $\mathfrak{H}_{k_0+1}$ lies on a line b whose image $\pi(b)$ has the number $n_{\pi(x_i),k_0+1}+1$ in $[\pi(x_i)]$—by adding it like indicated in the last paragraph, if necessary. Similarly for all lines x_i with $i \leq k_0+1$ add pre-images of $\pi(x_i)$ in that copy of $\mathfrak{A}$. Repeat until we have $n_{\pi(x_i),k_0+1} > k_0$ for all $i \leq k_0$, and do the same procedure for x_{k_0+1} to obtain $n_{\pi(x_{k_0+1},k_0+1)} > k_0$.

Note that now

$$k_0+1 \leq \min(\{n_{k_0+1}\} \cup \{n_{\pi(x_i),k_0+1} : x_i \in \mathfrak{H}_{k_0+1} \text{ and } 1 \leq i \leq k_0+1\})$$

holds and we have increased k by at least one. If necessary define further $\mathfrak{H}_i := \mathfrak{H}_{k_0+1}$, $i > k_0+1$, if that would increase k to an even higher number.

Induction gives a generalized hexagon $\cup_{k\in\mathbb{N}_0}\mathfrak{H}_k$; the diameter is 6 and there exist no circuits of length smaller than 12; hence the girth has to be 12, since there are no points or lines which are incident with only one line or point respectively, by the numbers $n_{\pi(x_i),k}$).

The fact that we at last reach every point of $\mathfrak{H}_0$ follows from the fact that the point with number z gets in the process of renumbering a number which is at most rz, where r is the limit of the sequence (r_i) defined by $r_0 = 1$ and $r_{i+1} = \frac{1+2^i}{2^i} r_i$. It is easy to check that this sequence converges. Hence also every new line will be reached in the process since it always has an number smaller than the number of some point.

Furthermore, the conditions maintained during the induction insure that we obtain a local isomorphism from $\mathfrak{H}$ onto $\mathfrak{P}$. Indeed, the map is a homomorphism between generalized polygons. Moreover, the induction starts with a map that is surjective on the points, hence any flag gets a pre-image (via the numbers $n_{\pi(x_i),k}$) and we have an epimorphism. Injectivity on point rows and line pencils is ensured by the conditions, whereas surjectivity follows from the fact that any point on any line l of $\mathfrak{P}$ has a pre-image on any pre-image of l in $\mathfrak{H}$ and the dual statement for any line through any point of $\mathfrak{P}$ (also via the numbers $n_{\pi(x_i),k}$). There exist points with the same image and by the Classification 2.1 of [7] we have a local isomorphism $(sssbb)$. □

Remark 3.3 (i) The above theorem actually holds for any generalized n-gon of countable infinite order, $n \geq 2$. Then there exists a generalized $2n$-gon admitting a local isomorphism onto the n-gon. It is also possible to construct the other classes of epimorphisms out of a given preimage. Moreover, similar to Remark 2.7 (i), we can also apply transfinite induction on polygons with point and line sets of other infinite cardinalities.

(ii) Unlike in group theory, the images of generalized polygons of gonality ≥ 3 under homomorphisms between polygons do not have to be generalized polygons. It is already known that the image of a gonality preserving homomorphism again is a generalized polygon if and only if the image contains an ordinary polygon; see, for example, [2]. However, given any generalized n-gon, $n \geq 3$, it is possible to construct a generalized m-gon, $m > n$, and a homomorphisms between the m-gon and the n-gon with precisely one point of the n-gon not contained in the image, hence the image contains an ordinary n-gon but is no generalized n-gon. The proof is omitted.

4. The finite case

In this section, we will disprove the existence of several types of epimorphisms from finite m-gons to n-gons. We concentrate on n-gons with $n > 2$, giving some brief comments on the case where the image is a digon at the end of the section. Recall that we exclusively deal with thick generalized polygons.

Some cases remain open, in particular, the case of a local isomorphism from a finite generalized $2n$-gon onto a generalized n-gon for $n = 3, 4$. We consider this as a very interesting open problem, since a solution would tell us at the same time something about spreads and ovoids in hexagons and octagons.

In what follows, we call an epimorphism *finite* if its pre-image is a finite generalized polygon (and hence so is its image).

All our proofs use the basic result of Feit and Higman [6].

Fact 4.1 *Finite (thick) generalized n-gons exist only for the five gonalities $n \in \{2, 3, 4, 6, 8\}$.*

We will frequently use the following inequalities between the parameters of a finite generalized polygon. These results are due to Higman [9] (case $n = 4, 8$) and Haemers and Roos [8] (case $n = 6$).

Fact 4.2 *For $n \in \{3,4,6,8\}$, let (s,t) be the order of a finite generalized n-gon. Then*

(i) $s = t$ *for* $n = 3$;

(ii) $s \leq t^2$ *and* $t \leq s^2$ *for* $n \in \{4,8\}$;

(iii) $s \leq t^3$ *and* $t \leq s^3$ *for* $n = 6$.

There are formulas for the numbers of points and lines of a finite generalized n-gon of order (s,t) in function of n, s and t. These can easily be derived by the reader or otherwise found in Lemma 1.5.4 of [13].

4.1. *bsbib* and *sbbbi*

Theorem 4.3 *There exists no finite bsbib epimorphism*

(i) *from a generalized hexagon to a projective plane;*

(ii) *from a generalized octagon to a projective plane;*

(iii) *from a generalized octagon to a generalized quadrangle.*

Hence, every finite bsbib epimorphism has as image a generalized digon.

Proof (i) Let (s,t) be the order of the generalized hexagon. By line pencil bijectivity the order of the projective plane is t. Hence, by point bijectivity,

$$(1+s)(1+st+s^2t^2) = t^2+t+1.$$

Consequently

$$t^2(s^3+s^2-1)+t(s^2+s-1)+s = 0,$$

which is impossible.

(ii) If (s,t) is the order of the octagon, then similarly to (i), one obtains

$$t^3(s^4+s^3)+t^2(s^3+s^2-1)+t(s^2+s-1)+s = 0,$$

which again is impossible.

(iii) Let (s,t) be the order of the generalized octagon and let (s^*,t^*) be the order of the generalized quadrangle. By line pencil bijectivity we have $t = t^*$. Moreover, the inverse image of a line of the quadrangle is, viewed as a set of points, the disjoint union of a set of lines of the octagon (by point and flag bijectivity). Hence there is a natural number n with $s^* + 1 = n(s + 1)$. By point bijectivity

$$\begin{aligned} 1 + st + s^2t^2 + s^3t^3 + s + s^2t + s^3t^2 + s^4t^3 &= \\ n + n^2st + n^2t - nt + ns + n^2s^2t + n^2st - nst. \end{aligned}$$

Furthermore, $t \leq s^2$ (see Fact 4.2) and $ns + n - 1 = n(s + 1) - 1 \leq t^2$ gives

$$\begin{aligned} \underbrace{s^4t^3}_{\geq t^5} + s^3t^3 + s^3t^2 + s^2t^2 + s^2t + st + s & \\ \underbrace{-n^2s^2t - 2n^2st - n^2t}_{\geq -t^5}\underbrace{-ns - n + 1}_{\geq -t^2} + nt + nst &= 0, \end{aligned}$$

which is impossible.

The conclusion now follows from Corollary 2.4 of [7] (see also the introduction above) and Fact 4.1. □

4.2. *ssbii*

Theorem 4.4 *There exists no finite ssbii epimorphism from a generalized hexagon to a projective plane.*

Proof Let (s, t) be the order of the generalized hexagon and let q be the order of the projective plane.

By *ssbii* we have $s+1|q+1$ (point row injectivity and flag bijectivity) and $t+1|q+1$ (line pencil injectivity and flag bijectivity), so let $s^* := \frac{q+1}{s+1}$ and $t^* := \frac{q+1}{t+1}$.

By flag bijectivity, we have

$$(1 + s)(1 + t)(1 + st + s^2t^2) = (1 + q)(1 + q + q^2);$$

hence

$$(1 + st + s^2t^2) = \frac{t^*(1 + q + q^2)}{1 + s},$$

which implies

$$(1 + st + s^2t^2) = \frac{t^*}{1 + s} + t^*s^*q \in \mathbb{Z}.$$

If we put $q^* := \frac{q+1}{(s+1)(t+1)}$, then, firstly, we have $(1 + st + s^2t^2) = q^*(1 + q + q^2)$, so $q^2 \leq 1 + st + s^2t^2$. Secondly, since $q^* \geq 1$, we see that $q + 1 \geq (s + 1)(t + 1)$, and this contradicts the previous inequality, as one easily calculates. □

It is an open question whether there exist finite *ssbii* epimorphisms with as preimage a generalized octagon, even if the image is a generalized digon.

4.3. *bssis* and *sbssi*

Here, most cases can be ruled out by some immediate inequalities. One case requires some further arguments, and we treat the latter in a separate theorem.

Theorem 4.5 *There exists no finite bssis epimorphism*

(i) *from a generalized quadrangle to a projective plane,*

(ii) *from a generalized hexagon to a projective plane,*

(iii) *from a generalized octagon to a projective plane,*

(iv) *from a generalized octagon to a generalized quadrangle.*

Proof (i) Let (s,t) be the order of the generalized quadrangle and let q be the order of the projective plane. Clearly, $s < q < t$. By point bijectivity, and using the inequality $t \leq s^2$, we have

$$\begin{aligned} 1+q+q^2 &= (1+s)(1+st) \\ &\geq (1+t^{\frac{1}{2}})(1+t^{\frac{3}{2}}) \\ &= 1+t^{\frac{1}{2}}+t^{\frac{3}{2}}+t^2, \end{aligned}$$

which implies $q > t$, a contradiction.

(ii) Let (s,t) be the order of the generalized hexagon and let q be the order of the projective plane. Again, $s < q < t$. But by point bijectivity, we have $(1+s)(1+st+s^2t^2) = 1+q+q^2$, implying $st < q$, thus contradicting $t > q$.

(iii) This is similar to (ii).

(iv) Let (s,t) be the order of the generalized octagon and let (s^*,t^*) be the order of the generalized quadrangle. Clearly, $s < s^*$ and $t^* < t$. Using the inequality $s^* \leq (t^*)^2$, the point bijectivity, and the inequality $t \leq s^2$, we obtain

$$\begin{aligned} \left(1+(t^*)^2\right)\left(1+(t^*)^3\right) &\geq (1+s^*)(1+s^*t^*) \\ &= (1+s)(1+st+s^2t^2+s^3t^3) \\ &\geq (1+t^{\frac{1}{2}})(1+t^{\frac{3}{2}}+t^3+t^{\frac{9}{2}}). \end{aligned}$$

Hence

$$1+(t^*)^2+(t^*)^3+(t^*)^5 \geq 1+t^{\frac{1}{2}}+t^{\frac{3}{2}}+t^2+t^3+t^{\frac{7}{2}}+t^{\frac{9}{2}}+t^5,$$

and so $t^* > t$, a contradiction. □

Theorem 4.6 *There exists no finite bssis epimorphism from a generalized octagon to a generalized hexagon.*

Proof Let (s,t) be the order of the generalized octagon $\mathfrak{O}$ and let (s^*,t^*) be the order of the generalized hexagon $\mathfrak{H}$. Let l be line of $\mathfrak{H}$ and let L be the set of points of $\mathfrak{O}$ which are mapped onto a point incident with l. If we view the lines of $\mathfrak{O}$ as sets of points, then clearly L is the union of lines, using line pencil surjectivity. Let m be a line contained in L and let $x \in L$. Let m' be any line contained in the shortest chain connecting x and m in $\mathfrak{O}$. Since this chain has length at most 7, the image gives rise to a closed path of length at most 8 in $\mathfrak{H}$. If m' is not mapped onto l, then the preimage of the point on the image of m' nearest to l contains at least two elements, a contradiction. Hence m' has to be mapped onto m and all points of m' belong to L. In particular, if L contains the points of an apartment, that is, an ordinary octagon, then L is the point set of a suboctagon.

First suppose that L contains the points of two lines m and m' at distance 6 from each other. Consider a point x on m at distance 7 from m' and let γ be a path of length 9 connecting x with m' (and not containing m). The image of γ gives rise to a closed path of length at most 10, hence, similarly as in the previous paragraph, all lines of γ must be mapped onto l. Hence L is the point set of a suboctagon $\mathfrak{O}'$ of $\mathfrak{O}$. But the element of γ incident with x can be chosen freely in the set of lines of $\mathfrak{O}$ through x, except for m. Since there are at least two choices for x on m, we infer from [13] (1.6.2) that $\mathfrak{O}'$ is thick, and Proposition 1.8.1 of [13] implies that $\mathfrak{O}'$ is *ideal*; that is, every line pencil in $\mathfrak{O}'$ coincides with some line pencil in $\mathfrak{O}$. Now Proposition 1.8.2 of [13] says that $\mathfrak{O}'$ coincides with $\mathfrak{O}$, a contradiction since, otherwise ($\mathfrak{H}$ would have only one line.

Suppose now that L contains all points of two lines m and m' at distance 4 from each other in $\mathfrak{O}$. Let m'' be the unique line of $\mathfrak{O}$ at distance 2 from both m and m'. Let x be any point of m not on m'' and let k be any line through x distinct from m. Finally, let k' be any line at distance 2 from k and at distance 4 from m. Suppose that the image of k and k' coincide. Since there is a path of length 8 connecting k' with m' and not containing m'', we obtain, by adding l, a closed path of length at most 10 in $\mathfrak{H}$. As before, this leads to a contradiction, because this time we know that k is not mapped onto l. Repeating this argument for all lines through x, we deduce that every line of the inverse image of the image K of k is incident with x. Suppose that there are i lines in the inverse image of K. Then, counting the points on K, we obtain $1+s^* = 1+is$, showing i is a constant. Hence the set of lines through x minus $\{m\}$ is divided into t/i inverse images of lines, and the latter are precisely the t^* lines of $\mathfrak{H}$ incident with the image of x and different from l. Hence $t/i = t^*$ and we infer that $s^*t^* = st$. Now, together with point bijectivity, this implies

$$(1+s)(1+s^*t^*)(1+s^{*2}t^{*2}) = (1+s^*)(1+s^*t^*+s^{*2}t^{*2}),$$

which, multiplying by t^* and separating the terms in $(s^*t^*)^3$, eventually gives

$$(s^*t^*)^3(st^*+t^*-1) = -(st^*-1)(1+s^*t^*+(s^*t^*)^2)-1<0,$$

a contradiction.

We conclude that L is contained in the line pencil of some point a of $\mathfrak{O}$, and this is true for every such set L. Hence, as in the previous paragraph, there is a constant $i = s^*/s$ giving the number of lines in the pre-image of any line of $\mathfrak{H}$. Now let a^* be the image of a. Let j be the number of lines l^* through a with the property that there is only one line of the inverse image of l^* incident with a. Now we count the number of pairs (x, m), where x is a point of $\mathfrak{O}$ and m is a line of $\mathfrak{O}$ through x and such that no other line through x has the same image as m. For given x, there are j such lines; for given m, there are s such points. This gives rise to the equality $(1+s)j = (1+t)s$. Hence $j = 1+t-\frac{1+t}{1+s}$. Counting the number of lines through a^*, we obtain $s^*(1+t^*-j) = s(1+t-j)$. Combining these two equalities, we compute

$$t^*-t+\frac{1+t}{1+s} = \frac{(1+t)s}{(1+s)s^*} \in \mathbb{Z}.$$

Now, by point bijectivity, and by the fact that the pre-image of every line of $\mathfrak{H}$ contains exactly $i = s^*/s$ lines, we have

$$\begin{aligned}(1+s)(1+st)(1+s^2t^2) &= (1+s^*)(1+s^*t^*+s^{*2}t^{*2}),\\ s(1+t)(1+st)(1+s^2t^2) &= s^*(1+t^*)(1+s^*t^*+s^{*2}t^{*2}),\end{aligned}$$

which, together with the previous equality, imply easily that $1+t^*$ is divisible by $1+s^*$. Hence $s^* \leq t^*$. Line pencil surjectivity implies $t^* < t$, while point row injectivity means $s < s^*$. Consequently we have

$$s < s^* \leq t^* < t.$$

Now using $t > t^*, s^*$, the point bijectivity, and the inequality $s \geq t^{1/2}$, we subsequently obtain

$$\begin{aligned}(1+t)(1+t^2+t^4) &> (1+s^*)(1+s^*t^*+s^{*2}t^{*2})\\ &= (1+s)(1+st)(1+s^2t^2)\\ &\geq (1+t^{1/2})(1+t^{3/2})(1+t^3).\end{aligned}$$

Cancelling out the equal terms in the extreme sides, and dividing everything by $t^{1/2}$, this implies

$$t^{1/2} + t^{3/2} + t^{7/2} > 1 + t + t^2 + t^3 + t^4.$$

clearly a contradiction. □

It is still open whether there is a finite *bssis* epimorphism from a generalized hexagon to a generalized quadrangle. By Fact 4.1, this is the only open case.

4.4. *sssbb*

Several attempts to rule the principle cases out did not work. Also, it seems hopeless to try to construct examples of finite *sssbb* epimorphisms from an octagon to a quadrangle, or from a finite hexagon to a projective plane, because nobody even knows whether the known finite octagons and hexagons have two disjoint suitable ovoids.

Theorem 4.7 *There exists no finite local isomorphism from a generalized octagon to a projective plane.*

Proof This would imply that $s = t = q$ for the order (s, t) of the octagon and the order q of the plane, contradicting the fact that $\sqrt{2st}$ must be an integer by [6]. □

Since the diameter of the pre-image is at least twice the diameter of the image, the only two open cases not involving digons are mentioned above.

4.5. Onto digons

4.5.1 *bsbib* and *sbbbi*. By Remark 2.7 (iii) of [7], the pre-images of the lines of a *bsbib* epimorphism of some generalized n-gon $\mathfrak{P}$ onto a generalized digon form a partition of the line set of $\mathfrak{P}$ in distance-2-spreads. In the finite case, there are examples of this for $n = 4$ (see Example 3.3 of [7]; for a general discussion of the existence problem of partitions of spreads in finite generalized quadrangles, see [10]). For $n = 6$, the only known distance-2-spread lives in the dual of the split Cayley hexagon $\mathsf{H}(2)$. But it is easy to prove that no partition of the line set into distance-2-spreads in this hexagon exists. For $n = 8$, no single distance-2-spread is known.

So, in conclusion, the problem is equivalent to the problem of partitioning the line set of generalized polygons into distance-2-spreads.

4.5.2 ***ssbii.*** Here, the problem is more general than in the previous paragraph. The inverse images of the elements are partial distance-2-ovoids and -spreads. For generalized hexagons and octagons, we did not succeed in finding nice such examples, but we did not do an extensive search in small polygons. For quadrangles however, here is an example. The idea is to refine a partition of distance-2-spreads and -ovoids. Let $T_2^*(O)$ be a generalized quadrangle obtained from the hyperoval O in the projective plane $\mathrm{PG}(2,q)$; for an explicit construction, see [10] and also [7]. Let p be a point of $\mathrm{PG}(2,q)$ not on O and let x be a point of $\mathrm{PG}(2,q)$ such that the line px does not meet O. The quadrangle $T_2^*(O)$ now lives in some projective space $\mathrm{PG}(3,q)$, which contains $\mathrm{PG}(2,q)$ as a (hyper)plane. The lines through x of $PG(3,q)$ not in $\mathrm{PG}(2,q)$ define a partition of the point set of $T_2^*(O)$ in partial ovoids; the sets of lines in $\mathrm{PG}(3,q)$ not in $\mathrm{PG}(2,q)$ through some common point of O and lying in the same plane through p define a partition of the line set of $T_2^*(O)$ into partial spreads. It is clear that, given any such partial ovoid, every partial spread contains a unique line incident with a unique point of the partial ovoid. Also the dual holds. Hence, by Theorem 2.4 (ii) of [7], we have an *ssbii* epimorphism onto a generalized digon.

4.5.3 ***bssis* and *sbssi*.** In the case of *sbssi*, it suffices to find a partition of the point set into at least three blocking sets which are not all distance-2-ovoids, where a *blocking set* is a subset of the point set such that every line has at least one point in common with the subset, see Theorem 2.5 of [7]; see Remark 2.7 (iv) of [7]). It is not so hard to construct such partitions for some member of a given class of polygons. Let us for example do this for the usual split Cayley hexagons $\mathsf{H}(q)$. We assume that q is odd. We choose a point p arbitrarily, and for each line at distance ≤ 3 from p, we can choose a point incident with that line and put it in some set S. We can do this in such a way that every line at distance ≤ 3 from p is incident with exactly one point of S. To construct the other points of our blocking set S, we need coordinates. We choose $p = (\infty)$ and consider the coordinates as given in [13] (3.5). We take the points (a,l,a',l',a'') satisfying $a = \pm a''$ and put them in S. Given any line $m = [k,b,k',b',k'']$, the points of S at distance 6 from p incident with m are obtained by solving the equation $a = \pm(ak+b)$. It is clear that this has always at least one and at most two solutions. Given a line $m' = [a,l,a',l']$, the points of S incident with m' are $(a,l,a',l',\pm a)$.

We now construct similarly a set S', choosing different points at distance ≤ 4 from p, using the condition $a = \pm a'' + 1$ and removing from S' the elements which are already in S. A routine check shows that every line of $\mathsf{H}(q)$ meets at least one element of S'. But $S \cup S'$ has at most

5 elements in common with every line; indeed, for lines at distance 5 from p, we have at most two times two points at distance 6 from p, and at most one at distance 4. Hence, if $q > 5$, the remaining points also form a blocking set. We thus have an *sbssi* epimorphism of $\mathsf{H}(q)$ onto a generalized digon of order $(2, q + q^2 + q^3 + q^4 + q^5)$.

For quadrangles, it suffices to take unions of ovoids. In conclusion we may say that finite *sbssi* epimorphisms onto digons are not rare, and that it is very likely that many examples exist. In view of the ease of finding the example with the hexagons, we did not try to find an explicit finite *sbssi* epimorphism involving a generalized octagon.

Acknowledgements

The authors would like to express their gratitude to Hans Cuypers for valuable help in the final stage of writing this article.

The second author is a Research Director of the Fund for Scientific Research – Flanders (Belgium).

References

[1] A. Blokhuis, personal communication.

[2] R. Bödi and L. Kramer, On homomorphisms between generalized polygons, *Geom. Dedicata* **58** (1995), 1–14.

[3] A.E. Brouwer, A non-degenerate generalized quadrangle with lines of size four is finite, *Advances in Finite Geometries and Designs*, Proceedings of the Third Isle of Thorns Conference (eds. J.W.P. Hirschfeld et al.), Oxford University Press, Oxford, 1991, 47–49.

[4] P.J. Cameron, Orbits of permutation groups on unordered sets, II., *J. London Math. Soc.* **23** (1981), 49–264.

[5] G. Cherlin, Notes on locally finite generalized quadrangles, 1996.

[6] W. Feit and G. Higman, The nonexistence of certain generalized polygons, *J. Algebra* **1** (1964), 114–131.

[7] R. Gramlich and H. Van Maldeghem, Epimorphisms of generalized polygons, Part 1: geometrical characterizations, *Des. Codes Cryptogr.* **21** (2000), 99–111.

[8] W.H. Haemers and C. Roos, An inequality for generalized hexagons, *Geom. Dedicata* **10** (1981), 219–222.

[9] D.G. Higman, Invariant relations, coherent configurations and generalized polygons, *Combinatorics* (eds. M. Hall and J.H. Van Lint), Proceedings ofthe Advanced Study Institute, Breukelen, 1974, *Part 3: Combinatorial Group Theory*, Reidel, Dordrecht, 1975, 347–363.

[10] S.E. Payne and J.A. Thas, *Finite Generalized Quadrangles*, Pitman, London, 1984.

[11] J. Tits, Sur la trialité et certains groupes qui s'en déduisent, *Inst. Hautes Études Sci. Publ. Math.* **2** (1959), 13–60.

[12] J. Tits, Endliche Spiegelungsgruppen, die als Weylgruppen auftreten, *Invent. Math.* **43** (1977), 283–295.

[13] H. Van Maldeghem, *Generalized Polygons*, Birkhäuser, Basel 1998.

THE PACKING PROBLEM IN STATISTICS, CODING THEORY AND FINITE PROJECTIVE SPACES: UPDATE 2001

J.W.P. Hirschfeld
School of Mathematical Sciences
University of Sussex
Brighton BN1 9QH
United Kingdom
http://www.maths.sussex.ac.uk/Staff/JWPH
jwph@sussex.ac.uk

L. Storme
Department of Pure Mathematics
Krijgslaan 281
Ghent University
B-9000 Gent
Belgium
http://cage.rug.ac.be/~ls
ls@cage.rug.ac.be

Abstract

This article updates the authors' 1998 survey [134] on the same theme that was written for the Bose Memorial Conference (Colorado, June 7-11, 1995). That article contained the principal results on the packing problem, up to 1995. Since then, considerable progress has been made on different kinds of subconfigurations.

1. Introduction

1.1. The packing problem

The packing problem in statistics, coding theory and finite projective spaces regards the determination of the maximal or minimal sizes of given

A. Blokhuis et al. (eds.), Finite Geometries, 201–246.

subconfigurations of finite projective spaces. This problem is not only interesting from a geometrical point of view; it also arises when coding-theoretical problems and problems from the design of experiments are translated into equivalent geometrical problems.

The geometrical interest in the packing problem and the links with problems investigated in other research fields have given this problem a central place in *Galois geometries*, that is, the study of finite projective spaces.

In 1983, a historical survey on the packing problem was written by the first author [127] for the 9th British Combinatorial Conference. A new survey article stating the principal results up to 1995 was written by the authors for the Bose Memorial Conference [134].

Since 1995, many interesting results have appeared: for instance, the result of Szőnyi [216] confirming the $1 \pmod{p}$ conjecture of Blokhuis on small minimal blocking k-sets in $PG(2,q)$, $q=p^h$, p prime; the classification of small Rédei-type blocking k-sets in $PG(2,q)$ by Blokhuis, Ball, Brouwer, Storme and Szőnyi [31]; the unified construction of Cherowitzo, O'Keefe and Penttila for hyperovals in $PG(2,q)$, q even [59]; greatly improved upper bounds on the sizes of caps in $PG(N,q)$ by Meshulam [161], and Bierbrauer and Edel [27]; and the determination of the maximal size of a cap in $PG(4,4)$ by Edel and Bierbrauer [72].

These recent results are the motivation for updating [134]. We collect in this article the known results up to 2001. For connections with Coding Theory and Statistics, see [134].

With the aim of making this survey both easily accessible and self-contained, in the next subsection the exact definition of the packing problem and the notation used in this article are repeated.

When comparing this survey to the one of 1995, it is a pleasure to see both the progress made on different subconfigurations and the new techniques. It is hoped that these results may motivate others to study the packing problem.

The authors also wish to thank all the colleagues for their suggestions and remarks in writing this update.

1.2. Subsets in projective spaces

Let $PG(N,q)$ be the projective space of N dimensions over the finite field $\mathbf{F}_q$ of q elements, $q=p^h$ with p prime, and let $|PG(N,q)| = \theta_N = (q^{N+1}-1)/(q-1)$.

In $PG(N,q)$, subspaces will be denoted by Π_l, where l is the dimension of the subspace. A Π_0 is a *point*, a Π_1 is a *line*, a Π_2 is a *plane*, a Π_3 is a *solid*, and a Π_{N-1} is a *hyperplane* or *prime*.

Let $e_i = (0, \dots, 0, 1, 0, \dots, 0)$, for $i = 0, \dots, N$, where 1 is in the $(i+1)$-th position and let $e = (1, \dots, 1)$.

For a unified notion of subsets of projective spaces, an $(n; r, s; N, q)$-set K is defined to be a set satisfying the following properties:

(a) the set K consists of n points of $PG(N, q)$ and is not contained in a proper subspace;

(b) some subspace Π_s contains r points of K, but no Π_s contains $r+1$ points of K;

(c) there is a subspace Π_{s+1} containing $r+2$ points of K.

An $(n; r, s; N, q)$-set K is called *complete* if it is not contained in an $(n+1; r, s; N, q)$-set.

Several natural problems on $(n; r, s; N, q)$-sets K arise immediately.

(I) Find the maximum value $m(r, s; N, q)$ of n for which a $(n; r, s; N, q)$-set exists.

(II) Characterize the sets with this size $m(r, s; N, q)$.

(III) Find the size $m'(r, s; N, q)$ of the second largest complete $(n; r, s; N, q)$-set.

(IV) Find the size $t(r, s; N, q)$ of the smallest complete $(n; r, s; N, q)$-set.

The following examples of these sets will be considered in subsequent sections.

(1) An $(n; 2, 1; N, q)$-set is an n-set with at most two points on any line of $PG(N, q)$ and is called an *n-cap* when $N \geq 3$ and a (plane) n-arc when $N = 2$. The numbers $m(2, 1; N, q)$, $m'(2, 1; N, q)$ and $t(2, 1; N, q)$ are respectively denoted by $m_2(N, q)$, $m'_2(N, q)$ and $t_2(N, q)$.

(2) An $(n; N, N-1; N, q)$-set is an n-set with at most N points in any hyperplane of $PG(N, q)$ and is called an *n-arc*. For $N = 2$, the two definitions of n-arc coincide. Here, $m(N, N-1; N, q)$, $m'(N, N-1; N, q)$ and $t(N, N-1; N, q)$ are denoted by $m(N, q)$, $m'(N, q)$ and $t(N, q)$.

(3) An $(n; r, 1; N, q)$-set K is a set of n points with at most r on a line. When $N = 2$, the set K is called a *(plane) (n, r)-arc*; when $N > 2$, the set K is called an (n, r)-cap. For these sets, the parameters $m(r, 1; N, q), m'(r, 1; N, q)$, $t(r, 1; N, q)$ are respectively denoted by $m_r(N, q), m'_r(N, q)$, $t_r(N, q)$.

(4) An $(n; r, r-1; N, q)$-set K is a set of n points at most r of which lie in a Π_{r-1} but some $r+2$ lie in a Π_r; that is, $r+1$ points are always

linearly independent, but some $r+2$ points are linearly dependent. Such a set K is called an (n,r)*-set* or n*-set of kind* r. The value $m(r,r-1;N,q)$ is denoted by $M_r(N,q)$. Hence $M_N(N,q) = m(N,q)$ and $M_2(N,q) = m_2(N,q)$.

For n-arcs, Problem (I) has been studied in detail. The main conjecture for this problem is given in Table 1.1.

$m(N,q)$	Conditions
$N+2$	$N \geq q-1$
$q+2$	q even, $N \in \{2, q-2\}$
$q+1$	in all other cases

Table 1.1: Main conjecture for $m(N,q)$

The results for this problem are gathered in Tables 2.1, 3.1, 3.3, 3.4, 3.5, and 3.7.

The sets above are discussed in the following sections: §2: n-arcs in $PG(2,q)$; §3: n-arcs in $PG(N,q)$, $N \geq 3$; §4: n-caps in $PG(N,q)$, $N \geq 3$; §5: (n,r)-arcs in $PG(2,q)$; §6: Multiple blocking sets in $PG(2,q)$; §7: Blocking sets in $PG(N,q)$; §8: n-tracks and almost MDS codes; §9: Minihypers.

Throughout the article, $\lfloor x \rfloor$ denotes the greatest integer smaller than or equal to x, while $\lceil x \rceil$ denotes the smallest integer greater than or equal to x.

2. n-arcs in $PG(2,q)$

q	$m(2,q)$	
q odd	$q+1$	[38]
q even	$q+2$	[38]

Table 2.1: $m(2,q)$

An $m(2,q)$-arc in $PG(2,q)$, q odd, is called an *oval* and an $m(2,q)$-arc in $PG(2,q)$, q even, is called a *hyperoval.* Formerly, the term *oval* was used in both cases.

Theorem 2.1 (Segre [194, 195]) *For q odd, an oval is the set of rational points of a conic.*

Bose [38] showed that, for q even, a conic plus its nucleus (the intersection point of its tangents) is a hyperoval. A hyperoval of this type is called *regular*. As shown by Segre [197], for $q = 2, 4, 8$, every hyperoval is regular.

For $q = 2^h, h \geq 4$, there exist *irregular* hyperovals, that is, hyperovals which are not the union of a conic and its nucleus. Several infinite classes of irregular hyperovals are known. In general, the following result is valid.

Theorem 2.2 (Segre [197], [130, §8.4]) *Any hyperoval of $PG(2,q)$, $q = 2^h$ and $h > 1$, is projectively equivalent to a hyperoval*

$$\mathcal{D}(F) = \{(1, t, F(t)) \parallel t \in \mathbf{F}_q\} \cup \{e_1, e_2\},$$

where F is a permutation polynomial over $\mathbf{F}_q$ of degree at most $q-2$, satisfying $F(0) = 0$, $F(1) = 1$, and such that $F_s(X) = (F(X+s) + F(s))/X$ is a permutation polynomial for each s in $\mathbf{F}_q$, satisfying $F_s(0) = 0$.

A polynomial $F(X)$ of this type is called an *o-polynomial* [57] and, conversely, every o-polynomial gives rise to a hyperoval. Table 2.2 presents the known infinite classes of hyperovals. For each infinite class, the simplest form of the corresponding o-polynomial is given [60, 185]. The references for these infinite classes of hyperovals are: (1) regular (Bose [38]), (2) translation (Segre [197]), (3) Segre [200], (4) Glynn I and Glynn II [99], (4) Payne [176], (5) Cherowitzo [58], (6) Subiaco (Cherowitzo, Penttila, Pinneri and Royle [60], Payne [177], Payne, Penttila and Pinneri [178]), (7) Adelaide (Cherowitzo, O'Keefe and Penttila [59]).

In Table 2.2,

$$\begin{aligned}
P(X) &= X^{1/6} + X^{3/6} + X^{5/6};\\
C(X) &= X^{\sigma} + X^{\sigma+2} + X^{3\sigma+4};\\
S_1(X) &= \frac{\omega^2(X^4+X)}{X^4+\omega^2X^2+1} + X^{1/2};\\
S_2(X) &= \frac{\delta^2X^4+\delta^5X^3+\delta^2X^2+\delta^3X}{X^4+\delta^2X^2+1} + \left(\frac{X}{\delta}\right)^{1/2};\\
S_3(X) &= \frac{(\delta^4+\delta^2)X^3+\delta^3X^3+\delta^2X}{X^4+\delta^2X^2+1} + \left(\frac{X}{\delta}\right)^{1/2};\\
S(X) &= \frac{T(\beta^m)(X+1)}{T(\beta)} + \frac{T((\beta X+\beta^q)^m)}{T(\beta)(X+T(\beta)X^{1/2}+1)^{m-1}} + X^{1/2}.
\end{aligned}$$

In fact, the derivation of the Subiaco and Adelaide hyperovals is more complicated [60, 177, 178, 59]. They come in a set of $q+1$, called a *herd*, which is defined by two o-polynomials f, g satisfying the property that

$$T_2\left(\frac{c(f(s)+f(t))(g(s)+g(t))}{s+t}\right)=1$$

for all $s,t \in \mathbf{F}_q$ with $s \neq t$, and a fixed c with $T_2(c)=1$, where T_2 is the trace function from $\mathbf{F}_q$ to $\mathbf{F}_2$.

Name	$F(X)$	$q=2^h$	Conditions
Regular	X^2		
Translation	X^{2^i}		$(h,i)=1$
Segre	X^6	h odd	
Glynn I	$X^{3\sigma+4}$	h odd	$\sigma=2^{(h+1)/2}$
Glynn II	$X^{\sigma+\lambda}$	h odd	$\sigma=2^{(h+1)/2}$; $\lambda=2^m$ if $h=4m-1$; $\lambda=2^{3m+1}$ if $h=4m+1$
Payne	$P(X)$	h odd	
Cherowitzo	$C(X)$	h odd	$\sigma=2^{(h+1)/2}$
Subiaco	$S_1(X)$	$h=4r+2$	$\omega^2+\omega+1=0$
Subiaco	$S_2(X)$	$h=4r+2$	$\delta=\zeta^{q-1}+\zeta^{1-q}$, ζ primitive in $\mathbf{F}_{q^2}$
Subiaco	$S_3(X)$	$h\neq 4r+2$	$T_2(1/\delta)=1$
Adelaide	$S(X)$	h even, $h\geq 4$	$\beta\in\mathbf{F}_{q^2}\setminus\{1\}$, $\beta^{q+1}=1$, $T(x)=x+x^q$, $m\equiv\pm(q-1)/3 \pmod{q+1}$

Table 2.2: Hyperovals in $PG(2,q), q$ even

Then, with

$$f_s(X)=\frac{f(X)+csg(X)+s^{1/2}X^{1/2}}{1+cs+s^{1/2}},$$

the herd is given by the o-polynomials g and f_s for $s \in \mathbf{F}_q$. For $q = 2^h$, $h \not\equiv 2 \pmod 4$, all Subiaco hyperovals are projectively equivalent to $\mathcal{D}(S_3)$, whereas for $q=2^h$, $h\equiv 2 \pmod 4$, there are two classes of hyperovals $\mathcal{D}(S_1)$, $\mathcal{D}(S_2)$.

Apart from the infinite classes of Table 2.2, there is one other sporadic hyperoval known, namely the O'Keefe–Penttila hyperoval in $PG(2,32)$ [167].

It is interesting to remark that the translation hyperoval with $F(X) = X^4$, the Subiaco hyperovals and Adelaide hyperovals also arise from a unified construction of Cherowitzo, O'Keefe and Penttila [59]. In a personal communication, T. Penttila [182] confirmed that the Adelaide hyperovals are distinct from the Subiaco hyperovals.

Except for the O'Keefe–Penttila hyperoval in $PG(2,32)$, every known hyperoval arises from an α-flock [58, 59].

In $PG(2,q)$, $q \leq 32$, the complete classification of hyperovals has been determined and is given in the next theorem.

Theorem 2.3 (i) (Segre [197]) *In $PG(2,q)$, $q = 2,4,8$, the only hyperovals are the regular hyperovals.*

(ii) (Hall [110], O'Keefe and Penttila [166]) *In $PG(2,16)$, there are exactly two distinct hyperovals. They are the regular hyperoval and the Subiaco hyperoval. This latter hyperoval is also called the* Lunelli-Sce *hyperoval* [152].

(iii) (Penttila and Royle [184]) *In $PG(2,32)$, there are exactly six distinct hyperovals. They are the regular hyperoval, the translation hyperoval $\mathcal{D}(X^4)$, the Segre hyperoval, the Payne hyperoval, the Cherowitzo hyperoval and the O'Keefe–Penttila hyperoval.*

Finding the values of n for which an n-arc is always contained in an oval, for q odd, or hyperoval, for q even, is relevant for solving problems in higher-dimensional spaces. Table 2.3 surveys the known results on this problem. In $PG(2,q)$, for q subject to the conditions in the second column, the third column gives an upper bound for $m'(2,q)$; the fourth column indicates when this upper bound is sharp in the case that equality does not hold in general. The fifth column gives the value of $m(2,q)$. So any n-arc with $n > m'(2,q)$ is contained in an $m(2,q)$-arc. This latter arc is a conic when q is odd and a hyperoval when q is even.

	q	$m'(2,q)$	Sharp	$m(2,q)$
(1)	$q = p^{2e}$, $p > 2,\ e \geq 1$	$\leq q - \sqrt{q}/4 + 25/16$		$q+1$
(2)	$q = p^{2e+1}$, $p > 2,\ e \geq 1$	$\leq q - \sqrt{pq}/4$ $+29p/16 + 1$		$q+1$
(3)	q prime, $q > 2$	$\leq 44q/45 + 8/9$		$q+1$
(4)	$q = p^h,\ p \geq 5$	$\leq q - \sqrt{q}/2 + 5$		$q+1$
(5)	$q = p^h,\ p \geq 3$, $q \geq 23^2,\ q \neq 5^5,\ 3^6$, h even for $p = 3$	$\leq q - \sqrt{q}/2 + 3$		$q+1$
(6)	$q = 2^{2e},\ e > 1$	$= q - \sqrt{q} + 1$		$q+2$
(7)	$q = 2^{2e+1},\ e \geq 1$	$\leq q - \sqrt{2q} + 2$	$q = 8$	$q+2$

Table 2.3: Upper bounds for $m'(2,q)$

Result (1) is due to Thas [223]; (2) and (7) to Voloch [228]; (3) to Voloch [226]; (4) and (5) to Hirschfeld and Korchmáros [131, 132]; (6) as an upper bound is due to Segre [201].

The sharpness of (6) was shown independently by Boros and Szőnyi [37], Fisher, Hirschfeld and Thas [88], and Kestenband [143]. That these arcs are unique follows from Hirschfeld [125, §14.8], [130, §8.7], Lisonek [147, p.193], Chao and Kaneta [55, 56].

Theorem 2.4 *In $PG(2,q)$, $q = p^{2e}$, $q > 4$, a complete $(q - \sqrt{q} + 1)$-arc exists. For $q = 9, 16, 25$, a complete $(q - \sqrt{q} + 1)$-arc is projectively unique.*

Conjecture $m'(2,q) = q - \sqrt{q} + 1$ for $q = p^{2e}$, $q > 9$.

For arcs in $PG(2,q)$, $q = 2^{2e}$, $q > 4$, of size smaller than $q - \sqrt{q} + 1$, there is the following result of Hirschfeld and Korchmáros [133].

Theorem 2.5 *A complete n-arc of $PG(2,q)$, $q = 2^{2e}$, $e > 2$, has size either $q + 2$, or $q - \sqrt{q} + 1$, or at most $q - 2\sqrt{q} + 6$.*

For $q = 16$, the list of sizes n of complete n-arcs in $PG(2,16)$ is $9, 10, 11, 12, 13, 18$.

For small q, the values $m'(2,q)$ for the size of the second largest complete arc are given in Table 2.4. For $q \leq 5$, there is no second largest complete arc, so we do not consider these values of q. For $q \leq 9$, see [125, 130, Ch.14]; for $q = 11$, see [192], [130, §14.8]; for $q = 13$, see [5], [101], [193], [6], [7], [130, §14.9]; for $q = 17, 19$, see [55, 78, 162]; for $q = 23, 25, 27$, see [56]; for $q = 29$, see [54]. See also [83], [147, p.193], [186], [215].

q	7	8	9	11	13	16	17	19	23	25	27	29
$m'(2,q)$	6	6	8	10	12	13	14	14	17	21	22	24

Table 2.4: $m'(2,q)$ in small planes

Similarly, for small q, the size $t(2,q)$ of the smallest complete arc in $PG(2,q)$ has been calculated and is given in Table 2.5; see [78], [83], [147, p.193], [186], [215], [156], [157].

q	2	3	4	5	7	8	9	11	13	16	17	19	23	25	27	29
$t(2,q)$	4	4	6	6	6	6	6	7	8	9	10	10	10	12	12	13

Table 2.5: $t(2,q)$ in small planes

General lower and upper bounds are given in Table 2.6. In the upper bound of Kim and Vu [144], d and c are absolute constants. Giulietti [97] has shown that, for q an odd square with $q \leq 1681$ or $q = 2401$, there exists a complete $4(\sqrt{q}-1)$-arc in $PG(2,q)$.

q	$t(2,q)$	
q	$> \sqrt{2q}+1$	[198]
$q=p^h$, p prime, $h=1,2,3$	$> \sqrt{3q}+\frac{1}{2}$	[28, 11, 188]
q	$\leq d\sqrt{q}\log^c q$	[144]

Table 2.6: Lower and upper bounds for $t(2,q)$

To conclude this section on arcs in $PG(2,q)$, the following theorem presents results on the spectrum of the values n for which a complete n-arc in $PG(2,q)$ exists.

Theorem 2.6 (i) (Bose [38]) *In $PG(2,q)$, a complete $(q+1)$-arc exists for q odd and a complete $(q+2)$-arc exists for q even.*

(ii) (Boros and Szőnyi [37], Fisher, Hirschfeld and Thas [88], Kestenband [143]) *In $PG(2,q)$, q square, $q>4$, there exists a complete $(q-\sqrt{q}+1)$-arc.*

(iii) (Szőnyi [212]) *Let $A=\{n/q \,\|$ there exists a complete n-arc in $PG(2,q)$, for some prime power $q\}$. Then $A\cap[0,1/2]$ is dense in $[0,1/2]$.*

(iv) (Voloch [225]) *If $E \subseteq PG(2,q)$ is an elliptic cubic curve with $2n$ rational points, then there exists an n-arc contained in E, which is complete if the modular (or j-)invariant $j(E)\neq 0$ ([118, Ch.IV, §4]) and $q\geq 175$ for q odd or $q\geq 256$ for q even.*

(v) (Giulietti [98]) *When $j(E)=0$ and $q>9887$, then the n-arc can be completed to an arc of size at most $n+3$.*

(vi) (Hadnagy [107], Voloch [227] and Szőnyi [213]) *Let $p\geq 5^{10}$ be a prime number. Then there exists a complete n-arc in $PG(2,p)$ for every integer n satisfying $\lfloor 2.46\cdot\log p\cdot p^{3/4}\rfloor \leq n < (\sqrt{p}+1)^2/2$.*

Many constructions of complete arcs are based on an idea of Segre [200] and Lombardo-Radice [148]; the points of the arc are chosen, with some exceptions, among the points of a conic or cubic curve.

This leads to complete arcs with approximately $q/2$ [125, §9.4], [130, §9.3], [201], $q/3$ [2, 145, 211, 227], $q/4$ [145], $2q^{9/10}$ [211] points.

3. n-arcs in $PG(N,q)$

In Tables 3.1–3.3, NRC stands for *normal rational curve*, that is, a $(q+1)$-arc in $PG(N,q)$, $2\leq N\leq q-2$, projectively equivalent to the $(q+1)$-arc $\{(1,t,\ldots,t^N)\,\|\,t\in\mathbf{F}_q^+\}$, where $\mathbf{F}_q^+=\mathbf{F}_q\cup\{\infty\}$ and where $t=\infty$ defines the point e_N. In $PG(3,q)$, $q=2^h$, $h>2$, $L_e=\{(1,t,t^e,t^{e+1})\,\|\,t\in\mathbf{F}_q^+\}$, with $e=2^v$, $(v,h)=1$, and with $t=\infty$ defining the point e_3.

Table 3.1 shows the value of $m(N,q)$ for small dimensions N. The characterization of the $m(N,q)$-arcs L in $PG(N,q)$ is given in the last column.

	q	N	$m(N,q)$	L
(1)	$q = p^h$	$N \geq q-1$	$N+2$	$\{e_0, \ldots, e_N, e\}$
(2)	q odd, $q > 3$	3	$q+1$	NRC
(3)	$q = 2^h$, $q > 4$	3	$q+1$	$L_e, e = 2^v, (v,h) = 1$
(4)	q odd, $q > 5$	4	$q+1$	NRC for $q > 83$
(5)	$q = 2^h$, $q > 4$	4	$q+1$	NRC
(6)	$q = 2^h$, $q \geq 8$	5	$q+1$	NRC for $q \geq 16$
(7)	$q = 2^h$, $q \geq 16$	6	$q+1$	NRC for $q > 64$

Table 3.1: $m(N,q)$ and $m(N,q)$-arcs

Result (1) is due to Bush [47] and (2) to Segre [196]. The value of $m(N,q)$ in (3) is due to Casse [49], as well as to Gulati and Kounias [106], while the characterization of the $(q+1)$-arcs is by Casse and Glynn [50]. In (4), the value of $m(N,q)$ is due to Segre [196]; the characterization of L follows from Table 3.2. The bound in (5) is by Casse [49], and by Gulati and Kounias [106], while the characterization of the $(q+1)$-arcs is due to Casse and Glynn [51]. Result (6) is by Maruta and Kaneta [158], and (7) by the same authors [142, 158] for the bound and by Storme and Thas [206] for the characterization of the $(q+1)$-arcs.

A problem that has been intensively studied is the extendability of n-arcs in $PG(N,q)$ to larger arcs. Table 3.2 summarizes the main results on this topic. An n-arc in $PG(N,q)$, satisfying the condition on n in the third column, can be extended uniquely to a $(q+1)$-arc L, whose description is given in column 4. Alternatively, the third column gives an upper bound for $m'(N,q)$. The results are respectively due to Thas [220, 223], Voloch [228], Voloch [226], Hirschfeld and Korchmáros for the next two formulas [131, 132], and Storme and Thas [206] for the latter two formulas.

Theorem 3.1 (Kaneta and Maruta [142]) *If every $(q+1)$-arc of the space $PG(N,q)$ is a normal rational curve, then $q+1$ is the maximum value of n for which n-arcs exist in $PG(N+1,q)$.*

N	q	$n >$	L
≥ 2	$q = p^{2e}$, p odd, $e \geq 1$	$q - \sqrt{q}/4 + N - 7/16$	NRC
≥ 2	$q = p^{2e+1}$, p odd, $e \geq 1$	$q - \sqrt{pq}/4 + 29p/16$ $+N - 1$	NRC
≥ 2	q prime, $q > 2$	$44q/45 + N - 10/9$	NRC
≥ 2	$q = p^h$, $p \geq 5$	$q - \sqrt{q}/2 + N + 3$	NRC
≥ 2	$q = p^h$, $p \geq 3$, $q \geq 23^2$, h even for $p = 3$, $q \neq 5^5$, 3^6	$q - \sqrt{q}/2 + N + 1$	NRC
3	$q = 2^h, h > 1$	$q - \sqrt{q}/2 + 9/4$	L_e
≥ 4	$q = 2^h$, $h > 2$	$q - \sqrt{q}/2 + N - 3/4$	NRC

Table 3.2: Upper bounds for $m'(N, q)$

Applying the results in Table 3.2 for $n = q + 1$ gives an upper bound on the dimension N for the spaces $PG(N, q)$, $q = p^h$, p prime, in which every $(q+1)$-arc is a normal rational curve. The upper bound is given in the second column of Table 3.3. Column 3 of Table 3.3 gives the upper bound, obtained from Theorem 3.1, on the dimension N of the spaces $PG(N, q)$ in which $m(N, q) = q + 1$.

As indicated in [134, §1.3], an n-arc in $PG(n - k - 1, q)$ defines an $[n, k, n-k+1]$ MDS code C. But the dual code $C^{\perp}$ of an $[n, k, n-k+1]$ MDS code is an $[n, n-k, k+1]$ MDS code [153, Ch.11, §2]. Hence, to $C^{\perp}$ also corresponds an n-arc $\hat{K}$ in $PG(k - 1, q)$.

This theorem, proved independently by Thas [221] and by Halder and Heise [109], makes it possible to link n-arcs in $PG(n-k-1, q)$ to n-arcs in $PG(k - 1, q)$. An n-arc K in $PG(n - k - 1, q)$ and an n-arc $\hat{K}$ in $PG(k - 1, q)$ are called C-*dual arcs* if and only if they define dual MDS codes.

q	$(q+1)$-arc = NRC	$m(N, q) = q + 1$
$q = p^{2e}$, p odd, $e \geq 1$	$2 \leq N <$ $\sqrt{q}/4 + 23/16$	$2 \leq N <$ $\sqrt{q}/4 + 39/16$
$q = p^{2e+1}$, p odd, $e \geq 1$	$2 \leq N <$ $\sqrt{pq}/4 - 29p/16 + 2$	$2 \leq N <$ $\sqrt{pq}/4 - 29p/16 + 3$
q prime, $q > 2$	$2 \leq N < q/45 + 19/9$	$2 \leq N < q/45 + 28/9$
$q = p^h$, $p \geq 5$	$2 \leq N < \sqrt{q}/2 - 2$	$2 \leq N < \sqrt{q}/2 - 1$
$q = p^h$, $p \geq 3$, h even for $p = 3$, $q \geq 23^2$, $q \neq 5^5$, 3^6	$2 \leq N < \sqrt{q}/2$	$2 \leq N < \sqrt{q}/2 + 1$
$q = 2^h$, $h > 2$	$4 \leq N < \sqrt{q}/2 + 7/4$	$4 \leq N < \sqrt{q}/2 + 11/4$

Table 3.3: $(q + 1)$-arcs in $PG(N, q)$

Combining this C-duality principle on arcs with the result stating that an n-arc of $PG(n - k - 1, q)$ is contained in a normal rational curve if and only if its C-dual n-arc in $PG(k-1,q)$ is contained in a normal rational curve [153, Ch.10, §8], the results in the preceding tables on arcs in $PG(N,q)$, with N small, immediately imply other results on arcs in $PG(N,q)$, where N is close to q.

In Table 3.4, in the spaces $PG(N,q)$, with N satisfying the bounds in the table, any n-arc, where n satisfies the bound in the second column, is contained in a normal rational curve. So, in these spaces, $m(N,q)=q+1$ and every $(q+1)$-arc is a normal rational curve.

Only n-arcs in $PG(N,q)$, with $n \geq N+4$, are considered since an $(N+3)$-arc of $PG(N,q)$, $2 \leq N \leq q-2$, is always contained in a unique normal rational curve [128, §21.1].

q	$n \geq$	$m(N,q)=q+1$
$q=p^{2e},\ p>2,\ e\geq 1$	$N+4$	$q-3 \geq N > q-\sqrt{q}/4-39/16$
$q=p^{2e+1},\ p>2,\ e\geq 1$	$N+4$	$q-3\geq N >$ $q-\sqrt{pq}/4+29p/16-3$
q prime, $q>2$	$N+4$	$q-3\geq N > 44q/45-28/9$
$q=p^h,\ p\geq 5$	$N+4$	$q-3\geq N > q-\sqrt{q}/2+1$
$q=p^h,\ p\geq 3$, h even for $p=3$, $q\geq 23^2,\ q\neq 5^5,\ 3^6$	$N+4$	$q-3\geq N > q-\sqrt{q}/2-1$
$q=2^h,\ h>2$	$N+6$	$q-5\geq N > q-\sqrt{q}/2-11/4$

Table 3.4: Arcs in $PG(N,q), N$ close to q

Theorem 3.2 (Glynn [100]) *In $PG(4,9)$, a 10-arc is one of two types; it is either a normal rational curve or is equivalent to $L=\{(1,t,t^2+\eta t^6,t^3,t^4)\,\|\, t\in \mathbf{F}_9\}\cup\{(0,0,0,0,1)\}$, where $\eta^4=-1$.*

The preceding result is of particular interest since this 10-arc L is the only known $(q+1)$-arc in $PG(N,q)$, q odd, $2\leq N\leq q-2$, which is not a normal rational curve.

There are some fundamental differences between arcs in spaces of even characteristic and odd characteristic; the existence and characterization of $(q+2)$-arcs are given by the next two results.

Theorem 3.3 (Thas [221]) *In $PG(q-2,q)$, q even, $m(q-2,q)=q+2$.*

Theorem 3.4 (Storme and Thas [207]) *A point $P=(a_0,\dots,a_{q-2})$ extends the normal rational curve $K=\{(1,t,\dots,t^{q-2})\,\|\, t\in\mathbf{F}_q^+\}$ to a $(q+2)$-arc if and only if $F(X)=\sum_{i=0}^{q-2}a_{q-2-i}X^{i+1}$ defines a $(q+2)$-arc $K'=\{(1,t,F(t))\,\|\, t\in\mathbf{F}_q\}\cup\{e_1,e_2\}$ in $PG(2,q)$; in this case, K' is a C-dual $(q+2)$-arc of $K\cup\{P\}$.*

This correspondence between the point P and the $(q+2)$-arc K' is one-to-one if the condition $F(1) = 1$ is added.

Table 3.5 presents the results by Storme and Thas [205] for the values of n for which there exist complete n-arcs in the respective spaces $PG(N,q)$. For more information on complete $(N+4)$-arcs and $(N+5)$-arcs, see [205].

q	N	$n \in$
$q = p^{2e}$, $e \geq 1$, $p > 2$	$q - 2 \geq N > q - \sqrt{q}/4 - 39/16$	$\{q+1\}$
$q = p^{2e+1}$, $e \geq 1$, $p > 2$	$q - 2 \geq N > q - \sqrt{pq}/4 + 29p/16 - 3$	$\{q+1\}$
q prime, $q > 2$	$q - 2 \geq N > 44q/45 - 28/9$	$\{q+1\}$
$q = p^h$, $p \geq 5$	$q - 2 \geq N > q - \sqrt{q}/2 + 1$	$\{q+1\}$
$q = p^h$, $p \geq 3$, h even for $p = 3$, $q \geq 23^2$, $q \neq 5^5$, 3^6	$q - 2 \geq N > q - \sqrt{q}/2 - 1$	$\{q+1\}$
$q = 2^h$, $q \geq 32$, $q \neq 64$	$q - 5 \geq N > q - \sqrt{q}/2 - 11/4$	$\{N+4, N+5, q+1\}$
$q = 64$	$N = 58$ or $N = 59$	$\{N+4, q+1\}$
$q = 2^h$, $q \geq 8$	$q - 4$	$\{q, q+1\}$
$q = 2^h$, $q \geq 8$	$q - 3$	$\{q+1\}$
$q = 2^h$, $q \geq 4$	$q - 2$	$\{q+2\}$

Table 3.5: Spectrum of complete arcs

Table 3.6 presents the known values for the size $m'(3,q)$ of the second largest complete n-arc in $PG(3,q)$. These results are by Kaneta [140], Chao and Kaneta [55], in combination with O'Keefe and Storme [170] for $q = 13$, and Kaneta [141].

q	8	9	11	13	16	17	19	25
$m'(3,q)$	7	9	9	10	12	13	14	18

Table 3.6: $m'(3,q)$

To conclude this section, Table 3.7 gives known results on arcs in small projective spaces. For the corresponding spaces $PG(N,q)$, $2 \leq N \leq q-2$, it is indicated if every $(q+1)$-arc is a normal rational curve, the values of N for which $m(N,q) = q+1$, and if there exist $(q+2)$-arcs. For $q \geq 11$, these results are by Chao and Kaneta [55, 56].

q	$(q+1)$-arc=NRC	$m(N,q)=q+1$	$(q+2)$-arcs
$5 \leq q \leq 27$, $q \neq 9$, q odd	$N \in \{2,\dots,q-2\}$	$N \in \{2,\dots,q-2\}$	
$8, 16$	$N \in \{3,\dots,q-4\}$	$N \in \{3,\dots,q-3\}$	$N \in \{2, q-2\}$
9	$N \in \{2,3\} \cup \{5,6,7\}$	$N \in \{2,\dots,7\}$	

Table 3.7: n-arcs in small spaces $PG(N,q)$

As a consequence of Tables 3.1, 3.2, 3.3, 3.4, 3.7,
(1) $m(5,q) = q+1$ for $q \geq 7$, except possibly for q odd and

$$q \in \{29,\dots,83\};$$

(2) $m(6,q) = q+1$ for $q \geq 9$, except possibly for q odd and

$$q \in \{29,\dots,127\} \cup \{169\}.$$

For an alternative summary on n-arcs, see [129].

4. n-caps in $PG(N,q)$

q	N	$m_2(N,q)$	
q odd	3	q^2+1	[38]
q even, $q>2$	3	q^2+1	[191]
$q=2$	N	2^N	[38]
$q=3$	4	20	[180]
$q=3$	5	56	[120]
$q=4$	4	41	[72]

Table 4.1: $m_2(N,q)$

The maximum size $m_2(N,q)$ of a cap is known only for the values given in Table 4.1. An $m_2(3,q)$-cap, $q>2$, is called an *ovoid* or *ovaloid*.

Theorem 4.1 (i) (Barlotti [21], Panella [175]) *For q odd or $q=4$, an ovoid in $PG(3,q)$ is an elliptic quadric.*

(ii) (Tits [224], [128, §16.4]) *There exists an ovoid in $PG(3,2^{2e+1})$, $e \geq 1$, which is not an elliptic quadric. This ovoid, the* Tits ovoid, *is projectively equivalent to the set $K = \{(0,1,0,0)\} \cup \{(1,z,y,x) \,\|\, z = xy + x^{\sigma+2} + y^{\sigma}$, with $(x,y) \in \mathbf{F}_q^2$ and $\sigma = 2^{e+1}\}$.*

(iii) (Fellegara [85], Penttila and Praeger [183]) *In $PG(3,8)$, an ovoid is an elliptic quadric or a Tits ovoid.*

(iv) (O'Keefe and Penttila [165, 168]) *In $PG(3,16)$, an ovoid always is an elliptic quadric.*

(v) (O'Keefe, Penttila and Royle [169]) *In $PG(3,32)$, an ovoid is an elliptic quadric or a Tits ovoid.*

(vi) (Segre [198]) *In $PG(N,2)$, a 2^N-cap is the complement of a hyperplane.*

(vii) (Hill [122]) *There are 9 projectively distinct 20-caps in $PG(4,3)$.*

(viii) (Hill [121]) *The 56-cap in $PG(5,3)$ is projectively unique.*

(ix) (Edel and Bierbrauer [72]) *There exist at least two types of 41-cap in $PG(4,4)$.*

Table 4.2 displays the values of n for which an n-cap in $PG(N,q)$ is contained in an $m_2(N,q)$-cap. The table is in two parts. The first part shows the exact values of $m_2'(N,q)$ for small q and upper bounds for q even. The second part of the table displays the upper bounds on $m_2'(3,q)$ for q odd, using the following relation between $m_2'(2,q)$ and $m_2'(3,q)$. For the corresponding value of $m_2'(2,q)$, see Table 2.3.

Theorem 4.2 (Nagy and Szőnyi [164]) *If, for q odd, $m_2'(2,q) \geq (5q+19)/6$, then*

$$m_2'(3,q) < qm_2'(2,q) + \tfrac{3}{4}(q + \tfrac{10}{3} - m_2'(2,q))^2 - q - 1.$$

N	q	$m_2'(N,q)$	$m_2(N,q)$-cap	
3	q even, $q \geq 8$	$\leq q^2 - q + 5$	ovoid	[52]
3	3	$= 8$	elliptic quadric	[79]
3	4	$= 14$	elliptic quadric	[136]
3	5	$= 20$	elliptic quadric	[4]
3	7	$= 32$	elliptic quadric	[76]
4	3	$= 19$	20-cap	[219]
5	3	$= 48$	56-cap	[19]
N	2	$= 2^{N-1} + 2^{N-3}$	complement of a hyperplane	[66]

Table 4.2(i): Upper bounds for and values of $m_2'(N,q)$

q odd, $q \geq 17$	$m_2'(3,q)$
$q = p^{2e}$, $p > 2$, $e \geq 1$	$< q^2 - q^{3/2}/4 + 39q/64 + O(q^{1/2})$
$q = p^{2e+1}$, $p > 2$, $e \geq 1$	$< q^2 - p^{3e+2}/4 + 119p^{2e+2}/64 - O(p^{e+2})$
q prime	$< 2641q^2/2700 - 4q/135 + 94/27$
$q = p^h$, $p \geq 5$	$< q^2 - q^{3/2}/2 + 67q/16 - O(q^{1/2})$
$q = p^h$, $p \geq 3$, h even for $p = 3$, $q \geq 23^2$, $q \neq 5^5$, 3^6	$< q^2 - q^{3/2}/2 + 35q/16 + O(q^{1/2})$

Table 4.2(ii): Upper bounds for $m_2'(3,q)$

The spectrum of the known values n for which there exists a complete n-cap in $PG(N,q)$ is given in the next result.

Theorem 4.3 (i) (Bose [38], Qvist [191]) *In $PG(3,q)$, $q > 2$, a complete (q^2+1)-cap exists.*

(ii) (Davydov and Tombak [66]) *In $PG(N,2)$, $N \geq 3$, a complete n-cap K, with $n \geq 2^{N-1}+1$, has order $n = 2^{N-1} + 2^{N-1-g}$ for some $g = 0, 2, 3, \ldots, N-1$. For each $g = 0, 2, 3, \ldots, N-1$, there exists a complete $(2^{N-1}+2^{N-1-g})$-cap in $PG(N,2)$.*

(iii) (Faina [77]) *Let $A = \{n/q^2 \parallel$ there exists a complete n-cap in $PG(3,q)$, for some prime power $q\}$. Then $A \cap [1/3, 1/2]$ is dense in $[1/3, 1/2]$.*

(iv) (Ferret and Storme [86]) *There does not exist a complete n-cap in $PG(3,q)$, q even, $q \geq 1024$, for which*

$$n \in \left[q^2 - (c-1)q + (2c^3 + c^2 - 5c + 6)/2,\ q^2 - (c-2)q - 2c^2 + 3c\right]$$

where $2 \leq c \leq \sqrt[4]{q}$.

Generalizing the idea of Segre [200] and Lombardo-Radice [148] for constructing plane arcs, complete caps containing a large number of points of an elliptic quadric have been constructed. This leads to complete caps with approximately $q^2/2$ [1, 77, 128, 181], $q^2/3$ [3, 77, 81] and $q\sqrt{q/2}$ [181] points.

In Tables 4.3(i) and 4.3(ii), the known upper bounds on $m_2(4,q)$ are presented. The bounds for $q = 9, 11, 13$ are obtained by applying the ideas of Gronchi [105] to the results of Heim [119]. The entries in Table 4.3(ii) are a consequence of the results in Theorem 4.4. Since the result by Storme, Thas and Vereecke is better than the bound by Nagy and Szőnyi for large values of q, we only state the bounds following from the second bound on $m_2(4,q)$, q odd.

Theorem 4.4 (i) (Nagy and Szőnyi [164]) *For q odd, let $m = \max\{(q^2 + 5q + 2)/2, m_2'(3,q)\}$. Then*

$$m_2(4,q) < qm + 2q^2.$$

(ii) (Storme, Thas and Vereecke [208]) *Let K be an n-cap in $PG(4,q)$, q odd, $n > (41q^3 + 202q^2 - 47q)/48$, and assume that every plane section of K, of size bigger than θ_K, with $\theta_K \geq (5q+25)/6$, is contained in a conic, then*

$$n < (q+1)\left(q \cdot \theta_K + \frac{3}{4}\left(q + \frac{10}{3} - \theta_K\right)^2 - q - 1 - \theta_K\right) + \theta_K.$$

In particular, when $m_2'(2,q) \geq (5q+25)/6$, this result can be applied.

(iii) (Hirschfeld and Thas [136]) *If $\gamma \geq q^2-q+4$ for q even and $q \geq 8$, and if every n-cap of $PG(3,q)$ with $n > \gamma$ is contained in an ovoid of $PG(3,q)$, then*

$$m_2(4,q) \leq (q-1)\gamma + q^2 + 2.$$

In particular, when $m_2'(3,q) \geq q^2-q+4$, this result can be applied.

q	$m_2(4,q)$	
$q=5$	≤ 96	[76]
$q=7$	≤ 285	[76]
$q=9$	≤ 703	
$q=11$	≤ 1266	
$q=13$	≤ 2107	
q odd, $q \geq 7$	$\leq q^3-q^2+8q-14$	[23]
q odd, $q \geq 67$	$< q^3-q^{5/2}/4+31q^2/16+O(q^{3/2})$	[126]
q even, $q \geq 8$	$\leq q^3-q^2+6q-3$	[52, 136]

Table 4.3(i): Upper bounds for $m_2(4,q)$

q odd	$m_2(4,q)$
$q=p^{2e}$, $p>2$, $e\geq 1$, $q \geq 49$	$< q^3-q^{5/2}/4+39q^2/64+O(q^{3/2})$
$q=p^{2e+1}$, $p>2$, $e \geq 1$	$< q^3-p^{e+1}q^2/4+119pq^2/64-O(p^{e+2}q)$
q prime, $q \geq 37$	$< 2641q^3/2700-79q^2/2700+O(q)$
$q=p^h$, $p \geq 5$, $q \geq 19$	$< q^3-q^{5/2}/2+67q^2/16-O(q^{3/2})$
$q=p^h$, $p\geq 3$, h even for $p=3$, $q \neq 5^5,\ 3^6$, $q \geq 23^2$	$< q^3-q^{5/2}/2+35q^2/16+O(q^{3/2})$

Table 4.3(ii): General upper bounds for $m_2(4,q)$, q odd

We now concentrate on $m_2(N,q)$. Tables 4.4(i) and (ii) give the upper bounds for $m_2(N,q)$, $N \geq 5$. The results in Table 4.4(ii) follow from those in Table 4.3(ii).

The latter formulas in Table 4.4(i), 4.4(ii) are of Bierbrauer and Edel [27]. This is an improvement to the result, for q odd, of Meshulam [161]. These are the best upper bounds, when the dimension N is large. More precisely, the larger the dimension N, the bigger the difference between the formula of Bierbrauer and Edel, and the other formulas of Table 4.4(i) and 4.4(ii). We remark that the bounds of Bierbrauer and Edel, and Meshulam, were originally bounds on caps in affine spaces $AG(N,q)$; bounds which have been rewritten into bounds on caps in projective spaces. In Table 4.4(i), the reference for $q=3$ is Barát *et al.*

[19], for $q = 5, 7$ is Edel, Storme and Sziklai [76], for q odd, $q > 7$, is Hill [121], and for q even, $q \geq 16$, is Storme, Thas and Vereecke [208].

Table 4.5 now gives general inequalities on $m_2(N, q)$. We first present formulas which are crucial in the determination of lower bounds on $m_2(N, q)$. This is then followed by Hill's recurrence relation [121], which makes it possible to calculate upper bounds on $m_2(N, q)$. To this formula of Hill, small improvements have been made by Storme, Thas and Vereecke. We refer to [208] for the exact improvements.

The references for the formulas in Table 4.5 are: Segre [198] for the first two formulas, Edel and Bierbrauer [73] for the third formula, and Hill [121] for the latter two formulas.

q, N	$m_2(N,q)$
$q = 3$, $N \geq 6$	$\leq 2 \cdot 3^{N-2} - 3^{N-3} + 2$
$q = 4$, $N \geq 4$	$\leq 4^{N-1} - 4^{N-2} - 2 \cdot 4^{N-3} - 2 \cdot 4^{N-4}/3 + 5/3$
$q = 5$, $N \geq 4$	$\leq 5^{N-1} - 5^{N-2} - 5^{N-3} - 5^{N-4}/2 + 3/2$
$q = 7$, $N \geq 4$	$\leq 7^{N-1} - 7^{N-2} - 4 \cdot 7^{N-3}/3 + 4/3$
$q = 9$, $N \geq 4$	$\leq 9^{N-1} - 109 \cdot 9^{N-4}/4 + 5/4$
$q = 11$, $N \geq 4$	$\leq 11^{N-1} - 331 \cdot 11^{N-4}/5 + 6/5$
$q = 13$, $N \geq 4$	$\leq 13^{N-1} - 547 \cdot 13^{N-4}/6 + 7/6$
q odd, $q > 7$	$\leq q^{N-1} - q^{N-2} + 8q^{N-3} - 15q^{N-4} - 2\theta_{N-5} + 1$
q even, $q \geq 16$	$m_2(5,q) < q^4 - q^3 + 5q^2 + 3q - 1$
q even, $q \geq 16$, $N \geq 6$	$< q^{N-1} - q^{N-2} + 5q^{N-3} + 2q^{N-4} + O(q^{N-5})$
$q = Q^h, Q > 2$	$(Nh+1)q^N/(Nh)^2 + m_2(N-1,q)$

Table 4.4(i): Upper bounds for $m_2(N, q)$

$q = p^h$ odd, p prime	$m_2(N,q)$
$q = p^{2e}$, $e \geq 1$, $q \geq 49$	$< q^{N-1} - q^{N-3/2}/4 + 39q^{N-2}/64 + O(q^{N-5/2})$
$q = p^{2e+1}$, $e \geq 1$	$< q^{N-4}(q^3 - p^{e+1}q^2/4 + 119pq^2/64 - O(p^{e+2}q))$
q prime, $q \geq 37$	$< 2641q^{N-1}/2700 - 79q^{N-2}/2700 + O(q^{N-3})$
$q = p^h$, $p \geq 5$, $q \geq 19$	$< q^{N-1} - q^{N-3/2}/2 + 67q^{N-2}/16 - O(q^{N-5/2})$
$q = p^h$, $p \geq 3$, h even for $p = 3$, $q \geq 23^2$, $q \neq 5^5$, 3^6	$< q^{N-1} - q^{N-3/2}/2 + 35q^{N-2}/16 + O(q^{N-5/2})$
$q = Q^h$	$(Nh+1)q^N/(Nh)^2 + m_2(N-1,q)$

Table 4.4(ii): General upper bounds for $m_2(N, q)$, q odd

q, N	$m_2(N,q)$
q	$m_2(N+3,q) \geq q^2 m_2(N,q)+1$
q	$m_2(N+3r,q) \geq q^{2r} m_2(N,q)+(q^{2r}-1)/(q^2-1)$
q	$m_2(N+6,q) \geq (q^4+q^2-1)m_2(N,q)+q^2+1$
$q>2, N\geq 4$	$m_2(N,q) \leq qm_2(N-1,q)-(q+1)$
$q>2, N\geq 5$	$m_2(N,q) \leq q^{N-4}m_2(4,q)-q^{N-4}-2\theta_{N-5}+1$

Table 4.5: General inequalities

We now present lower bounds on $m_2(N,q)$. The lower bounds for arbitrary values of the dimension N follow from an inductive application of the formulas of Table 4.5. To obtain such a lower bound, in general, we start with the cardinality of a known n-cap in a particular projective space of small dimension. These cardinalities are listed in Table 4.6(i).

q	N	$m_2(N,q) \geq$	
q even, $q>4$	4	$2q^2+q+9$	[74]
$q\equiv 1 \pmod 8$	4	$(5q^2-2q-7)/2$	[26]
$3<q\equiv 3 \pmod 8$	4	$(5q^2-8q-13)/2$	[26]
$q\equiv 5 \pmod 8$	4	$(5q^2-6q-11)/2$	[26]
$q\equiv 7 \pmod 8$	4	$(5q^2-4q-9)/2$	[26]
q odd	5	q^3+q^2+3q+3	[73]
q even	5	q^3+2q^2+q+2	[163]
q	6	q^4+2q^2	[73]

Table 4.6(i): Lower bounds on $m_2(N,q)$, $N=4,5,6$

In Tables 4.6(ii) and 4.6(iii), we use the general inequalities to obtain lower bounds for $m_2(N,q)$ for arbitrary dimensions N. The bound for $N=9$ follows from Edel and Bierbrauer [73].

q	N	$m_2(N,q) \geq$	Condition
$q>4$	7	$aq^4+bq^3+cq^2+1$	$m_2(4,q) \geq aq^2+bq+c$
q odd	8	$q^5+q^4+3q^3+3q^2+1$	
q even	8	$q^5+2q^4+2q^3+4q^2+q+2$	
q	9	$q^6+2q^4+q^2$	

Table 4.6(ii): Lower bounds on $m_2(N,q)$, $N=7,8,9$

The bounds in Table 4.6(iii) follow from an inductive application of the bounds of Edel and Bierbrauer. Namely, for $q>3$, the formulas for $N=1+6l$ and $N=4+6l$ start from $m_2(4,q)\geq aq^2+bq+c$, the one for $N=2+6l$ from $m_2(8,q)\geq aq^5+bq^4+cq^3+dq^2+eq+f$, and the one for $N = 5 + 6l$ from $m_2(5, q) \geq aq^3 + bq^2 + cq + d$. For exact values of the parameters a, b, c, d, e, f, we refer to Tables 4.6(i) and 4.6(ii).

In the first formula for $q = 3$, the corresponding lower bounds on $m_2(v, 3)$ can be found in Table 4.7. The second formula for $q = 3$ follows from Calderbank and Fishburn [48].

Table 4.7 presents the sizes of the largest known caps in $PG(N, q)$, for N and q small. This table is from Edel and Bierbrauer [75] and from Edel [71].

For the size $t_2(N, q)$ of the smallest complete n-cap in $PG(N, q)$, the trivial lower bound is $t_2(N, q) > \sqrt{2q^{N-1}}$. Upper bounds on this parameter are given in Table 4.8. Other small caps are given by Östergard [171].

q	N	$m_2(N, q) \geq$
3	$6l + v$, $v = 7, \ldots, 12$	$112^l m_2(v, 3)$
3	$6l$	$32^l((3l/8)(7/2)^{l-1} + 1)$
> 3	$6l$	$q^{4l} + (1 + l)q^{4l-2} + (l^2 - 2l + 1)q^{4l-4}$
> 3	$6l + 1$	$aq^{4l} + bq^{4l-1} + (c - a + al)q^{4l-2} + (bl - l)q^{4l-3}$
> 3	$6l + 2$	$aq^{4l+1} + bq^{4l} + (c - a + al)q^{4l-1} + (d - b + bl)q^{4l-2}$
> 3	$6l + 3$	$q^{4l+2} + (l + 1)q^{4l} + (l^2 - 2l + 3)q^{4l-2}/2$
> 3	$6l + 4$	$aq^{4l+2} + bq^{4l+1} + (al + c)q^{4l} + blq^{4l-1}$
> 3	$6l + 5$	$aq^{4l+3} + bq^{4l+2} + (c + al)q^{4l+1} + (d + bl)q^{4l}$

Table 4.6(iii): Lower bounds on $m_2(N, q)$, $N \geq 10$

N \ q	3	4	5	7	8	9	11	13
2	4	6	6	8	10	10	12	14
3	10	17	26	50	65	82	122	170
4	20	41	66	132	208	212	316	388
5	56	126	186	434	695	840		
6	112	288	675	2499	4224	6723		
7	248	756	1715	6472	13520	17220		
8	532	2110	4700	21555	45174	68070		
9	1216	4938	17124	122500	270400	544644		
10	2744	15423	43876	323318	878800	1411830		
11	6464	34566	120740	1067080	2812160	5580100		
12	13312							
13	30464							

Table 4.7: Large caps in small projective spaces

N	q	$t_2(N,q)$	
3	2	$= 5$	[128, §18.2]
3	3	$= 8$	[128, §18.4]
3	4	$= 10$	[82], [186]
3	5	$= 12$	[81], [186]
3	7	$= 17$	[171]
3	8	≤ 20	[171]
3	9	≤ 24	[171]
3	11	≤ 30	[171]
3	13	≤ 37	[171]
3	16	≤ 41	[173]
3	17	≤ 52	[171]
4	2	$= 9$	[91]
4	3	$= 11$	[79], [186]
4	4	≤ 20	[173]
4	5	≤ 31	[171]
4	7	≤ 57	[171]
4	8	≤ 72	[171]
4	9	≤ 88	[171]
5	2	$= 13$	[91]
5	3	≤ 22	[173]
5	4	≤ 50	[171]
5	5	≤ 83	[171]
$2m-1 \geq 9$	2	$\leq 15 \cdot 2^{m-3} - 3$	[91]
$2m-2 \geq 10$	2	$\leq 23 \cdot 2^{m-4} - 3$	[91]
3	q even	$\leq 2q + t_2(2,q)$	[174]
$2r$	q even, $q > 2$	$\leq q^r + 3(q^{r-1} + \cdots + q) + 2$	[173]
$2r+1$	q even, $q > 2$	$\leq 3(q^r + \cdots + q) + 2$	[173]
$4k+2$	q odd, $q \geq 5$	$\leq q^{2k+1} + t_2(2k,q)$	[65]

Table 4.8: Upper bounds for $t_2(N,q)$

5. (n,r)-arcs in $PG(2,q)$

To begin this section on (plane) (n,r)-arcs K, Table 5.1 presents the known exact values of $m_r(2,q)$, while Tables 5.2 and 5.3 give upper and lower bounds on $m_r(2,q)$.

Theorem 5.1 (i) (Ball and Blokhuis [15]) *In $PG(2,q)$, q even, a (k,r)-arc, with $r|q$ and with $k \geq rq - q + r/2$ when $r < q/2$ or with $k > rq - r + 523r/1000$ when $r = q/2$, can be extended in a unique way to an $(rq-q+r,r)$-arc.*

(ii) (Hadnagy and Szőnyi [108]) *An $(qn - q + n - \varepsilon, n)$−arc can be embedded into a maximal arc, provided that $\varepsilon \leq c \cdot n$, c is a constant satisfying $0 < c < 2/3$, n divides q, and $K = q/n$ is large enough. More* *precisely,* $K = 2$, *when* $0 \le c \le 1/3$, $K = 3$, *when* $0 < c \le 0.449$, $K = 4$, *when* $0.449 < c \le 1/2$, $K \ge (1+2c)(1-c)/(1-c-c^2)$, *when* $1/2 < c \le 3/5$ *and* $K \ge (1+2c)(4-5c)/(4-6c)$, *when* $3/5 < c < 2/3$.

(iii) (Szőnyi [217]) *If* $\varepsilon \le q^{1/4}/2$, *then any* (k,p)*-arc of* $PG(2,q)$, $q = p^h$, p *prime, with* $k = qp - q + p - \varepsilon$, *can be embedded in a maximal arc.*

r	$q = p^h$, p prime	$m_r(2,q)$	
2	q odd	$q+1$	[38]
2^e	2^h	$(2^e-1)q+2^e$	[69]
$\sqrt{q}+1$	q square, $q \ge 25$	$q\sqrt{q}+1$	[53]
$(q+1)/2$	q odd prime	$(q^2-q)/2+1$	[10]
$(q+3)/2$	q odd prime	$(q^2+q)/2+1$	[10]
q	q	q^2	
$q-1$	q square, $q > 4$	$q^2-q-2\sqrt{q}-1$	[13]
$q-2$	q odd square, $q > 121$	$q^2-2q-3\sqrt{q}-2$	[9]
$q+1-t$, $t > 1$	$q = p^{2e}$, $p > 3$, $t < \min(q^{1/6}, q^{1/4}/2)$	$q^2+q+1-t(q+\sqrt{q}+1)$	[34]
$q+1-t$, $t > 1$	$q = p^{2e}$, $p = 2,3$, $t < \min(2^{-1/3}q^{1/6}, q^{1/4}/2)$	$q^2+q+1-t(q+\sqrt{q}+1)$	[34]

Table 5.1: Exact values of $m_r(2,q)$

The second row of Table 5.2 is implied by the result of Ball, Blokhuis and Mazzocca [16], which states that a maximal arc, that is, an (n,r)-arc with $n = (r-1)q + r$, does not exist for q odd. This had previously been shown for $q = 3^h$ with $r = 3$ by Thas [222] and for $q = 9$ by Cossu [64]. The proof of [16] has been simplified by Ball and Blokhuis [14].

In Table 5.3, the integer $q = p^h$ is *exceptional* if h is odd, $h \ge 3$ and p divides $\lfloor 2\sqrt{q} \rfloor$.

For small r and q, the known values and bounds on $m_r(2,q)$ are enumerated in Table 5.4. The values $m_3(2,11)$ and $m_3(2,13)$ were determined by Marcugini, Milani and Pambianco ([154] and [155], respectively). See Ball [10] for the exact references for the other values.

For recent results on *multi-arcs* in $PG(2,q)$, that is, (n,r)-arcs in $PG(2,q)$ which are allowed to have multiple points, we refer to Ball [12] and Ball *et al.* [17]. In the latter article, a detailed study of $(q^2+q+2, q+2)$-arcs in $PG(2,q)$ is made.

r	Conditions	$m_r(2,q) \leq$	
r		$(r-1)q+r$	[22]
r	q odd	$(r-1)q+r-2$	[16]
r	q odd, $r\|q$	$(r-1)q+r/2$	[15]
r	$4 \leq r < q$, $r \not\| q$	$(r-1)q+r-3$	[152]
r	$9 \leq r < q$, $r \not\| q$	$(r-1)q+r-4$	[152]
r	q prime, $r \geq (q+3)/2$	$(r-1)q+r-(q+1)/2$	[10]
r	$r \leq 2q/3$	$(r-1)q+q-r$	[123]
r	$r \geq q/2$, $(r,q) > 1$, K has a skew line	$(r-1)q+q-r$	[10]
r	$r > 2q/3$	$(r-1)q+(q+r-r/q)/2-$ $\sqrt{(q+r-r/q)^2/4-(r^2-r)}$	[123]
r	K does not have a skew line	$rq-\sqrt{(q+1-r)q}$	[10]
r	K does not have a skew line	$m_r(2,q) < (r-1)q+\lfloor r^2/q \rfloor$	[12]
r	$(r,q)=1$, $r < \sqrt{q}+1$	$(r-1)q+1$	[8]
r	$(r,q)=1$, $6 \leq r \leq \sqrt{2q}+1 \leq q-1$	$(r-1)q+1$	[53]
r	$(r,q)=p^e$, K has a skew line	$(r-1)q+p^e$	[12]
r	q prime, $r \leq (q+1)/2$	$(r-1)q+1$	[10]
$q-1$	$q=17,19$	$q^2-3q/2-5/2$	[10, 13]
$q-1$	$q=p^3$, p prime, $q \geq q_0$	q^2-q-2p^2+5p-1	[150]
$q-1$	$q=p^{2e+1}$, $q>19$	$q^2-q-p^e\lceil \frac{p^{e+1}+1}{p^e+1} \rceil -1$	[10, 13]
$q-2$	$q=p^{2e+1}$, $q>17$	$q^2-2q-p^e\lceil \frac{p^{e+1}+1}{p^e+1} \rceil -2$	[9, 10]
$q-2$	$q=2^{2e}$, $q>4$, or $q \in \{25,49,81,121\}$	$q^2-2q-2\sqrt{q}-2$	[9, 10]
$q-2$	$q=17$	$q^2-5q/2-7/2$	[9, 10]

Table 5.2: Upper bounds for $m_r(2,q)$

r	q	$m_r(2,q) \geq$	
$r=3$	q non-exceptional	$q+1+\lfloor 2\sqrt{q} \rfloor$	[229]
$r=3$	q exceptional	$q+\lfloor 2\sqrt{q} \rfloor$	[229]
$r=\sqrt{q}+1$	q square	$(r-1)q+1$	[39]
$r=(q+1)/2$	q odd	$(r-1)q+1$	[23]
$r=(q+3)/2$	q odd	$(r-1)q+1$	[23]
$r=q-1$	q	$(r-1)q+1$	[124]
$r=q-2$	q even	$(r-1)q+2$	[124]
r	q square	$(r-1)q+r-\sqrt{q}+$ $\sqrt{q}(r-q)$	[124]
$r=q-\sqrt{q}$	q square	$(r-1)q+\sqrt{q}$	[159]
$(q-r)\|q$	q	$(r-1)q+q-r$	[160]

Table 5.3: Lower bounds for $m_r(2,q)$

r \ q	3	4	5	7	8	9	11	13
2	4	6	6	8	10	10	12	14
3		9	11	15	15	17	21	23
4			16	22	28	28	31 − 34	34 − 40
5				29	33	37	43 − 45	46 − 53
6				36	42	48	56	61 − 66
7					49	55	67	79
8						65	77 − 78	92
9							89 − 90	105
10							100 − 102	116 − 119
11								131 − 133
12								144 − 147

Table 5.4: Small values of $m_r(2,q)$

$q = p^h$, p prime	$k \geq$	Sharpness
$q = p$, p odd	$3(p+1)/2$	sharp
$q = p^{2e}$, $e > 0$	$q + \sqrt{q} + 1$	sharp
$q = p^{2e+1}$, $e > 0$	$q + p^{e+1} + 1$	sharp for $e = 1$
$q = p^{2e+1}$, $e > 0$, $p > 3$	$q + q^{2/3} + 1$	sharp for q cube
$q = p^{2e+1}$, $e > 0$, $p = 2, 3$	$q + 2^{-1/3}q^{2/3} + 1$	

Table 6.1: Lower bounds for the size of a blocking set

6. Multiple blocking sets in $PG(2,q)$

A *t-blocking k-set* S in $PG(2,q)$ is a set of k points such that every line of $PG(2,q)$ intersects S in at least t points, and there is a line intersecting S in exactly t points.

For $t = 1$, a 1-blocking set is simply called a *blocking set*. A *trivial* blocking set S is a blocking set containing a line of $PG(2,q)$. A t-blocking set is called *minimal* or *irreducible* when no proper subset of it is a t-blocking set. A 2-blocking set is also called a *double* blocking set and a 3-blocking set a *triple* blocking set.

A subset S of $PG(2,q)$ is a t-blocking set if and only if its complement $S' = PG(2,q) \setminus S$ is a $(k', q+1-t)$-arc. So the study of t-blocking sets is equivalent to the study of $(n, q+1-t)$-arcs.

First of all, lower bounds on the sizes of minimal non-trivial blocking sets S are given in Table 6.1. The third column indicates when this bound is sharp. These results are respectively of Blokhuis [28], Bruen [43, 44], Blokhuis [30], and Blokhuis, Storme and Szőnyi [34].

Theorem 6.1 (Bruen [43]) *In $PG(2,q)$, q square, a non-trivial blocking set of cardinality $q+\sqrt{q}+1$ is a Baer subplane.*

The following two theorems present results on the spectrum of the values k for which there exists a minimal blocking k-set in $PG(2,q)$.

Theorem 6.2 (i) (Bruen and Thas [46]) *Let S be a minimal blocking k-set in $PG(2,q)$. Then $k \leq q\sqrt{q}+1$, with equality holding if and only if S is a unital in $PG(2,q)$, q square.*

(ii) (Blokhuis and Metsch [32], Innamorati [138]) *In $PG(2,q)$, with q square and $q \geq 25$ or $q=9$, there is no minimal blocking set of size $q\sqrt{q}$.*

(iii) (Ball and Blokhuis [13]) *For q square, $q \geq 16$, there is no minimal blocking k-set S with $q+\sqrt{q}+1 < k < q+2\sqrt{q}+1$.*

(iv) (Innamorati and Maturo [139]) *In $PG(2,q)$, $q \geq 4$, there exists a blocking k-set for every k with $2q-1 \leq k \leq 3q-3$.*

Let S be a minimal non-trivial blocking k-set in $PG(2,q)$. Then $|S \setminus L| \geq q$ for every line L of $PG(2,q)$. If $|S \setminus L| = q$ for some line L, then the blocking set S is *of Rédei type* with respect to L.

A blocking k-set S in $PG(2,q)$ is called *small* when $k < 3(q+1)/2$. We note that the number e described in Theorem 6.3 (i)(2) is called the *exponent* of the minimal blocking k-set.

Theorem 6.3 (i) (Polverino [188], Szőnyi [216]) *Let S be a small minimal blocking k-set in $PG(2,q)$, $q=p^h$.*

(1) *Then*

$$q+1+p^e\left\lceil \frac{q/p^e+1}{p^e+1}\right\rceil \leq k \leq \frac{1+(p^e+1)(q+1)-\sqrt{\Delta}}{2},$$

where $\Delta = (1+(p^e+1)(q+1))^2-4(p^e+1)(q^2+q+1)$, for some integer e, $1 \leq e \leq n/2$.

(2) *If k lies in the interval belonging to e and $p^e \neq 4,8$, then each line intersects S in $1 \pmod{p^e}$ points.*

(ii) (Szőnyi [216]) *Let $q=p^2$, p prime, and let S be a minimal blocking k-set which is not a Baer subplane. Then $k \geq 3(q+1)/2$.*

(iii) (Polverino [188], Szőnyi [216]) *For p prime, $p \geq 7$, let S be a small minimal blocking k-set in $PG(2,p^3)$. Then $k = p^3+p^2+1$ or $k = p^3+p^2+p+1$.*

In $PG(2,q)$, q odd, there exists a minimal blocking k-set of Rédei type, the *projective triangle*, of cardinality $3(q+1)/2$. This is a blocking k-set projectively equivalent to the set containing the points $(1,0,-c)$, $(0,-c,1)$, $(-c,1,0)$, where c is a square of $\mathbf{F}_q$. In $PG(2,7)$, there also exists a sporadic example of a blocking k-set of cardinality $3(7+1)/2=12$;

this is the dual of the Hesse configuration $(9_4, 12_3)$ of the nine inflexions of a cubic curve.

Theorem 6.4 (i) (Blokhuis, Ball, Brouwer, Storme, and Szőnyi [31]) *Let S be a minimal blocking k-set of Rédei type in $PG(2,q)$, with $q = p^h$, p prime. Suppose that e, with $0 \le e < h$, is the largest integer such that each line intersects S in $1 \pmod{p^e}$ points. Then $k \equiv 1 \pmod{p^e}$ and one of the following holds:*

(1) *$e = 0$ and $3(q+1)/2 \le k \le 2q$;*

(2) *$e = 1$, $p = 2$ and $(4q+5)/3 \le k \le 2q-1$;*

(3) *$p^e > 2$, $e|h$, and $q + q/p^e + 1 \le k \le q + (q-1)/(p^e-1)$.*

Also, if $p^e > 3$ or $(p^e, k) = (3, q + q/3 + 1)$, and L is a line such that $|S \setminus L| = q$, then the subset $U = S \setminus L$ of the affine plane $AG(2,q) \cong PG(2,q) \setminus L \cong \mathbf{F}_{q^2}$ is an $\mathbf{F}_{p^e}$-linear subspace of $\mathbf{F}_{q^2}$.

(ii) (Polito and Polverino [187]) *There exist small minimal blocking k-sets, not of Rédei type, in every projective plane $PG(2,q)$, $q = p^t$, p prime, $t \ge 4$.*

(iii) (Polverino [189]) *In $PG(2,p^3)$, p prime, $p \ge 7$, every small minimal blocking k-set is of Rédei type.*

(iv) (Polverino and Storme [190]) *In $PG(2,q^3)$, $q = p^h$, $h \ge 1$, p prime, $p \ge 7$, the non-trivial minimal blocking k-sets with exponent $e \ge h$ are as follows:*

(1) *A Baer subplane $PG(2,q^{3/2})$ of cardinality $q^3 + q^{3/2} + 1$ when q is a square.*

(2) *A minimal blocking k-set of size q^3+q^2+1, projectively equivalent to the set $K = \{(x, \mathrm{T}(x), 1) || x \in \mathbf{F}_{q^3}\} \cup \{(x, \mathrm{T}(x), 0) || \ x \in \mathbf{F}_{q^3} \setminus \{0\}\}$, with $\mathrm{T} : \mathbf{F}_{q^3} \to \mathbf{F}_q : x \mapsto x + x^q + x^{q^2}$.*

(3) *A minimal blocking k-set of size $q^3 + q^2 + q + 1$, projectively equivalent to the set $K = \{(x, x^q, 1) || x \in \mathbf{F}_{q^3}\} \cup \{(x, x^q, 0) || x \in \mathbf{F}_{q^3} \setminus \{0\}\}$.*

(v) (Lovász and Schrijver [149]) *Let S be a blocking k-set of cardinality $3(p+1)/2$ in $PG(2,p)$, p odd prime, of Rédei type. Then S is projectively equivalent to the projective triangle.*

(vi) (Sziklai and Szőnyi [210]) *Let S be a minimal non-trivial blocking k-set of cardinality $3(p+1)/2$ in $PG(2,p)$, $p > 2$ prime. Then every point of S lies on exactly $(p-1)/2$ tangents and there are at least $(p+1)/2$ lines intersecting S in exactly 2 points.*

(vii) (Gács [93]) *For a Rédei type blocking k-set S in $PG(2,p)$, p odd prime, which is not a projective triangle, $|S| \ge p + 2(p-1)/3 + 1$.*

(viii) (Gács [94]) *In $PG(2,11)$, every minimal blocking 18-set is a projective triangle.*

(ix) (Gács [94]) *In $PG(2,7)$, a minimal blocking 12-set is a projective triangle or is the sporadic example.*

A blocking k-set S of $PG(2,q)$ is of *almost Rédei type* when there exists a line L of $PG(2,q)$ such that $|S \setminus L|$ is close to q; more precisely, $|S \setminus L| = q + m$ with $m \leq \sqrt{q}/2$.

Theorem 6.5 (i) (Blokhuis, Pellikaan and Szőnyi [33]) *Let S be a minimal non-trivial blocking k-set in $PG(2,q)$, $q = p^h$, p prime, such that $|S \cap L| = k - q - m$, with $0 < m \leq \sqrt{q}/2$. Then one of the following possibilities occurs:*

(1) *if $k < 3(q+1)/2$, then $m \equiv 0 \pmod{p}$;*
(2) *if $m = 1$, then $k \geq 3(q+1)/2$;*
(3) *if $m = 2$ and $p \neq 2$, then $k \geq 3(q+1)/2$;*
(4) *if $m > 2$ and $m \not\equiv 0 \pmod{p}$, then $k \geq q + (q+1)/2 + m$.*

(ii) *Let $q = p$ be a prime, let S be a blocking k-set, not of Rédei type, with $k = 3(p+1)/2$ in $PG(2,p)$, and let L be a line of $PG(2,p)$. Then, there are three cases:*

(1) (Blokhuis, Pellikaan and Szőnyi [33]) *when $|S \cap L| \leq (p+3)/2 - 3$, then $|S \cap L| \leq (9p-15)/20$;*

(2) (Gács, Sziklai and Szőnyi [95]) *if $|S \setminus L| = p + 1$, then $p = 3$ or $p = 5$ and S is of Rédei type with respect to an other line, or $p = 7$ and S is the sporadic blocking 12-set in $PG(2,7)$;*

(3) (Gács [92]) *if $|S \setminus L| = p + 2$, then $p \leq 7$ and S is of Rédei type with respect to an other line.*

Table 6.2 shows the values of n for which there exists a minimal blocking k-set in $PG(2,q)$ (Blokhuis [30]). The non-existence of a minimal blocking 14-set in $PG(2,8)$ was proved by Barát, Del Fra, Innamorati and Storme [18], and the non-existence of minimal blocking k-sets in $PG(2,8)$ for $k = 22, 23$ was proved by Del Fra and Innamorati [68].

q	$n \in$
4	$\{7, 8, 9\}$
5	$\{9, \ldots, 12\}$
7	$\{12, \ldots, 19\}$
8	$\{13, 15, 16, \ldots, 21\}$
9	$\{13, 15, 16, \ldots, 26, 28\}$

Table 6.2: The sizes of a blocking k-set in small planes

To give an idea of how good Theorem 6.2(i) is in general, there are asymptotic results on large minimal blocking k-sets.

Theorem 6.6 (i) (Szőnyi [214]) *In $PG(2,q)$, $q > q_0$, there exist minimal blocking k-sets of size $cq \log q$.*

(ii) (Hirschfeld and Szőnyi [135]) *In $PG(2,q)$, q square, $q > q_0(\lambda)$, there exist minimal blocking k-sets of size $cq^{1+\lambda}$, $1/4 < \lambda \leq 1/2$.*

$q = p^h$, p prime	t	$\|S\| \geq$	Sharp	
$q = p^{2e}$, $p > 3$, $e \geq 2$	$t < \min(q^{1/6}, q^{1/4}/2)$	$t(q+\sqrt{q}+1)$	yes	[34]
$q = p^{2e}$, $p = 2, 3$, $e \geq 2$	$t < \min(2^{-1/3}q^{1/6}, q^{1/4}/2)$	$t(q+\sqrt{q}+1)$	yes	[34]
$q = p^2$	$t < q^{1/4}/2$	$t(q+\sqrt{q}+1)$	yes	[34]

$q = p^h$, p prime	t	$\|S\| \geq$	Sharp for	
$q = p$ prime, $p > 3$	$t < p/2$	$(2t+1)(p+1)/2$	$t = 1$, $(p-1)/2$	[10]
$q = p$ prime, $p > 3$	$t > p/2$	$(t+1)p$	$t = (p+1)/2$	[10]
$q < 9$	2	$3q$	$t = 2$	[25, 61] [152]
$q = 11, 13, 17, 19$	2	$(5q+7)/2$		[10, 13]
q square, $q > 4$	2	$2q + 2\sqrt{q} + 2$	$t = 2$	[10, 13]
$q = p^3$, $q \geq q_0$	2	$2q + 2p^2 - 5p + 2$		[150]
$q = p^{2e+1} > 19$	2	$2q + p^e\lceil \frac{p^{e+1}+1}{p^e+1} \rceil + 2$		[10, 13]
$q = 5, 7, 9$	3	$4q$	$t = 3$	[9, 61] [152]
$q = 8$	3	31	$t = 3$	[124]
$q = 11, 13, 17$	3	$(7q+9)/2$		[9, 10]
q odd square, $q > 121$	3	$3q + 3\sqrt{q} + 3$	$t = 3$	[9, 10]
$q = p^{2e+1}$, $q > 17$	3	$3q + p^e\lceil \frac{p^{e+1}+1}{p^e+1} \rceil + 3$		[9, 10]
q even square, $q > 4$, or $q \in \{25, 49, 81, 121\}$	3	$3q + 2\sqrt{q} + 3$		[9, 10]

q	t	Condition	$\|S\| \geq$	
q	t	S does not contain a line	$tq + \sqrt{tq} + 1$	[10]
q	t	S contains a line and $(t-1, q) = 1$	$q(t+1)$	[29]
q	t	S contains a line and $(t-1, q) > 1$, $t \leq q/2 + 1$	$tq + q - t + 2$	[45]
q	t	S contains a line and $(t-1, q) > 1$, $t \geq q/2 + 1$	$t(q+1)$	[10]

Table 6.3: Lower bounds for t-blocking k-sets

Now, Table 6.3 gives lower bounds on the number of points in a t-blocking k-set S of $PG(2,q)$. The table is subdivided into three parts. In the first and second part, lower bounds on t-blocking k-sets, with no condition on the t-blocking k-set, are given. The third part gives lower bounds on t-blocking k-sets with conditions on the t-blocking k-sets.

There are various characterizations of the smallest t-blocking k-sets.

Theorem 6.7 (i) (Gács and Szőnyi [96]) *In $PG(2,q)$, with q an odd square and $q \geq 169$, a double blocking k-set containing $2q + 2\sqrt{q} + 2$ points is the union of two disjoint Baer subplanes.*

(ii) (Blokhuis, Storme and Szőnyi [34]) *Let S be a t-blocking k-set in $PG(2,q)$, $q = p^h$, p prime, of size $t(q+1) + c$. Let $c_2 = c_3 = 2^{-1/3}$ and $c_p = 1$ for $p > 3$.*

(1) *If $q = p^{2d+1}$ and $t < \frac{1}{2}(q - c_p q^{2/3})$, then $c \geq c_p q^{2/3}$.*

(2) *If q is a square with $q > 4$, and $t < \min(c_p q^{1/6}, q^{1/4}/2)$, $c < c_p q^{2/3}$, then $c \geq t\sqrt{q}$ and S contains the union of t disjoint Baer subplanes.*

(3) *If $q = p^2$, $t < q^{1/4}/2$ and $c < p\left\lceil \frac{1}{4} + \sqrt{\frac{p+1}{2}} \right\rceil$, then $c \geq t\sqrt{q}$ and S contains the union of t disjoint Baer subplanes.*

A t-blocking k-set is called *small* when $k < tq + (q+3)/2$. Similarly to Theorem 6.3, there are the following results on small minimal t-blocking k-sets. These results are based upon the abstract [150].

Theorem 6.8 (Lovász and Szőnyi [150]) (i) *Let S be a small minimal t-blocking k-set in $PG(2,q)$, $q = p^h$, p prime, $t < p$, with $q > q_0(t)$.*

(1) *Then*

$$tq + t + p^e \left\lceil \frac{q/p^e + 1}{p^e + 1} \right\rceil \leq k \leq \frac{1 + (q+1)(2t - 1 + p^e) - \sqrt{\Delta}}{2},$$

where $\Delta = (1 + (q+1)(2t - 1 + p^e))^2 - 4(q^2 + q + 1)(t^2 + tp^e)$, for some integer $1 \leq e \leq h/2$.

(2) *If k lies in the interval belonging to e, then each line intersects S in $t \pmod{p^e}$ points.*

(ii) *Let S be a small minimal t-blocking k-set in $PG(2,p^2)$, p prime, $t < \sqrt{p}/2$, $p > p_0(t)$. Then S is the disjoint union of t Baer subplanes.*

7. Blocking sets in $PG(N,q)$

The concepts of Section 6 can be extended to $PG(N,q)$. A *blocking set in $PG(N,q)$ with respect to k-dimensional subspaces of $PG(N,q)$* is a set of points K such that any k-dimensional subspace intersects K in at least one point.

Theorem 7.1 (Bose and Burton [41]) *The smallest blocking sets of $PG(N,q)$ with respect to k-dimensional subspaces are the $(N-k)$-dimensional subspaces of $PG(N,q)$.*

A blocking set with respect to k-dimensional subspaces in $PG(N,q)$ is called *non-trivial* when no $(N-k)$-dimensional subspace is contained in K. *Minimality* is defined in the usual way.

We now present a characterization result on the smallest such non-trivial minimal blocking sets.

Theorem 7.2 (Beutelspacher [24], Heim [119]) *In $PG(N,q)$, the smallest non-trivial minimal blocking sets with respect to k-dimensional subspaces are cones with an $(N-k-2)$-dimensional vertex and a non-trivial 1-blocking set of minimum cardinality in a plane of $PG(N,q)$ as base curve.*

Improvements to this characterization for q square have been obtained by Bokler and Metsch.

Theorem 7.3 (Bokler and Metsch [36], Bokler [35]) *Let B be a minimal non-trivial blocking set with respect to k-dimensional subspaces of $PG(N,q)$, $N \geq 2$, q square, $q \geq 16$, $N \geq k$. If $|B| \leq \theta_{N-k} + \sqrt{q}\theta_{N-k-1}$, then B contains a cone with t-dimensional vertex and with base a subgeometry $PG(2(N-k-t-1), \sqrt{q})$, for some t with $\max\{-1, N-2k-1\} \leq t \leq N-k-1$.*

A significant result on small minimal non-trivial 1-blocking sets of $PG(2,q)$ was shown by Szőnyi (Theorem 6.3(i)(2)); this states that every line intersects S in 1 (mod p^e) points. This result gives precise disjoint intervals in which the sizes of small minimal non-trivial 1-blocking sets of $PG(2,q)$ must lie.

These results were extended to blocking sets with respect to k-dimensional subspaces of $PG(N,q)$ by Szőnyi and Weiner [218]. In the following theorem, let $S(q)$ denote the set of sizes k of the small minimal blocking k-sets in $PG(2,q)$; see also Theorem 6.3(i)(1).

Theorem 7.4 (Szőnyi and Weiner [218]) *Let B be a minimal blocking set B of $PG(N,q)$ with respect to k-dimensional subspaces, $q = p^h$, $p > 2$ prime, and assume that $|B| < 3(q^{N-k}+1)/2$. Then any subspace that intersects B intersects it in 1 (mod p) points.*

Regarding the sizes of minimal blocking sets of $PG(N,q)$ with respect to k-dimensional subspaces, the following results are known.

Theorem 7.5 (Szőnyi and Weiner [218]) *Let B be a minimal blocking set of $PG(N,q)$ with respect to k-dimensional subspaces, $q = p^h$, p prime. If $p = 2$ let $|B| < \sqrt{2}q^{N-k}$, otherwise let $|B| < 3(q^{N-k}+1)/2$. Then*

(i) $|B| \in S(q^{N-k})$;
(ii) *if* $p > 2$, *then* $((|B|-1)(q^{N-k})^{N-2}+1) \in S((q^{N-k})^{N-1})$.

In the following theorems, let $l(q,e)$ and $u(q,e)$ denote respectively the smallest and largest value so that for any minimal non-trivial blocking k-set B in $PG(2,q)$, $q=p^h$, p prime, and of size $l(q,e) \leq |B| \leq u(q,e)$, e is the largest integer such that each line intersects B in 1 (mod p^e) points; see also Theorem 6.3(i).

Theorem 7.6 (Szőnyi and Weiner [218]) *Let B be a minimal blocking set in $PG(N,q)$, $q=p^h$, $p>2$ prime, with respect to k-dimensional subspaces. Assume that $|B| < 3(q^{N-k}+1)/2$ and let e be the integer for which*

$$l((q^{N-k})^{N-1}, e) \leq (|B|-1)(q^{N-k})^{N-2}+1 \leq u((q^{N-k})^{N-1}, e).$$

Then each subspace that intersects B in at least one point intersects it in 1 (mod p^e) *points.*

For characterizations of small minimal blocking sets in $PG(N,q)$, the following results have been obtained. The first result is on *Rédei-type* blocking sets with respect to k-dimensional subspaces of $PG(N,q)$. A blocking set with respect to k-dimensional subspaces of $PG(N,q)$, of size q^k+d, is called a *Rédei-type* blocking set with respect to k-dimensional subspaces, when it intersects at least one hyperplane Π_{N-1} in exactly d points. We describe the theorem via the corresponding affine space $AG(N,q) = PG(N,q) \setminus \Pi_{N-1}$.

Theorem 7.7 (Storme and Sziklai [203]) *Let $U \subset AG(N,q)$, $N \geq 3$, $|U| = q^k$. Suppose U determines $|D| \leq \frac{q+3}{2}q^{k-1}+q^{k-2}+q^{k-3}+\ldots+q^2+q$ directions and suppose that U is a $\mathbf{F}_p$-linear set of points, where $q = p^h$, $p>3$. If $N-1 \geq (N-k)h$, then U is a cone with an $(N-1-h(N-k))$-dimensional vertex in Π_{N-1} and with base a $\mathbf{F}_p$-linear point set $U_{(N-k)h}$ of size $q^{(N-k)(h-1)}$, contained in some affine $(N-k)h$-dimensional subspace of $AG(N,q)$.*

Theorem 7.8 (Szőnyi and Weiner [218]) (i) *Let B be a minimal blocking set of $PG(N,q)$ with respect to k-dimensional subspaces, $q=p^h$ and $p>2$ prime. Suppose that $|B| < 3(q^{N-k}+1)/2$ and $h(N-k) \leq N$. Assume that B is not contained in an $(h(N-k)-1)$-dimensional subspace of $PG(N,q)$, then B is projectively equivalent to a subgeometry $PG(h(N-k),p)$.*

(ii) *Let B be a minimal blocking set of $PG(N,q)$ with respect to k-dimensional subspaces, $q=p^h$ and $p>2$ prime. Assume that e is* *an integer,* $1 \leq e \leq h/2$, *and* $h(N-k)/e \leq N$. *Suppose also that* $(|B|-1)(q^{N-k})^{N-2}+1 \leq u((q^{N-k})^{N-1}, e)$. *Then* B *is contained in an* $(\lceil h(N-k)/e \rceil -1)$*-dimensional subspace of* $PG(N,q)$ *or* B *is projectively equivalent to* $PG(h(N-k)/e, p^e)$.

The following theorem presents more specific characterizations. Parts (i), (iii) and (iv) are from Storme and Weiner [209], and Part (ii) from Szőnyi and Weiner [218].

Theorem 7.9 (i) *A non-trivial minimal blocking set of* $PG(N,p^2)$, $p > 3$ *prime, with respect to hyperplanes and of size less than or equal to* $3(p^2+1)/2$, *is a Baer subplane or a planar blocking set of size* $3(p^2+1)/2$.

(ii) *A non-trivial minimal blocking set of* $PG(N,p^3)$, $p > 2$ *prime, with respect to hyperplanes and of size less than* $3(p^3+1)/2$, *is a planar blocking set or a subgeometry* $PG(3,p)$.

(iii) *Let* $s(q^2)$ *be the size of the smallest non-trivial blocking set in* $PG(2,q^2)$ *not equal to a Baer subplane. Let* B *be a minimal blocking set in* $PG(N,q^2)$, $q = p^h$, $h \geq 1$, $p > 3$ *prime,* $N \geq 3$, *with respect to hyperplanes and with* $|B| \leq s(q^2)$. *Then* B *is a line or a minimal planar blocking set of* $PG(N,q^2)$.

(iv) *In* $PG(N,q^3)$, $q = p^h$, $h \geq 1$, p *prime,* $p \geq 7$, $N \geq 3$, *a minimal blocking set* B, *with respect to hyperplanes, of cardinality at most* q^3+q^2+q+1 *is one of the following:*

(1) *a line;*

(2) *a Baer subplane when* q *is square;*

(3) *a minimal blocking set of cardinality* q^3+q^2+1 *in a plane of* $PG(N,q^3)$;

(4) *a minimal blocking set of cardinality* q^3+q^2+q+1 *in a plane of* $PG(N,q^3)$;

(5) *a subgeometry* $PG(3,q)$ *in a solid of* $PG(N,q^3)$.

For an article containing general information on a particular class of blocking sets in $PG(N,q)$ with respect to k-dimensional subspaces, called *linear* blocking sets, we refer to Lunardon [151].

8. n-tracks and almost MDS codes

An *n-track* in $PG(N,q)$ is an n-set of kind $N-1$. In other words, it is a set of n points such that every N of them are linearly independent, but some $N+1$ of them are linearly dependent.

An $[n,k,d]$ code C over $\mathbf{F}_q$ is an *almost MDS code* if its minimum distance d is equal to $n-k$.

N	q	$M_{N-1}(N,q)$
2		q^2+q+1
3	$q>2$	q^2+1
$2q-2$	$q>3$	$2q+1$
$2q-1$	$q>3$	$2q+2$
$\geq 2q$		0

Table 8.1: Exact values of $M_{N-1}(N,q)$

N	q	$M_{N-1}(N,q)\geq$	Conditions
5		$m_3(2,q)$	
	$q=p^h$	$q+\lfloor 2\sqrt{q}\rfloor$	q exceptional
	$q=p^h$	$q+\lfloor 2\sqrt{q}\rfloor+1$	q non-exceptional

Table 8.2: Lower bounds for $M_{N-1}(N,q)$

$N,\ q$	$M_{N-1}(N,q)\leq$	Conditions
	$M_{N-2}(N-1,q)+1$	
$q=p^h$	$q(q-N+3)+1$	$m(N,q)=q+1$ and $N=q-p^l+3$ for some $l\leq h$
$q=p^h$	$q(q-N+3)$	$m(N,q)=q+1$ but $N\neq q-p^l+3$ for all $l\leq h$
$N\geq q$	$2q+N-1$	$N\geq 4$ when $q=2$ and $N\geq 6$ when $q=3$

Table 8.3: Upper bounds for $M_{N-1}(N,q)$

N \ q	2	3	4	5	7	8	9	11
2	7	13	21	31	57	73	91	133
3	8	10	17	26	50	65	82	122
4		11	11	$12-20$	$16-30$	$14-36$	$16-43$	$22-57$
5		12	12	$12-14$	$15-31$	$15-37$	$17-44$	$23-58$
6			9	$10-15$	$13-28$	$14-34$	$17-39$	$18-49$
7			10	$11-16$	$13-20$	$14-35$	$18-40$	$18-50$
8				11	$13-21$	$14-23$	$19-36$	$19-50$
9				12	$13-22$	$14-24$	$20-26$	$20-51$
10					$14-23$	$14-25$	$16-27$	$18-44$
11					$15-24$	$15-26$	$16-28$	$18-32$
12					15	$15-27$	$16-29$	$18-33$
13					16	$16-28$	$17-30$	$18-34$

Table 8.4: Small values of $M_{N-1}(N,q)$

Almost MDS codes and n-tracks are equivalent objects since an n-track in the space $PG(n-k-1,q)$ defines a parity check matrix of an $[n,k,n-k]$ code over $\mathbf{F}_q$.

These two equivalent structures have been studied in detail by De Boer [67] and by Dodunekov and Landjev [70], resulting in Tables 8.1, 8.2, 8.3, which give the known exact values, as well as upper and lower bounds on the maximum number $M_{N-1}(N,q)$ of points of an n-track in $PG(N,q)$. The integer $q=p^h$ is *exceptional* if h is odd, $h \geq 3$ and p divides $\lfloor 2\sqrt{q} \rfloor$.

Table 8.4 gives values for $M_{N-1}(N,q)$ for small N and q; see De Boer [67].

9. Minihypers

A subconfiguration closely related to (multiple) blocking sets with respect to k-dimensional subspaces of $PG(N,q)$, is the subconfiguration *minihyper*.

An $\{n,m;N,q\}$-*minihyper* K is the complement of a $(\theta_N - n; \theta_{N-1} - m, N-1; N,q)$-set; that is, K is a set of n points in $PG(N,q)$, $N \geq 2, n \geq 1$, such that (a) $|K \cap \Pi_{N-1}| \geq m$ for every hyperplane Π_{N-1}, and (b) $|K \cap \Pi_{N-1}| = m$ for some hyperplane Π_{N-1}.

When constructing a code, from an economical point of view, it is desirable to obtain an $[n,k,d]$ code C over $\mathbf{F}_q$ whose length n is minimal for given values of k,d, and q. The Griesmer bound [104, 202] shows that if there is an $[n,k,d]$ code C over $\mathbf{F}_q$ for given k,d, and q, then $n \geq \sum_{i=0}^{k-1} \lceil d/q^i \rceil$.

Theorem 9.1 (Hamada [111, 112]) *For $k \geq 3$ and $1 \leq d < q^{k-1}$, there is a one-to-one correspondence between the set of all non-equivalent $[n,k,d]$ codes C over $\mathbf{F}_q$ meeting the Griesmer bound and the set of all $\{\theta_{k-1}-n, \theta_{k-2}-n+d; k-1,q\}$-minihypers.*

There are a number of articles studying this correspondence. For surveys on recent work, see Hamada [113, 114, 115]. The general results are presented in the following theorems. We also wish to remark that this notion of minihypers is linked to that of *anticodes*, introduced by Farrell [84].

Theorem 9.2 (Hamada and Helleseth [116], Hamada and Maekawa [117]) *Let t,q,h and λ_i, $i=1,\ldots,h$, be any integers such that $t \geq 2, h \geq 1$, $q > (h-1)^2$ and $0 \leq \lambda_1 \leq \cdots \leq \lambda_h < t$.*

(i) *If $t < \lambda_{h-1} + \lambda_h + 1$, there is no $\{\sum_{i=1}^h \theta_{\lambda_i}, \sum_{i=1}^h \theta_{\lambda_i - 1}; t,q\}$-minihyper.*

(ii) *If* $t \geq \lambda_{h-1} + \lambda_h + 1$, *then* F *is a* $\{\sum_{i=1}^{h} \theta_{\lambda_i}, \sum_{i=1}^{h} \theta_{\lambda_i - 1}; t, q\}$*-minihyper if and only if* F *is the pairwise disjoint union of a subspace* $PG(\lambda_1, q)$, *a subspace* $PG(\lambda_2, q), \ldots, PG(\lambda_h, q)$.

Theorem 9.3 (Barát and Storme [20]) *If* F *is a* $\{\varepsilon_0 + \varepsilon_1(q+1), \epsilon_1; t, q\}$*-minihyper of* $PG(t,q)$, $t \geq 3$, $q = p^h$, p *prime*, $q \geq 2^{12}$, $c_2 = c_3 = 2^{-1/3}$ *and* $c_p = 1$ *when* $p > 3$, *with* $\varepsilon_1 < c_p q^{1/6}$ *and with* $\varepsilon_0 + \varepsilon_1 < c_p q^{2/3} - (\varepsilon_1 - 1)(\varepsilon_1 - 2)/2$, *then* F *contains the union of a collection of* s *disjoint lines and/or Baer subplanes.*

Theorem 9.4 (Ferret and Storme [87]) *Assume that the set* F *is a* $\{\sum_{i=0}^{s} \epsilon_i \theta_i, \sum_{i=0}^{s} \epsilon_i \theta_{i-1}; t, q\}$*-minihyper, where* $\sum_{i=0}^{s} \epsilon_i < 2\sqrt{q}$, q *square*, $q = p^f$, p *prime, with* $q > 2^{18}$ *when* $p > 3$ *and* $q > 2^{20}$ *when* $p = 2, 3$. *Then* F *consists of the disjoint union of one of the following:*

(i) ϵ_s *spaces* $PG(s,q), \epsilon_{s-1}$ *spaces* $PG(s-1,q), \ldots, \epsilon_0$ *points;*

(ii) *one subgeometry* $PG(2l+1, \sqrt{q})$, *for some integer* l *with* $1 \leq l \leq s$, ϵ_s *spaces* $PG(s,q), \ldots, \epsilon_{l+1}$ *spaces* $PG(l+1,q), \epsilon_l - \sqrt{q} - 1$ *spaces* $PG(l,q), \epsilon_{l-1}$ *spaces* $PG(l-1,q), \ldots, \epsilon_0$ *points;*

(iii) *one subgeometry* $PG(2l, \sqrt{q})$, *for some integer* l *with* $1 \leq l \leq s$, ϵ_s *spaces* $PG(s,q), \ldots, \epsilon_{l+1}$ *spaces* $PG(l+1,q), \epsilon_l - 1$ *spaces* $PG(l,q), \epsilon_{l-1} - \sqrt{q}$ *spaces* $PG(l-1,q), \epsilon_{l-2}$ *spaces* $PG(l-2,q), \ldots, \epsilon_0$ *points.*

Theorem 9.5 (Ferret and Storme [87]) *Assume that the set* F *is a* $\{\sum_{i=0}^{s} \epsilon_i \theta_i, \sum_{i=0}^{s} \epsilon_i \theta_{i-1}; t, q\}$*-minihyper,* $t \geq 3$, *where*

(a) $\sum_{i=0}^{s} \epsilon_i \leq q^{6/9}/(1 + q^{1/9})$, $q = p^f$, f *odd*, p *prime*, $p > 3$, $q \geq 2^{12}$;

(b) $\sum_{i=0}^{s} \epsilon_i \leq c_p q^{5/9}$, $q = p^f$, f *odd*, $p = 2, 3$, $q > 2^{12}$, $c_p = 2^{-1/3}$.

Then F *is the disjoint union of* ϵ_s *spaces* $PG(s,q)$, ϵ_{s-1} *spaces* $PG(s-1,q)$, $\ldots$, ϵ_0 *points.*

Theorem 9.6 (Govaerts and Storme [102]) *Let* $q > 2$ *and* $\delta < \epsilon$, *where* $q + \epsilon$ *is the size of the smallest non-trivial blocking sets in* $PG(2,q)$. *If* F *is a* $\{\delta\theta_\mu, \delta\theta_{\mu-1}; N, q\}$*-minihyper satisfying* $\mu \leq N - 1$, *then* F *is the disjoint union of* δ *subspaces* $PG(\mu, q)$.

Theorem 9.7 (Govaerts and Storme [103]) *A* $\{\delta\theta_\mu, \delta\theta_{\mu-1}; N, q\}$*-minihyper* F, $q > 16$ *square*, $\delta < q^{5/8}/\sqrt{2} + 1$, $2\mu + 1 \leq N$, *is a unique disjoint union of subspaces* $PG(\mu, q)$ *and subgeometries* $PG(2\mu+1, \sqrt{q})$.

References

[1] L.M. Abatangelo, Calotte complete in uno spazio di Galois $PG(3,q)$, q pari, *Atti Sem. Mat. Fis. Univ. Modena* **29** (1980), 215–221.

[2] V. Abatangelo, A class of complete $((q+8)/3)$-arcs in $\mathrm{PG}(2, 2^h)$ ($h \geq 6$ even), *Ars Combin.* **16** (1983), 103–111.

[3] V. Abatangelo, Un nuovo procedimento per la costruzione di calotte di $PG(3,q)$, q pari, *Rend. Sem. Mat. Brescia* **7** (1984), 19–25.

[4] V. Abatangelo, G. Korchmáros and B. Larato, Classification of maximal caps in $PG(3,5)$ different from elliptic quadrics, *J. Geom.* **57** (1996), 9–19.

[5] A.H. Ali, Classification of arcs in the plane of order 13, Ph.D. Thesis, University of Sussex, 1993.

[6] A.H. Ali, J.W.P. Hirschfeld and H. Kaneta, The automorphism group of a complete $(q-1)$-arc in $PG(2,q)$, *J. Combin. Des.* **2** (1994), 131–145.

[7] A.H. Ali, J.W.P. Hirschfeld and H. Kaneta, On the size of arcs in projective spaces, *IEEE Trans. Inform. Theory* **41** (1995), 1649–1656.

[8] S. Ball, On sets of points in finite planes, Ph.D. Thesis, University of Sussex, 1994.

[9] S. Ball, On the size of a triple blocking set in $PG(2,q)$, *European J. Combin.* **17** (1996), 427–435.

[10] S. Ball, Multiple blocking sets and arcs in finite planes, *J. London Math. Soc.* **54** (1996), 581–593.

[11] S. Ball, On small complete arcs in a finite plane, *Discrete Math.* **174** (1997), 29–34.

[12] S. Ball, Some notes on arcs and multi-arcs in projective and affine planes, unpublished manuscript (2000).

[13] S. Ball and A. Blokhuis, On the size of a double blocking set in $PG(2,q)$, *Finite Fields Appl.* **2** (1996), 125–137.

[14] S. Ball and A. Blokhuis, An easier proof of the maximal arcs conjecture, *Proc. Amer. Math. Soc.* **126** (1998), 3377–3380.

[15] S. Ball and A. Blokhuis, On the incompleteness of (k,n)-arcs in Desarguesian planes of order q where n divides q, *Geom. Dedicata* **74** (1999), 325–332.

[16] S. Ball, A. Blokhuis and F. Mazzocca, Maximal arcs in Desarguesian planes of odd order do not exist, *Combinatorica* **17** (1997), 31–47.

[17] S. Ball, R. Hill, I.N. Landjev and H.N. Ward, On $(q^2+q+2, q+2)$-arcs in the projective plane $PG(2,q)$, in preparation.

[18] J. Barát, A. Del Fra, S. Innamorati and L. Storme, Minimal blocking sets in $PG(2,8)$ and maximal partial spreads in $PG(3,8)$, in preparation.

[19] J. Barát, Y. Edel, R. Hill, C. Jones, I.N. Landjev and L. Storme, Caps in $PG(5,3)$ and $PG(6,3)$, in preparation.

[20] J. Barát and L. Storme, Multiple blocking sets in $PG(n,q)$, $n \geq 3$, submitted.

[21] A. Barlotti, Un' estensione del teorema di Segre–Kustaanheimo, *Boll. Un. Mat. Ital.* **10** (1955), 96–98.

[22] A. Barlotti, Su $\{k;n\}$-archi di un piano lineare finito, *Boll. Un. Mat. Ital.* **11** (1956), 553–556.

[23] A. Barlotti, Some topics in finite geometrical structures, *Institute of Statistics Mimeo Series* **439**, Univ. of North Carolina, 1965.

[24] A. Beutelspacher, Blocking sets and partial spreads in finite projective spaces, *Geom. Dedicata* **9** (1980), 130–157.

[25] J. Bierbrauer, (k,n)-arcs of maximal size in the plane of order 8, Unpublished manuscript (1988).

[26] J. Bierbrauer and Y. Edel, A family of caps in projective 4-space in odd characteristic, *Finite Fields Appl.* **6** (2000), 283–293.

[27] J. Bierbrauer and Y. Edel, Bounds on affine caps, *J. Combin. Des.*, to appear.

[28] A. Blokhuis, Note on the size of a blocking set in $PG(2,p)$, *Combinatorica* **14** (1994), 111–114.

[29] A. Blokhuis, On multiple nuclei and a conjecture of Lunelli and Sce, *Bull. Belg. Math. Soc.* **3** (1994), 349–353.

[30] A. Blokhuis, Blocking sets in Desarguesian planes, *Paul Erdős is Eighty*, Volume 2 (eds. D. Miklós, V.T. Sós and T. Szőnyi), Bolyai Soc. Math. Studies **2** (1996), 133–155.

[31] A. Blokhuis, S. Ball, A.E. Brouwer, L. Storme and T. Szőnyi, On the number of slopes of the graph of a function defined on a finite field, *J. Combin. Theory Ser. A* **86** (1999), 187–196.

[32] A. Blokhuis and K. Metsch, Large minimal blocking sets, strong representative systems, and partial unitals, *Finite Geometry and Combinatorics* (eds. F. De Clerck *et al.*), London Math. Soc. Lecture Note Series **191**, Cambridge University Press, Cambridge, 1993, 37–52.

[33] A. Blokhuis, R. Pellikaan and T. Szőnyi, Blocking sets of almost Rédei type, *J. Combin. Theory Ser. A* **78** (1997), 141–150.

[34] A. Blokhuis, L. Storme and T. Szőnyi, Lacunary polynomials, multiple blocking sets and Baer subplanes, *J. London Math. Soc.* **60** (1999), 321–332.

[35] M. Bokler, Minimal blocking sets in projective spaces of square order, *Des. Codes Cryptogr.*, to appear.

[36] M. Bokler and K. Metsch, On the smallest minimal blocking sets in projective spaces generating the whole space, *Contrib. Algebra Geom.*, to appear.

[37] E. Boros and T. Szőnyi, On the sharpness of a theorem of B. Segre, *Combinatorica* **6** (1986), 261–268.

[38] R.C. Bose, Mathematical theory of the symmetrical factorial design, *Sankhyā* **8** (1947), 107–166.

[39] R.C. Bose, On the application of finite projective geometry for deriving a certain series of balanced Kirkman arrangements, Golden Jubilee Commemoration Volume (1958–1959), *Calcutta Math. Soc.* (1959), 341–354.

[40] R.C. Bose, On some connections between the design of experiments and information theory, *Bull. Inst. Internat. Statist.* **38** (1961), 257–271.

[41] R.C. Bose and R.C. Burton, A characterization of flat spaces in a finite projective geometry and the uniqueness of the Hamming and the MacDonald codes, *J. Combin. Theory* **1** (1966), 96–104.

[42] R.C. Bose and J.N. Srivastava, On a bound useful in the theory of factorial designs and error correcting codes, *Ann. Math. Statist.* **35** (1964), 408–414.

[43] A.A. Bruen, Baer subplanes and blocking sets, *Bull. Amer. Math. Soc.* **76** (1970), 342–344.

[44] A.A. Bruen, Blocking sets in finite projective planes, *SIAM J. Appl. Math.* **21** (1971), 380–392.

[45] A.A. Bruen, Polynomial multiplicities over finite fields and intersection sets, *J. Combin. Theory Ser. A* **60** (1992), 19–33.

[46] A.A. Bruen and J.A. Thas, Blocking sets, *Geom. Dedicata* **6** (1977), 193–203.

[47] K.A. Bush, Orthogonal arrays of index unity, *Ann. Math. Statist.* **23** (1952), 426–434.

[48] A.R. Calderbank and P.C. Fishburn, Maximal three-independent subsets of $\{0,1,2\}^n$, *Des. Codes Cryptogr.* **4** (1994), 203–211.

[49] L.R.A. Casse, A solution to Beniamino Segre's 'Problem $I_{r,q}$' for q even, *Atti Accad. Naz. Lincei Rend.* **46** (1969), 13–20.

[50] L.R.A. Casse and D.G. Glynn, The solution to Beniamino Segre's problem $I_{r,q}, r=3, q=2^h$, *Geom. Dedicata* **13** (1982), 157–164.

[51] L.R.A. Casse and D.G. Glynn, On the uniqueness of $(q+1)_4$-arcs of $PG(4,q), q = 2^h, h \geq 3$, *Discrete Math.* **48** (1984), 173–186.

[52] J.M. Chao, On the size of a cap in $PG(n,q)$ with q even and $n \geq 3$, *Geom. Dedicata* **74** (1999), 91–94.

[53] J.M. Chao, On the maximum size of a $(k, \sqrt{q}+1)$-arc in $PG(2,q)$, with q square, *Atti Sem. Mat. Fis. Univ. Modena* **47** (1999), 529–531.

[54] J.M. Chao and H. Kaneta, A complete 24-arc in $PG(2,29)$ with automorphism group $PSL(2,7)$, *Rend. Mat. Appl.* **16** (1996), 537–544.

[55] J.M. Chao and H. Kaneta, Classical arcs in $PG(r,q)$ for $11 \leq q \leq 19$, *Discrete Math.* **174** (1997), 87–94.

[56] J.M. Chao and H. Kaneta, Classical arcs in $PG(r,q)$ for $23 \leq q \leq 27$, submitted.

[57] W.E. Cherowitzo, Hyperovals in Desarguesian planes of even order, *Ann. Discrete Math.* **37** (1988), 87–94.

[58] W.E. Cherowitzo, α-flocks and hyperovals, *Geom. Dedicata* **72** (1998), 221–246.

[59] W.E. Cherowitzo, C.M. O'Keefe and T. Penttila, A unified construction of finite geometries in characteristic two, submitted.

[60] W.E. Cherowitzo, T. Penttila, I. Pinneri and G.F. Royle, Flocks and ovals, *Geom. Dedicata* **60** (1996), 17–37.

[61] Z. Ciechanowicz, Ph.D. Thesis, University of London, 1980.

[62] G.D. Cohen, M.G. Karpovsky, H.F. Mattson, and J.R. Schatz, Covering radius: survey and recent results, *IEEE Trans. Inform. Theory* **31** (1985), 328–343.

[63] G.D. Cohen, S.N. Litsyn, H. Honkala and H.F. Mattson, Weighted coverings and packings, *IEEE Trans. Inform. Theory* **41** (1995), 1856–1867.

[64] A. Cossu, Su alcune proprietà dei $\{k,n\}$-archi di un piano proiettivo sopra un corpo finito, *Rend. Mat. e Appl.* **20** (1961), 271–277.

[65] A.A. Davydov and P.R.J. Östergard, Recursive constructions of complete caps, *J. Statist. Planning Infer.*, to appear.

[66] A.A. Davydov and L.M. Tombak, Quasiperfect linear binary codes with distance 4 and complete caps in projective geometry, *Problems Inform. Transmission* **25** (1990), 265–275.

[67] M.A. de Boer, Almost MDS codes, *Des. Codes Cryptogr.* **9** (1996), 143–155.

[68] A. Del Fra and S. Innamorati, private communication (2000).

[69] R.H.F. Denniston, Some maximal arcs in finite projective planes, *J. Combin. Theory* **6** (1969), 317–319.

[70] S.M. Dodunekov and I.N. Landjev, On near-MDS codes, *J. Geom.* **54** (1995), 30–43.

[71] Y. Edel, Private communication.

[72] Y. Edel and J. Bierbrauer, 41 is the largest size of a cap in $PG(4,4)$, *Des. Codes Cryptogr.* **16** (1999), 51–60.

[73] Y. Edel and J. Bierbrauer, Recursive constructions for large caps, *Bull. Belg. Math. Soc. Simon Stevin* **6** (1999), 249–258.

[74] Y. Edel and J. Bierbrauer, A family of caps in projective 4-space in characteristic 2, *Congr. Numer.* **141** (1999), 191–202.

[75] Y. Edel and J. Bierbrauer, Large caps in small spaces, *Des. Codes Cryptogr.*, to appear.

[76] Y. Edel, L. Storme and P. Sziklai, New upper bounds for the sizes of caps in $PG(N,5)$ and $PG(N,7)$, in preparation.

[77] G. Faina, Complete caps having less than $(q^2+1)/2$ points in common with an elliptic quadric of $PG(3,q)$, q odd, *Rend. Mat.* **8** (1988), 277–281.

[78] G. Faina, S. Marcugini, A. Milani and F. Pambianco, The spectrum of the values k for which there exists a complete k-arc in $PG(2,q)$ for $q \leq 23$, *Ars Combin.* **47** (1997), 3–11.

[79] G. Faina, S. Marcugini, A. Milani and F. Pambianco, The sizes k of the complete k-caps in $PG(n,q)$, for small q and $3 \leq n \leq 5$, *Ars Combin.* **50** (1998), 225–243.

[80] G. Faina and F. Pambianco, Fasci di coniche e archi completi, preprint (1995).

[81] G. Faina and F. Pambianco, A class of complete k-caps in $PG(3,q)$ for q an odd prime, *J. Geom.* **57** (1996), 93–105.

[82] G. Faina and F. Pambianco, Small complete k-caps in $PG(r,q), r \geq 3$, *Discrete Math.* **174** (1997), 114–123.

[83] G. Faina and F. Pambianco, On some 10-arcs for deriving the minimum order for complete arcs in small projective planes, *Discrete Math.* **208/209** (1999), 261–271.

[84] P.G. Farrell, Linear binary anticodes, *Electronics Letters* **6** (1970), 419–421.

[85] G. Fellegara, Gli ovaloidi di uno spazio tridimensionale di Galois di ordine 8, *Atti Accad. Naz. Lincei Rend.* **32** (1962), 170–176.

[86] S. Ferret and L. Storme, unpublished manuscript (2001).

[87] S. Ferret and L. Storme, Minihypers and linear codes meeting the Griesmer bound: improvements to results of Hamada, Helleseth and Maekawa, submitted.

[88] J.C. Fisher, J.W.P. Hirschfeld and J.A. Thas, Complete arcs in planes of square order, *Ann. Discrete Math.* **30** (1986), 243–250.

[89] R.A. Fisher, The theory of confounding in factorial experiments in relation to the theory of groups, *Ann. Eugen. London* **11** (1942), 341–353.

[90] R.A. Fisher, A system of confounding for factors with more than two alternatives giving completely orthogonal cubes and higher powers, *Ann. Eugen. London* **12** (1945), 283–290.

[91] E.M. Gabidulin, A.A. Davydov and L.M. Tombak, Linear codes with covering radius 2 and other new covering codes, *IEEE Trans. Inform. Theory* **37** (1991), 219–224.

[92] A. Gács, A remark on blocking sets of almost Rédei type, *J. Geom.* **60** (1997), 65–73.

[93] A. Gács, On a generalization of Rédei's theorem, submitted.

[94] A. Gács, The Rédei method applied to finite geometry, Ph.D. Thesis, Eötvös Loránd University, Budapest, Hungary, 1997.

[95] A. Gács, P. Sziklai and T. Szőnyi, Two remarks on blocking sets and nuclei in planes of prime order, *Des. Codes Cryptogr.* **10** (1997), 29–39.

[96] A. Gács and T. Szőnyi, Double blocking sets and Baer-subplanes of odd square order, unpublished manuscript (1995).

[97] M. Giulietti, Small complete caps in $PG(2,q)$, for q an odd square, *J. Geom.* **69** (2000), 110–116.

[98] M. Giulietti, On plane arcs contained in cubic curves, submitted.

[99] D.G. Glynn, Two new sequences of ovals in finite Desarguesian planes of even order, *Combinatorial Mathematics X* (ed. L.R.A. Casse), Lecture Notes in Mathematics **1036**, Springer, 1983, 217–229.

[100] D.G. Glynn, The non-classical 10-arc of $PG(4,9)$, *Discrete Math.* **59** (1986), 43–51.

[101] C.E. Gordon, Orbits of arcs in projective spaces, *Finite Geometry and Combinatorics*, London Math. Soc. Lecture Note Series **191**, Cambridge University Press, Cambridge, 1993, 161–174.

[102] P. Govaerts and L. Storme, On a particular class of minihypers of $PG(N,q)$ and its applications. I: The result for general q, submitted.

[103] P. Govaerts and L. Storme, On a particular class of minihypers of $PG(N,q)$ and its applications. II: Improvements for q square, submitted.

[104] J.H. Griesmer, A bound for error-correcting codes, *IBM J. Res. Develop.* **4** (1960), 532–542.

[105] P. Gronchi, I blocking sets ed il packing problem, *Boll. Un. Mat. Ital.* **7-A** (1993), 227–236.

[106] B.R. Gulati and E.G. Kounias, On bounds useful in the theory of symmetrical factorial designs, *J. Roy. Statist. Soc. Ser. B* **32** (1970), 123–133.

[107] É. Hadnagy, Small complete arcs in $PG(2,p)$, *Finite Fields Appl.* **5** (1999), 1–12.

[108] É. Hadnagy and T. Szőnyi, On embedding partial unitals and large (k,n)-arcs, *Ars Combin.*, to appear.

[109] H.-R. Halder and W. Heise, On the existence of finite chain-m-structures and k-arcs in finite projective space, *Geom. Dedicata* **3** (1974), 483–486.

[110] M. Hall, Ovals in the Desarguesian plane of order 16, *Ann. Mat. Pura Appl.* **102** (1975), 159–176.

[111] N. Hamada, Characterization resp. nonexistence of certain q-ary linear codes attaining the Griesmer bound, *Bull. Osaka Women's Univ.* **22** (1985), 1–47.

[112] N. Hamada, Characterization of minihypers in a finite projective geometry and its applications to error-correcting codes, *Bull. Osaka Women's Univ.* **24** (1987), 1–24.

[113] N. Hamada, A characterization of some $[n, k, d; q]$-codes meeting the Griesmer bound using a minihyper in a finite projective geometry, *Discrete Math.* **116** (1993), 229–268.

[114] N. Hamada, The nonexistence of some quaternary linear codes meeting the Griesmer bound and the bounds for $n_4(5, d)$, $1 \leq d \leq 256$, *Math. Japonica* **43** (1996), 7–21.

[115] N. Hamada, The nonexistence of some ternary linear codes of dimension 6 and the bounds for $n_3(6, d)$, $1 \leq d \leq 243$, *Math. Japonica* **43** (1996), 577–593.

[116] N. Hamada and T. Helleseth, A characterization of some q-ary codes ($q > (h-1)^2$, $h \geq 3$) meeting the Griesmer bound, *Math. Japonica* **38** (1993), 925–940.

[117] N. Hamada and T. Maekawa, A characterization of some q-ary codes ($q > (h-1)^2, h \geq 3$) meeting the Griesmer bound: Part 2, *Math. Japonica* **46** (1997), 241–252.

[118] R. Hartshorne, *Algebraic Geometry*, Springer, New York, 1977.

[119] U. Heim, On t-blocking sets in projective spaces, unpublished manuscript (1994).

[120] R. Hill, On the largest size of cap in $S_{5,3}$, *Atti Accad. Naz. Lincei Rend.* **54** (1973), 378–384.

[121] R. Hill, Caps and codes, *Discrete Math.* **22** (1978), 111–137.

[122] R. Hill, On Pellegrino's 20-caps in $S_{4,3}$, *Combinatorial Geometries and their Applications (Rome 1981)*, *Ann. Discrete Math.* **18** (1983), 443–448.

[123] R. Hill, Some problems concerning (k, n)-arcs in finite projective planes, *Rend. Sem. Mat. Brescia* **7** (1984), 367–383.

[124] R. Hill and J.R.M. Mason, On (k, n)-arcs and the falsity of the Lunelli–Sce conjecture, *Finite Geometries and Designs*, London Math. Soc. Lecture Note Series **49**, Cambridge University Press, Cambridge, 1981, 153–168.

[125] J.W.P. Hirschfeld, *Projective Geometries over Finite Fields*, Oxford University Press, Oxford, 1979.

[126] J.W.P. Hirschfeld, Caps in elliptic quadrics, *Ann. Discrete Math.* **18** (1983), 449–466.

[127] J.W.P. Hirschfeld, Maximum sets in finite projective spaces, *Surveys in Combinatorics,* London Math. Soc. Lecture Note Series **82**, Cambridge University Press, Cambridge, 1983, 55–76.

[128] J.W.P. Hirschfeld, *Finite Projective Spaces of Three Dimensions,* Oxford University Press, Oxford, 1985.

[129] J.W.P. Hirschfeld, Complete arcs, *Discrete Math.* **174** (1997), 177–184.

[130] J.W.P. Hirschfeld, *Projective Geometries over Finite Fields* (Second Edition), Oxford University Press, Oxford, 1998.

[131] J.W.P. Hirschfeld and G. Korchmáros, On the embedding of an arc into a conic in a finite plane, *Finite Fields Appl.* **2** (1996), 274-292.

[132] J.W.P. Hirschfeld and G. Korchmáros, On the number of rational points on an algebraic curve over a finite field, *Bull. Belg. Math. Soc. Simon Stevin* **5** (1998), 313–340.

[133] J.W.P. Hirschfeld and G. Korchmáros, Arcs and curves over a finite field, *Finite Fields Appl.* **5** (1999), 393–408.

[134] J.W.P. Hirschfeld and L. Storme, The packing problem in statistics, coding theory and finite projective spaces, *J. Statist. Planning Infer.* **72** (1998), 355–380.

[135] J.W.P. Hirschfeld and T. Szőnyi, Constructions of large arcs and blocking sets in finite planes, *European J. Combin.* **12** (1991), 499–511.

[136] J.W.P. Hirschfeld and J.A. Thas, Linear independence in finite spaces, *Geom. Dedicata* **23** (1987), 15–31.

[137] J.W.P. Hirschfeld and J.A. Thas, *General Galois Geometries*, Oxford University Press, Oxford, 1991.

[138] S. Innamorati, The non-existence of certain large minimal blocking sets, *Mitt. Math. Sem. Giessen* **235** (1998), 1–23.

[139] S. Innamorati and A. Maturo, On irreducible blocking sets in projective planes, *Ratio Math.* **2** (1991), 151–155.

[140] H. Kaneta, private communication (1995).

[141] H. Kaneta, private communication (1998).

[142] H. Kaneta and T. Maruta, An elementary proof and an extension of Thas' theorem on k-arcs, *Math. Proc. Cambridge Philos. Soc.* **105** (1989), 459–462.

[143] B.C. Kestenband, A family of complete arcs in finite projective planes, *Colloq. Math.* **57** (1989), 59–67.

[144] J.H. Kim and V.H. Vu, Small complete arcs in projective planes, submitted.

[145] G. Korchmáros, New examples of complete k-arcs in $PG(2,q)$, *European J. Combin.* **4** (1983), 329–334.

[146] P. Kustaanheimo, A note on a finite approximation of the Euclidean plane geometry, *Soc. Sci. Fenn. Comment. Phys.-Math.* **15**, No. 19 (1950).

[147] P. Lisonek, *Computer-assisted studies in algebraic combinatorics*, Ph.D. Thesis, Research Institute for Symbolic Computation, J. Kepler University Linz, 1994.

[148] L. Lombardo-Radice, Sul problema dei k-archi completi di $S_{2,q}$, *Boll. Un. Mat. Ital.* **11** (1956), 178–181.

[149] L. Lovász and A. Schrijver, Remarks on a theorem of Rédei, *Studia Scient. Math. Hungar.* **16** (1981), 449–454.

[150] L. Lovász and T. Szőnyi, Multiple blocking sets and algebraic curves, abstract from *Finite Geometry and Combinatorics* (Third International Conference at Deinze (Belgium), May 18-24, 1997).

[151] G. Lunardon, Normal spreads, *Geom. Dedicata* **75** (1999), 245–261.

[152] L. Lunelli and M. Sce, Considerazione aritmetiche e risultati sperimentali sui $\{K;n\}_q$-archi, *Ist. Lombardo Accad. Sci. Rend. A* **98** (1964), 3–52.

[153] F.J. MacWilliams and N.J.A. Sloane, *The Theory of Error-Correcting Codes*, North-Holland, Amsterdam, 1977.

[154] S. Marcugini, A. Milani and F. Pambianco, Maximal $(n,3)$-arcs in $PG(2,11)$, *Discrete Math.* **208/209** (1999), 421–426.

[155] S. Marcugini, A. Milani and F. Pambianco, Maximal $(n,3)$-arcs in $PG(2,13)$, in preparation.

[156] S. Marcugini, A. Milani and F. Pambianco, Complete arcs in $PG(2,25)$: the spectrum of the sizes and the classification of the smallest complete arcs, in preparation.

[157] S. Marcugini, A. Milani and F. Pambianco, Minimal complete arcs in $PG(2,q)$, $q \leq 29$, in preparation.

[158] T. Maruta and H. Kaneta, On the uniqueness of $(q+1)$-arcs of $PG(5,q)$, $q = 2^h, h \geq 4$, *Math. Proc. Cambridge Philos. Soc.* **110** (1991), 91–94.

[159] J.R.M. Mason, On the maximum sizes of certain (k,n)-arcs in finite projective geometries, *Math. Proc. Cambridge Philos. Soc.* **91** (1982), 153–170.

[160] J.R.M. Mason, A class of $((p^n-p^m)(p^n-1), p^n-p^m)$-arcs in $PG(2,p^n)$, *Geom. Dedicata* **15** (1984), 355–361.

[161] R. Meshulam, On subsets of finite abelian groups with no 3-term arithmetic progression, *J. Combin. Theory Ser. A* **71** (1995), 168–172.

[162] A. Milani, S. Marcugini and F. Pambianco, Computer aided determination of very small arcs in $PG(2,q)$, $q \leq 67$, preprint.

[163] A.C. Mukhopadhyay, Lower bounds on $m_t(r,s)$, *J. Combin. Theory Ser. A* **25** (1978), 1–13.

[164] G.P. Nagy and T. Szőnyi, Caps in finite projective spaces of odd order, *J. Geom.* **59** (1997), 103–113.

[165] C.M. O'Keefe and T. Penttila, Ovoids of $PG(3,16)$ are elliptic quadrics, *J. Geom.* **38** (1990), 95–106.

[166] C.M. O'Keefe and T. Penttila, Hyperovals in $PG(2,16)$, *European J. Combin.* **12** (1991), 51–59.

[167] C.M. O'Keefe and T. Penttila, A new hyperoval in $PG(2,32)$, *J. Geom.* **44** (1992), 117–139.

[168] C.M. O'Keefe and T. Penttila, Ovoids of $PG(3,16)$ are elliptic quadrics, II, *J. Geom.* **44** (1992), 140–159.

[169] C.M. O'Keefe, T. Penttila and G.F. Royle, Classification of ovoids in $PG(3,32)$, *J. Geom.* **50** (1994), 143–150.

[170] C.M. O'Keefe and L. Storme, Arcs fixed by A_5 and A_6, *J. Geom.* **55** (1996), 123–138.

[171] P.R.J. Östergard, Computer search for small complete caps, *J. Geom.* **69** (2000), 172–179.

[172] S. Packer, Sets of odd type and caps in Galois geometries of order four, D.Phil. Thesis, University of Sussex, 1995.

[173] F. Pambianco and L. Storme, Small complete caps in spaces of even characteristic, *J. Combin. Theory Ser. A* **75** (1996), 70–84.

[174] F. Pambianco and L. Storme, unpublished manuscript (1995).

[175] G. Panella, Caratterizzazione delle quadriche di uno spazio (tridimensionale) lineare sopra un corpo finito, *Boll. Un. Mat. Ital.* **10** (1955), 507–513.

[176] S.E. Payne, A new infinite family of generalized quadrangles, *Congr. Numer.* **49** (1985), 115–128.

[177] S.E. Payne, A tensor product action on q-clan generalized quadrangles with $q = 2^e$, *Linear Algebra Appl.* **226–228** (1995), 115–137.

[178] S.E. Payne, T. Penttila and I. Pinneri, Isomorphisms between Subiaco q-clan geometries, *Bull. Belg. Math. Soc.* **2** (1995), 197–222.

[179] S.E. Payne, T. Penttila and G.F. Royle, Constructing a cyclic q-clan, *Mostly Finite Geometries* (ed. N.L. Johnson), Marcel Dekker, New York, 1997, 365–378.

[180] G. Pellegrino, Sul massimo ordine delle calotte in $S_{4,3}$, *Matematiche* (Catania) **25** (1970), 1–9.

[181] G. Pellegrino, Sulle calotte complete, non ovaloidi, dello spazio $PG(3,q)$, q dispari, *Rend. Circ. Mat. Palermo* **47** (1998), 141–168.

[182] T. Penttila, private communication (2000).

[183] T. Penttila and C.E. Praeger, Ovoids and translation ovals, *Proc. London Math. Soc.* **56** (1997), 607–624.

[184] T. Penttila and G.F. Royle, Classification of hyperovals in $PG(2,32)$, *J. Geom.* **50** (1994), 151–158.

[185] T. Penttila and G.F. Royle, On hyperovals in small projective planes, *J. Geom.* **54** (1995), 91–104.

[186] T. Penttila and G.F. Royle, private communication (1995).

[187] P. Polito and O. Polverino, On small blocking sets, *Combinatorica* **18** (1998), 133–137.

[188] O. Polverino, Small blocking sets and complete k-arcs in $PG(2,p^3)$, *Discrete Math.* **208/209** (1999), 469–476.

[189] O. Polverino, Small blocking sets in $PG(2,p^3)$, *Des. Codes Cryptogr.* **20** (2000), 319–324.

[190] O. Polverino and L. Storme, Small blocking sets in $PG(2,q^3)$, submitted.

[191] B. Qvist, Some remarks concerning curves of the second degree in a finite plane, *Ann. Acad. Sci. Fenn. Ser. A* **134** (1952).

[192] A.R. Sadeh, The classification of k-arcs and cubic surfaces with twenty-seven lines over the field of eleven elements, D.Phil. Thesis, University of Sussex, 1984.

[193] M. Scipioni, Sugli archi completi nei piani desarguesiani, Tesi di laurea, University of Rome "La Sapienza", 1990.

[194] B. Segre, Sulle ovali nei piani lineari finiti, *Atti Accad. Naz. Lincei Rend.* **17** (1954), 1–2.

[195] B. Segre, Ovals in a finite projective plane, *Canad. J. Math.* **7** (1955), 414–416.

[196] B. Segre, Curve razionali normali e k-archi negli spazi finiti, *Ann. Mat. Pura Appl.* **39** (1955), 357–379.

[197] B. Segre, Sui k-archi nei piani finiti di caratteristica due, *Rev. Math. Pures Appl.* **2** (1957), 289–300.

[198] B. Segre, Le geometrie di Galois, *Ann. Mat. Pura Appl.* **48** (1959), 1–97.

[199] B. Segre, *Lectures on Modern Geometry*, Cremonese, Rome, 1961.

[200] B. Segre, Ovali e curve σ nei piani di Galois di caratteristica due, *Atti dell' Accad. Naz. Lincei Rend.* (8) **32** (1962), 785–790.

[201] B. Segre, Introduction to Galois geometries (ed. J.W.P. Hirschfeld), *Atti Accad. Naz. Lincei Mem.* **8** (1967), 133–236.

[202] G. Solomon and J.J. Stiffler, Algebraically punctured cyclic codes, *Inform. and Control* **8** (1965), 170–179.

[203] L. Storme and P. Sziklai, Linear point sets and Rédei type k-blocking sets in $PG(n,q)$, submitted.

[204] L. Storme and T. Szőnyi, Caps in $PG(n,q)$, q even, $n \geq 3$, *Geom. Dedicata* **45** (1993), 163–169.

[205] L. Storme and J.A. Thas, Complete k-arcs in $PG(n,q)$, q even, *Discrete Math.* **106/107** (1992), 455–469.

[206] L. Storme and J.A. Thas, M.D.S. codes and arcs in $PG(n,q)$ with q even: an improvement of the bounds of Bruen, Thas and Blokhuis, *J. Combin. Theory Ser. A* **62** (1993), 139–154.

[207] L. Storme and J.A. Thas, k-arcs and dual k-arcs, *Discrete Math.* **125** (1994), 357–370.

[208] L. Storme, J.A. Thas and S.K.J. Vereecke, New upper bounds for the sizes of caps in finite projective spaces, *J. Geom.*, to appear.

[209] L. Storme and Z. Weiner, Minimal blocking sets in $PG(n,q)$, $n \geq 3$, *Des. Codes Cryptogr.* **21** (2000), 235–251.

[210] P. Sziklai and T. Szőnyi, Blocking sets and algebraic curves, *Rend. Circ. Mat. Palermo* **51** (1998), 71–86.

[211] T. Szőnyi, Small complete arcs in Galois planes, *Geom. Dedicata* **18** (1985), 161–172.

[212] T. Szőnyi, Note on the order of magnitude of k for complete k-arcs in $PG(2,q)$, *Discrete Math.* **66** (1987), 279–282.

[213] T. Szőnyi, Arcs in cubic curves and 3-independent sets of abelian groups, *Combinatorics*, Eger, Colloquia Mathematica Societatis János Bolyai, North-Holland, Amsterdam **52** (1987), 499–508.

[214] T. Szőnyi, Note on the existence of large minimal blocking sets in Galois planes, *Combinatorica* **12** (1992), 227–235.

[215] T. Szőnyi, Arcs, caps, codes, and 3-independent subsets, *Giornate di Geometrie Combinatorie* (eds. G. Faina and G. Tallini), Università degli Studi di Perugia, Perugia, 1993, 57–80.

[216] T. Szőnyi, Blocking sets in Desarguesian affine and projective planes, *Finite Fields Appl.* **3** (1997), 187–202.

[217] T. Szőnyi, On the embedding of (k,p)-arcs in maximal arcs, *Des. Codes Cryptogr.* **18** (1999), 235–246.

[218] T. Szőnyi and Z. Weiner, Small blocking sets in higher dimensions, *J. Combin. Theory Ser. A*, to appear.

[219] G. Tallini, Calotte complete di $S_{4,q}$ contenenti due quadriche ellittiche quali sezioni iperpiane, *Rend. Mat. e Appl.* **23** (1964), 108–123.

[220] J.A. Thas, Normal rational curves and k-arcs in Galois spaces, *Rend. Mat.* **1** (1968), 331–334.

[221] J.A. Thas, Connection between the Grassmannian $G_{k-1;n}$ and the set of the k-arcs of the Galois space $S_{n,q}$, *Rend. Mat.* **2** (1969), 121–134.

[222] J.A. Thas, Some results concerning $\{(q+1)(n-1); n\}$-arcs and $\{(q+1)(n-1)+1; n\}$-arcs in finite projective planes of order q, *J. Combin. Theory Ser. A* **19** (1975), 228–232.

[223] J.A. Thas, Complete arcs and algebraic curves in $PG(2, q)$, *J. Algebra* **106** (1987), 451–464.

[224] J. Tits, Ovoides et groupes de Suzuki, *Arch. Math.* **13** (1962), 187–198.

[225] J.F. Voloch, On the completeness of certain plane arcs. *European J. Combin.* **8** (1987), 453–456.

[226] J.F. Voloch, Arcs in projective planes over prime fields, *J. Geom.* **38** (1990), 198–200.

[227] J.F. Voloch, On the completeness of certain plane arcs II, *European J. Combin.* **11** (1990), 491–496.

[228] J.F. Voloch, Complete arcs in Galois planes of non-square order, *Advances in Finite Geometries and Designs* (eds. J.W.P. Hirschfeld, D.R. Hughes, J.A. Thas), Oxford University Press, Oxford, 1991, 401–406.

[229] W.C. Waterhouse, Abelian varieties over finite fields, *Ann. Sci. École Norm. Sup.* **2** (1969), 521–560.

THE GEOMETRIC APPROACH TO LINEAR CODES

I.N. Landjev
Institute of Mathematics
8 Acad. G. Bonchev str.
1113 Sofia
Bulgaria
ivan@moi.math.bas.bg

Abstract The geometric nature of certain optimality problems in coding theory has been long known. The geometric approach to such problems is based on the equivalence between q-ary linear codes with no coordinate identically zero and multisets of points in projective geometries over finite fields. In this paper we give geometric proofs for several results on optimal linear codes and discuss generalizations that arise from the geometric approach.

1. Preliminaries

Let $\mathcal{P}$ be the set of points of the projective geometry $\mathrm{PG}(k-1,q)$. A *multiset* in $\mathrm{PG}(k-1,q)$ is a mapping $\mathfrak{k}\colon \mathcal{P} \to \mathbb{N}_0$. For any subset $\mathcal{Q}$ of $\mathcal{P}$ we set $\mathfrak{k}(\mathcal{Q}) = \sum_{P\in\mathcal{Q}} \mathfrak{k}(P)$. A multiset in $\mathrm{PG}(k-1,q)$ is called an $(n,w;k-1,q)$-*arc* or (n,w)-arc, for short, if

(a) $\mathfrak{k}(\mathcal{P}) = n$;

(b) $\mathfrak{k}(H) \leq w$ for any hyperplane H;

(c) there exists a hyperplane H_0 with $\mathfrak{k}(H_0) = w$.

The integer $\mathfrak{k}(P)$ is called the *multiplicity* of the point P and n is called the *cardinality* of $\mathfrak{k}$. Given an $(n,w;k-1,q)$-arc, denote by a_i the number of hyperplanes H with $\mathfrak{k}(H) = i$. The sequence $\{a_i\}_{i\geq 0}$ is called the *spectrum* of $\mathfrak{k}$.

There exists a familiar correspondence between the linear block codes of full length, that is, codes with no coordinate position identically zero, and the arcs in coordinate projective geometries. We refer for the basic notions in coding theory to any standard book, such as [7, 8, 12, 13].

A. Blokhuis et al. (eds.), Finite Geometries, 247–256.

Let $\mathcal{C}$ be an $[n,k]_q$ linear code of full length and let

$$\mathbf{G} = [\boldsymbol{c}_1, \dots, \boldsymbol{c}_k]^t = [\boldsymbol{g}_1, \dots, \boldsymbol{g}_n], \quad \boldsymbol{c}_i \in \mathbb{F}^n, \ \boldsymbol{g}_i \in \mathbb{F}^k.$$

be a generator matrix of $\mathcal{C}$. We define the arc $\mathfrak{k}_S$ *induced by the sequence* $S = \{\boldsymbol{c}_1, \dots, \boldsymbol{c}_k\}$ of $\mathcal{C}$ by

$$\mathfrak{k}_S \colon \left\{ \begin{array}{lll} \mathcal{P} & \to & \mathbb{N}_0, \\ P & \mapsto & |\{j \mid P = \langle g_j \rangle\}|. \end{array} \right. \tag{1.1}$$

The code $\mathcal{C}$ and the arc $\mathfrak{k}_S$ are said to be *associated* to each other. Let us note that an arc associated with an $[n,k,d]_q$ code has parameters $(n, n-d; k-1, q)$. Clearly, a linear code can be associated with different arcs but we have the following theorem.

Theorem 1.1 *For every multiset $\mathfrak{k}$ of cardinality n in $\mathrm{PG}(k-1,q)$ there exist both a linear code $\mathcal{C} < \mathbb{F}_q^n$ and a generating sequence S of $\mathcal{C}$ that induce $\mathfrak{k}$. Two arcs $\mathfrak{k}_1$ and $\mathfrak{k}_2$ in $\mathrm{PG}(k-1,q)$ associated with the linear codes $\mathcal{C}_1$ and $\mathcal{C}_2$ are equivalent if and only if the codes $\mathcal{C}_1$ and $\mathcal{C}_2$ are semilinearly isomorphic.*

A canonical way of associating an arc to a linear code has been presented in [4]. Theorem 1.1 can be generalized for linear codes over finite chain rings and arcs in projective Hjelmslev geometries [11].

A central bound in the theory of optimal linear codes is the so-called Griesmer bound [5]. It says that for any $[n,k,d]_q$ code

$$n \geq g_q(k,d) \stackrel{def}{=} \sum_{i=0}^{k-1} \left\lceil \frac{d}{q^i} \right\rceil. \tag{1.2}$$

Linear codes meeting the Griesmer bound are called *Griesmer codes.* Arcs associated with Giresmer codes are called *Griesmer arcs.* A linear code is called *divisible* if there is an integer larger than 1 dividing the weights of all its words. An (n,w)-arc in $\mathrm{PG}(k-1,q)$ is *divisible* if there is an integer $\ell > 1$ such that $\mathfrak{k}(H) \equiv n \bmod \ell$ for all hyperplanes H.

Let $\mathfrak{k}$ be an $(n,w;k-1,q)$-arc. Denote by γ_i the maximum number of points contained in an i-dimensional flat of $\mathrm{PG}(k-1,q)$, that is,

$$\gamma_i = \gamma_i(\mathfrak{k}) = \max_{\Delta} \mathfrak{k}(\Delta),$$

where Δ runs over all i-dimensional flats of $\mathrm{PG}(k-1,q)$. If $n = t + g_q(k,d)$, where $d = n - w$, then

$$\gamma_i(\mathfrak{k}) = t + \sum_{k-1-i}^{k-1} \left\lceil \frac{d}{q^i} \right\rceil. \tag{1.3}$$

Let π and σ be disjoint flats with $\dim \pi + \dim \sigma = k - 2$. We define the *projection* $\varphi_{\pi,\sigma}$ from π onto σ by

$$\varphi_{\pi,\sigma} \colon \begin{cases} \mathcal{P} \setminus \pi & \to \quad \sigma, \\ Q & \mapsto \quad \sigma \cap \langle \pi, Q \rangle, \end{cases}$$

where $\langle \pi, Q \rangle$ denotes the flat generated by π and Q. Flats of dimension $(s + \dim \pi)$ containing π are mapped via $\varphi_{\pi,\sigma}$ into $(s-1)$-dimensional flats in σ.

Let $\mathfrak{k}$ be an (n, w)-arc in $\mathrm{PG}(k-1, q)$ and let $\varphi = \varphi_{\pi,\sigma}$ be a projection from the i-dimensional flat π onto the $(k-2-i)$-dimensional flat σ. The *induced arc* $\mathfrak{k}^{\varphi}$ is defined on the set of points of σ by

$$\mathfrak{k}^{\varphi}(P) = \mathfrak{k}(\Delta) - \mathfrak{k}(\pi),$$

where Δ is an $(i+1)$-dimensional flat through π with $\varphi(\Delta) = P$.

In this paper we give geometric proofs and generalize some classical results on linear codes over finite fields. These include Hill and Lizak's extension theorem (Section 2), Dodunekov's theorem on the rank of the small weight codewords (Section 3) and Ward's theorem on divisibility of Griesmer codes (Section 4).

2. The extension theorem of Hill and Lizak

The following theorem has been proved in [10, 9] as a generalization of the parity check extension for binary codes of odd minimum distance.

Theorem 2.1 *Let C be an $[n, k, d]_q$ code with $\gcd(d, q) = 1$ and such that all nonzero weights are congruent to 0 or d modulo q. Then C can be extended to an $[n+1, k, d+1]_q$ code.*

The geometric version of this result is given below. Interestingly, it can be proved in this form as a corollary of a result by Bose and Burton on blocking sets [2].

Theorem 2.2 *Let $\mathfrak{k}$ be an $(n, w; k-1, q)$-arc with $\gcd(n-w, q) = 1$. Assume that the multiplicities of all hyperplanes are congruent to n or w modulo q. Then $\mathfrak{k}$ can be extended to an $(n+1, w)$-arc.*

Proof Fix a hyperplane H_0 in $\mathrm{PG}(k-1, q)$ with $\mathfrak{k}(H_0) = w$. For any subspace of codimension 2 $\delta \subset H_0$ consider the hyperplanes H_i, $i = 0, \ldots, q$, containing δ. Let α of them be of multiplicity congruent to n modulo q. Then

$$\begin{aligned} n &= \sum_{i=0}^{q} \mathfrak{k}(H_i) - q\mathfrak{k}(\delta) \\ &\equiv \alpha n + (q+1-\alpha) w \pmod{q}, \end{aligned}$$

whence $(\alpha - 1)(n - w) \equiv 0 \pmod{q}$ and $\alpha = 1$. Hence the number of hyperplanes of multiplicity $\equiv n \pmod{q}$ equals the number of subspaces of codimension 1 in H_0 and forms a blocking set in the dual space. By a result by Bose and Burton [2], a blocking set with respect to the lines having cardinality $(q^{k-1} - 1)/(q - 1)$ consists of the points of a hyperplane. By duality, this implies that all hyperplanes of multiplicity $\equiv n \pmod{q}$ pass through a fixed point P. Moreover, these are all hyperplanes through P. Hence we can get a $(n+1, w)$-arc by increasing the multiplicity of P by 1. □

Recently, Simonis [14] proved a generalization of this result allowing words of more weights modulo q.

Theorem 2.3 *Let $\mathcal{C}$ be a $[n, k, d]$-code over a finite field $\mathbb{F}_q$ of characteristic p with spectrum $\{A_i\}$. If $d \not\equiv 0 \pmod{p}$ and*

$$\sum_{i \not\equiv d \pmod{p}} A_i = q^{k-1},$$

then $\mathcal{C}$ can be extended to an $[n+1, k, d+1]$-code.

We can generalize on this result using a special case of a theorem by Heim [6], which says that if a blocking set contains "not too many" points then it contains a hyperplane.

Theorem 2.4 *Let $\mathcal{B}$ be a blocking set in $PG(k, q)$ with respect to $(k-t)$-dimensional subspaces. If*

$$|\mathcal{B}| < q^t + q^{t-1} + \ldots + 1 + q^{t-1} \cdot r(q),$$

where $q + r(q) + 1$ is the minimal size of a non-trivial blocking set of $\mathrm{PG}(2, q)$*, then $\mathcal{B}$ contains a t-dimensional subspace.*

This result suggests a new proof and a generalization of Simonis' result.

Theorem 2.5 *Let $\mathfrak{k}$ be a (n, w)-arc in* $\mathrm{PG}(k-1, q)$*, $q = p^s$, with spectrum $\{a_i\}_{i \geq 0}$. If $w \not\equiv n \pmod{p}$ and*

$$\sum_{i \not\equiv w \pmod{p}} a_i \leq q^{k-2} + q^{k-3} + \ldots + 1 + q^{k-3} \cdot r(q),$$

where $q + r(q) + 1$ is the minimal size of a non-trivial blocking set of $\mathrm{PG}(2, q)$*, then there exists an $(n+1, w)$-arc.*

Proof Fix a hyperplane H_0 of multiplicity w. Let δ and H_i be the same as in Theorem 2.2 and let α be the number of H_i with $\mathfrak{k}(H_i) \not\equiv w$

(mod p). Once again, counting the multiplicities of the points on the hyperplanes through δ, we get $(\alpha-1)(n-w) \pmod p$. Hence $\alpha \neq 0$ for any choice of δ and the set of hyperpanes of multiplicity $\not\equiv w \pmod p$ is a blocking set in the dual space. By Theorem 2.4, there is a point P which lies only on hyperplanes H with $\mathfrak{k}(H) \not\equiv w \pmod p$, that is, hyperplanes that are not of maximal multiplicity. Now we can increase the multiplicity of P by 1, without changing the maximal number of points on a hyperplane. □

3. Dodunekov's theorem on the rank of the small weight words

The following theorem has been proved by Dodunekov [3]. A weaker version of the theorem has been also proved by Simonis.

Theorem 3.1 *Let C be an $[n,k,d]_q$ code with $n = t + g_q(k,d)$. Then C has a basis of words whose weights are at most $t+d$.*

A very elegant proof by induction of Dodunekov's theorem has been given by Ward [15]. Below we state the geometric version of this result.

Theorem 3.2 *Let $\mathfrak{k}$ be an $(n, n-d)$-arc in $\mathrm{PG}(k-1,q)$ with $n = t + g_q(k,d)$. Then the hyperplanes of multiplicity $n-d, n-d-1, \ldots, n-d-t$ do not have a common point.*

Proof Assume otherwise and let the common point be P. Define a new arc $\mathfrak{h}$ in $\mathrm{PG}(k-1,q)$ by

$$\mathfrak{h}(Q) = \begin{cases} \mathfrak{k}(Q) + t + 1 & \text{for } Q \neq P, \\ \mathfrak{k}(Q) & \text{for } Q = P. \end{cases} \tag{3.1}$$

The arc $\mathfrak{h}$ has parameters

$$(g_q(k,d) + (t+1)\frac{q^k-1}{q-1} - 1, g_q(k,d) + (t+1)\frac{q^{k-1}-1}{q-1} - d - 1).$$

Now the parameters of the associated code violate the Griesmer bound. □

4. Divisible codes and Ward's theorem

One of the remarkable results of the past few years is Ward's theorem on the divisibility of Griesmer codes over a prime field [15]. In this section, we give an alternative proof of his result.

Lemma 4.1 *Let $\mathfrak{k}$ be an (n,w)-arc in* $\mathrm{PG}(k-1,p)$, $k \geq 3$. *A subspace Δ of codimension 2 is said to have the property $(*)$ if, for all hyperplanes H through Δ, the congruence $\mathfrak{k}(H) \equiv n \pmod{p^e}$ holds.*

(a) *Let $e = 1$ and let there exist a subspace δ of codimension 3 containing at least $p-1$ subspaces of codimension 2 with the property $(*)$. Then $\mathfrak{k}(H) \equiv n \pmod{p}$ for every hyperplane H through δ.*

(b) *Let $e \geq 2$ and let there exist a subspace δ of codimension 3 with the following properties:*

  (i) $\mathfrak{k}(\delta) \equiv n \pmod{p^{e-2}}$;

  (ii) $\mathfrak{k}(\Delta) \equiv n \pmod{p^{e-1}}$ *for every subspace Δ of codimension 2 through δ;*

  (iii) *there exist at least $p-1$ subspaces Δ of codimension 2 through δ with property $(*)$.*

  Then $\mathfrak{k}(H) \equiv n \pmod{p^e}$ for every hyperplane H containing δ.

Proof We give a proof of (b) only. The proof of (a) is similar and simpler.

Consider a projection φ from δ onto some plane σ in $\mathrm{PG}(k-1,p)$ disjoint from δ. The induced arc $\mathfrak{k}^\varphi$ on σ has parameters $(n-\mathfrak{k}(\delta), w-\mathfrak{k}(\delta))$. Set $n' = \sum_{P\in\sigma} \mathfrak{k}^\varphi(P) = n - \mathfrak{k}(\delta)$. Now define an arc on the points of σ by

$$\mathfrak{h}(P) = \frac{n' - \mathfrak{k}^\varphi(P)}{p^{e-1}}. \tag{4.1}$$

Note that the right-hand side in (4.1) is an integer by (ii). Let $P \in \sigma$ be the image of a subspace Δ of codimension 2 with property $(*)$. For every line $l \in \sigma$ incident with P we have

$$\begin{aligned}\mathfrak{h}(l) &= \frac{(p+1)n' - \sum_{Q\in l} \mathfrak{k}^\varphi(Q)}{p^{e-1}} \\ &= \frac{(p+1)n' - \mathfrak{k}(H) + \mathfrak{k}(\delta)}{p^{e-1}}\end{aligned}$$

On the other hand, by (4.1),

$$N = \mathfrak{h}(\sigma) = \sum_{Q\in\sigma} \mathfrak{h}(Q) = \frac{p(p+1)n'}{p^{e-1}},$$

whence $\mathfrak{h}(l) \equiv N \pmod{p}$.

Consider a hyperplane H_0 with $\delta \subset H_0$ and set $l_\infty = \varphi(H_0)$. Identify the points of $\sigma \setminus l_\infty$ with the elements of $\mathbb{F}_{p^2}$. Two lines $\langle X, Y\rangle$, and $\langle X', Y'\rangle$ are parallel if and only if $(X-Y)^{p-1}$ and $(X'-Y')^{p-1}$ are equal. Hence we can identify the points on l_∞ with the set $\mathcal{U}_{p+1}$ of all $(p+1)-st$ root of unity in $\mathbb{F}_{p^2}$.

Without loss of generality we can assume that $\varphi(\Delta) \neq 0$ for all subspaces Δ of codimension 2 through δ having the property $(*)$. Consider the polynomial

$$\begin{aligned} F(x,y) &= \prod_{P\in\mathbb{F}_{p^2}} \left(1-(1-Px)^{p-1}y\right)^{\mathfrak{h}(P)} \cdot \prod_{\zeta\in\mathcal{U}_{p+1}} (1-\zeta x^{p-1}y)^{\mathfrak{h}(\zeta)} \\ &= \sum_{i=0}^{N} F_i(x)y^i. \end{aligned}$$

Let $T \in \mathbb{F}_{p^2}$ and $T = \varphi(\Delta)$ for a subspace Δ of codimension 2 through δ with property $(*)$. If $P \neq T$ we have $(1-PT^{-1})^{p-1} = UT^{1-p}$, where $U \in \mathcal{U}_{p+1}$ corresponds to the point $\langle P, T\rangle \cap l_\infty$. If l_ζ denotes the line through T and $\zeta \in l_\infty$ then

$$F(T^{-1},y) = \prod_{\zeta\in\mathcal{U}_{p+1}} (1-\zeta y)^{\mathfrak{h}(l_\zeta)-\mathfrak{h}(T)}. \tag{4.2}$$

Since the numbers $\mathfrak{h}(l_\zeta)$, $\zeta \in \mathcal{U}_{p+1}$, are the same modulo p, we can set $\mathfrak{h}(l_\zeta)-\mathfrak{h}(T) = \alpha+\beta_\zeta p$ for some integers $0 \le \alpha < p$ and $\beta_\zeta \ge 0$, $\zeta \in \mathcal{U}_{p+1}$. This implies

$$\begin{aligned} F(T^{-1},y) &= \prod_{\zeta\in\mathcal{U}_{p+1}} (1-\zeta y)^{\alpha+\beta_\zeta p} \\ &= (1-y)^{\alpha(p+1)} \prod_{\zeta\in\mathcal{U}_{p+1}} (1-\zeta y)^{\beta_\zeta p} \\ &= (1-y)^{\alpha(p+1)} \prod_{\zeta\in\mathcal{U}_{p+1}} \left(1-(\zeta y)^{\beta_\zeta p}\right), \end{aligned}$$

whence $F_i(T^{-1}) = 0$ for all $i = 1, \dots, p-1$. Since $\deg F \le p-1$ and since $F_1(x) = 0$ for at least $p-1$ different values of x we have $F_1(x) \equiv 0$.

Furthermore,

$$\begin{aligned} F(0,y) &= (1-y)^{N-\mathfrak{h}(l_\infty)} \\ &= 1-\binom{N-\mathfrak{h}(l_\infty)}{1}y+\binom{N-\mathfrak{h}(l_\infty)}{2}y^2-\dots. \end{aligned}$$

The coefficient of y vanishes; so $\mathfrak{h}(l_\infty) \equiv N \pmod{p}$. This implies

$$\frac{(p+1)n' - \mathfrak{k}(H_0) + \mathfrak{k}(\delta)}{p^{e-1}} \equiv \frac{p(p+1)n'}{p^{e-1}} \pmod{p}. \qquad (4.3)$$

By (i) we have $n' \equiv 0 \pmod{p^{e-2}}$, which in conjunction with (4.3) implies

$$\frac{n' - \mathfrak{k}(H_0) + \mathfrak{k}(\delta)}{p^{e-1}} \equiv 0 \pmod{p}.$$

This gives $\mathfrak{k}(H_0) \equiv n \pmod{p^e}$. □

Theorem 4.2 *Let $\mathfrak{k}$ be a Griesmer (n,w)-arc in* $\mathrm{PG}(k-1,p)$, *with $n - w \equiv 0 \pmod{p^e}$, $e \geq 1$. Then $\mathfrak{k}(H) \equiv n \pmod{p^e}$ for every hyperplane H.*

Proof Set $d = n - w = a_0q^{k-1} + a_1q^{k-2} + \ldots + a_{k-2}q$, where a_i are non-negative integers and $0 \leq a_1, a_2, \ldots, a_{k-2} < q$. We also set $\phi_i = q^i + q^{i-1} + \ldots + 1$. We use induction on $m = k - e$ and k.

If $m \leq 1$ and k is free the proof is trivial. In this case, $d = a_0p^{k-1}$, $a_0 > 0$, whence $g_q(k,d) = a_0\phi_{k-1}$ and $\gamma_0 = a_0$. Therefore, $\mathfrak{k}(P) = a_0$ for any P and

$$n - \mathfrak{k}(H) = a_0(\phi_{k-1} - \phi_{k-2}) = a_0p^{k-1} \equiv 0 \pmod{p^e}.$$

In particular, the theorem holds for $k = 2$ and any $e \geq 1$.

Let $e = 1$ and assume we have proved the result for all Griesmer arcs in geometries of dimension up to $k - 1$. By the induction hypothesis, we have $\mathfrak{k}(H) \equiv n \pmod{p}$ for all hyperplanes H containing points of maximal multiplicity. Let Δ be a subspace of codimension 2 with $\mathfrak{k}(\Delta) = \gamma_{k-3}$. Denote by $H_0, \ldots, H_q$ all hyperplanes through Δ. Clearly, $\mathfrak{k}(H_i) = \gamma_{k-2}$ for all i. Consider an arbitrary hyperplane H different from the H_i's. We are going to prove that $\delta = H \cap \Delta$ satisfies the condition of Lemma 4.1(a).

Assume that none of the subspaces in H_0 through δ has $(*)$. This means that for any point from $H_0 \setminus \Delta$ we have $\mathfrak{k}(P) \leq \gamma_0 - 1 = a_0$. Denote by Δ_i, $i = 1, \ldots, q$, the codimension 2 subspaces in H_0 through

δ other than Δ. We have

$$\begin{aligned}
\mathfrak{k}(H_0) &= \mathfrak{k}(\delta) + \mathfrak{k}(\Delta \setminus \delta) + \sum_{i=1}^{q} \mathfrak{k}(\Delta_i \setminus \delta) \\
&\leq \gamma_{k-3} + a_0 q^{k-2} \\
&= a_0\phi_{k-3} + a_1\phi_{k-4} + \ldots + a_{k-3}\phi_0 + (k-2) + a_0 q^{k-2} \\
&= a_0\phi_{k-2} + a_1\phi_{k-4} + \ldots + a_{k-3}\phi_0 + (k-2) \\
&< a_0\phi_{k-2} + a_1\phi_{k-3} + \ldots + a_{k-3}\phi_1 + (k-2) \\
&= \gamma_{k-2},
\end{aligned}$$

a contradiction. Therefore each one of the hyperplanes H_i contains a subspace with $(*)$ through δ. This implies that δ satisfies the condition of Lemma 4.1(a) and hence $\mathfrak{k}(H) \equiv n \pmod{p}$.

Now assume we have proved the theorem for all Griesmer arcs with $m = k - e \leq \mu - 1$ and all Griesmer arcs with $m = \mu$ in the projective geometry of dimension $k < \kappa$. We are going to prove that the result holds for all Griesmer arcs with $k - e = \mu$ and $k = \kappa$.

By the induction hypothesis we have the following:

(1) $\mathfrak{k}(H) \equiv n \pmod{p^e}$ for all hyperplanes H containing points of maximum multiplicity;

(2) $\mathfrak{k}(\Delta) \equiv w \equiv n \pmod{p^{e-1}}$ for any subspace Δ of codimension 2 contained in a maximal hyperplane;

(3) $\mathfrak{k}(\delta) \equiv \gamma_{k-2} \equiv w \equiv n \pmod{p^{e-2}}$ for any subspace δ of codimension 3 contained in a maximal subspace of codimension 2.

Now let Δ, $H_0, \ldots, H_q$, H, δ and Δ_i be the same as above. The subspace $\delta = H \cap \Delta$ has properties (*i*) and (*ii*) from Lemma 4.1(b) by parts (2) and (3) of the induction hypothesis. If we assume that none of the subspaces of codimension 2 through δ in H_0 we get contradiction using the above counting argument. Now each of the hyperplanes H_i contains at least one hyperline different from Δ having $(*)$, which shows that δ has property (iii) from Lemma 4.1(b). Now using the lemma we get $\mathfrak{k}(H) \equiv n \pmod{p^e}$. □

Lemma 4.1 enables us to prove more than divisibility for just Griesmer arcs. It has been proved in [1] that any $(p^2+p+2, p+2)$-arc in PG$(2,p)$, p prime, which is not a Griesmer arc, with at least p points of multiplicity 2 is divisible. This can be viewed as a corollary of Lemma 4.1(a). More generally, any (n,w)-arc in PG$(2,p)$ with $n = t + g_p(2,d)$, $d = n - w = a_0p^2 + a_1p$, $0 < a_1 < q$, with at least p points of multiplicity $a_0 + t + 1$ is divisible.

Acknowledgements

This research has been supported by the Bulgarian NSF under Contract No. MM-901/1999.

References

[1] S. Ball, R. Hill, I.N. Landjev and H.N. Ward, On $(q^2+q+2, q+2)$-arcs in the projective plane $\mathrm{PG}(2,q)$, *Des. Codes Cryptogr.*, to appear.

[2] R.C. Bose and R.C. Burton, A characterization of flat spaces in a finite geometry and the uniqueness of the Hamming and the MacDonald codes, *J. Combin. Theory* **1** (1966), 96–104.

[3] S.M. Dodunekov, Optimal linear codes, Doctor of Math. Sciences Thesis, Institute of Mathematics, Sofia, 1985.

[4] S. Dodunekov and J. Simonis, Codes and projective multisets, *Electron. J. Combin.* **5** (1998), no. #R37.

[5] J.H. Griesmer, A bound for error-correcting codes, *IBM J. Res. Develop.* **4** (1960), 532–542.

[6] U. Heim, On t-blocking sets in projective spaces, preprint.

[7] W. Heise and P. Quattrocchi, *Informations- und Codierungstheorie*, 3rd Edition, Springer, Berlin, 1995.

[8] R. Hill, *A First Course in Coding Theory*, Oxford University Press, Oxford, 1986.

[9] R. Hill, An extension theorem for linear codes, *Des. Codes Cryptogr.* **17** (1999), 151-157.

[10] R. Hill and P. Lizak, Extensions of linear codes, Proc. Intern. Symposium on Inform. Theory, Canada, 1995, 345.

[11] T. Honold and I.N. Landjev, Linear codes over finite chain rings, *Electron. J. Combin.* **7** (2000), no. #R11.

[12] F.J. MacWilliams and N.J.A. Sloane, *The Theory of Error-correcting Codes*, North Holland, Amsterdam, 1977.

[13] V.S. Pless, W.C. Huffman and R.A. Brualdi, An introduction to algebraic codes, *Handbook of Coding Theory* (eds. V.S. Pless and W.C. Huffman), North Holland, Amsterdam, 1998, 3–139.

[14] J. Simonis, Adding a parity-check bit, *IEEE Trans. Inform. Theory* **46** (2000), 1544–1545.

[15] H.N. Ward, Divisibility of codes meeting the Griesmer bound, *J. Combin. Theory Ser. A* **83** (1998), 79–93.

FLOCKS AND LOCALLY HERMITIAN 1-SYSTEMS OF $Q(6,q)$

D. Luyckx
Ghent University
Dept. of Pure Mathematics and Computer Algebra
Galglaan 2
B 9000 Ghent
Belgium
dluyckx@cage.rug.ac.be

J. A. Thas
Ghent University
Dept. of Pure Mathematics and Computer Algebra
Galglaan 2
B 9000 Ghent
Belgium
jat@cage.rug.ac.be

Abstract

A connection between locally hermitian 1-systems of $Q(6,q)$ and a generalization of flocks of quadratic cones in $\mathsf{PG}(3,q)$ is given. This connection will be used to describe geometrically the semi-classical non-hermitian spread $\mathcal{S}_{[9]}$ of the hexagon $\mathsf{H}(q)$. The construction yields several examples of locally hermitian 1-systems of $Q(6,q)$ which are shown to be pairwise non-isomorphic. Finally, all semi-classical non-hermitian 1-systems of $Q(6,q)$, q odd, are determined and classified.

1. Introduction

A flock of a quadratic cone K in $\mathsf{PG}(3,q)$ is a set of q mutually disjoint conics on K. Flocks of quadratic cones in $\mathsf{PG}(3,q)$ have been thoroughly studied by various authors; we refer for instance to [1], [4],

A. Blokhuis et al. (eds.), Finite Geometries, 257–275.

[7], [8] and [11]. In particular, it is known that flocks of quadratic cones are closely related to ovoids of $Q^+(5,q)$ in the following way.

Let $\mathcal{F}$ be a flock of a quadratic cone K in $\mathsf{PG}(3,q)$; then $|\mathcal{F}| = q$. We now embed the cone K in a nonsingular hyperbolic quadric $Q^+(5,q)$. For each conic $\mathcal{C}_i \in \mathcal{F}$, the plane $\langle \mathcal{C}_i \rangle$ will be denoted by π_i. It is clear that $\pi_i^\perp$, with $\perp$ the polarity of $Q^+(5,q)$, intersects $Q^+(5,q)$ in a conic $\mathcal{C}_i^*$ and, as every $\mathcal{C}_i$ is contained in $p^\perp$, the point p belongs to all conics $\mathcal{C}_i^*$. Now it is easily seen that no two points of $\cup_{i=1}^q \mathcal{C}_i^*$ are collinear on $Q^+(5,q)$ and hence these points form an ovoid of $Q^+(5,q)$.

By using the Klein correspondence, this ovoid of $Q^+(5,q)$ yields a spread of $\mathsf{PG}(3,q)$ consisting of q reguli sharing one line, which in turn gives rise to a translation plane of order q^2. This construction of translation planes is due independently to Thas and Walker; see, for example, [11]. The process can also be reversed: starting from a spread of $\mathsf{PG}(3,q)$ consisting of q reguli through a common line, one can construct a flock of a quadratic cone, see [5].

In the following we will study a similar connection between some 1-systems of $Q(6,q)$ and flocks of a quadratic cone with a line vertex.

2. The connection between locally hermitian 1-systems of $Q(6,q)$ and flocks

A 1-system $\mathcal{M}$ of the parabolic quadric $Q(6,q)$ in $\mathsf{PG}(6,q)$ is a set $\{L_0, L_1, \dots, L_{q^3}\}$ consisting of q^3+1 lines on $Q(6,q)$ having the property that the tangent space of $Q(6,q)$ at L_i has no point in common with $(L_0 \cup L_1 \cup \dots \cup L_{q^3}) \setminus L_i$, $i = 0, 1, \dots, q^3$. A 1-system $\mathcal{M}$ of the parabolic quadric $Q(6,q)$ is said to be *locally hermitian* at a line L of $\mathcal{M}$ if for any line $M \in \mathcal{M} \setminus \{L\}$, the regulus on $Q(6,q)$ defined by L and M, that is, the regulus containing L of the hyperbolic quadric $\langle L, M \rangle \cap Q(6,q)$, is entirely contained in $\mathcal{M}$. Such 1-systems consist of q^2 reguli through the line L; the union of the points on the elements of $\mathcal{M}$ will be denoted by $\tilde{\mathcal{M}}$.

For more details on the theory of m-systems, we refer to Shult and Thas [10]. Here, we only mention the following theorem we will need.

Theorem 2.1 *If $\mathcal{M}$ is an m-system of the finite classical polar space P, then, for any generator G of P,*

$$|G \cap \tilde{\mathcal{M}}| = \frac{q^{m+1}-1}{q-1}.$$

From now on, we assume that q is odd, which implies that a polarity $\perp$ is defined by $Q(6,q)$. Let $\mathcal{M}$ be a 1-system of $Q(6,q)$ which is locally hermitian at some line L. The q^2 hyperbolic quadrics, the reguli through L of which constitute $\mathcal{M}$, will be denoted by $Q_i^+(3,q)$, $i = 1,2,\ldots,q^2$. We now have a closer look at the induced structure in $L^\perp \cap Q(6,q) = LQ(2,q)$. As each $Q_i^+(3,q)$ contains the line L, it follows that $\mathcal{C}_i := \langle Q_i^+(3,q)\rangle^\perp \cap Q(6,q)$ is a conic lying on $LQ(2,q)$ for $i = 1,2,\ldots,q^2$. The fact that $\mathcal{M}$ is a 1-system implies that, for $i \neq j$, the hyperbolic quadrics $Q_i^+(3,q)$ and $Q_j^+(3,q)$ only share the line L and that $\langle Q_i^+(3,q), Q_j^+(3,q)\rangle$ is 5-dimensional. Let $\langle Q_i^+(3,q), Q_j^+(3,q)\rangle \cap Q(6,q) = Q$. Consider an arbitrary line N of $Q_i^+(3,q)$ disjoint from L. If Q did contain planes, there would exist a plane of Q on N having at least one point in common with $Q_j^+(3,q)$. But such a plane would intersect $\tilde{\mathcal{M}}$ in at least $q+2$ points, a contradiction to Theorem 2.1. Hence $\langle Q_i^+(3,q), Q_j^+(3,q)\rangle \cap Q(6,q)$ is an elliptic quadric $Q_{ij}^-(5,q)$. It now immediately follows that $\langle \mathcal{C}_i\rangle \cap \langle \mathcal{C}_j\rangle = \langle Q_{ij}^-(5,q)\rangle^\perp$ is an internal point of the parabolic quadric $Q(6,q)$, and hence also of the cone $LQ(2,q)$.

These observations justify the following definitions.

Definitions (i) A *flock* of a quadratic cone $LQ(2,q)$, with line vertex L, of $\mathsf{PG}(4,q)$ is a partition of $LQ(2,q) \setminus L$ into q^2 mutually disjoint conics, with the extra condition that any two disjoint conics of the partition generate $\mathsf{PG}(4,q)$.

(ii) If q is odd and the planes $\pi_1, \pi_2, \ldots, \pi_{q^2}$ of the elements of the flock mutually intersect in internal points of the cone, then we speak of an *i-flock*; if the planes mutually intersect in external points, then we speak of an *e-flock*.

(iii) If q is even and the nuclei of the elements of the flock are all distinct, that is, together with L these nuclei form the nucleus plane of the cone, then the flock is called *general*.

Conversely, if we start from an i-flock of a cone $LQ(2,q)$ in $\mathsf{PG}(4,q)$, q odd, and embed it in a parabolic quadric $Q(6,q)$, then we obtain a 1-system of this $Q(6,q)$ consisting of q^2 reguli through the line L, simply by reversing the above construction. Hence we have the following theorem.

Theorem 2.2 *For q odd, locally hermitian* 1*-systems of* $Q(6,q)$ *and* i-*flocks of* $LQ(2,q)$ *are equivalent objects.*

If we assume q even, then any locally hermitian 1-system of $Q(6,q)$ also yields a flock of a quadratic cone $LQ(2,q)$. However, all planes of the elements of the flock share the nucleus of $Q(6,q)$, and, moreover, starting from such a flock, the construction cannot be reversed.

We observe that locally hermitian 1-systems of $Q(6,q)$ are known to exist. The simplest example is certainly the classical 1-system of

$Q(6,q)$, which is the classical or hermitian spread of some elliptic quadric $Q^-(5,q) \subseteq Q(6,q)$. This 1-system even has a stronger property than being locally hermitian at L: it is hermitian, which means that the regulus defined by any two distinct lines of the 1-system is completely contained in the 1-system or, equivalently, that it is locally hermitian at any of its lines. It is clear that in this case $Q^-_{ij}(5,q) = Q^-(5,q)$ for all $i \neq j$ and consequently, for q odd, all conic planes of the associated i-flock share one point, which is internal to $Q(6,q)$.

Other examples are given in [2], where Bloemen, Thas and Van Maldeghem study locally hermitian spreads of the hexagon $\mathsf{H}(q)$. They describe for instance the unique semi-classical non-hermitian spread $\mathcal{S}_{[9]}$ of $\mathsf{H}(q)$, where q is odd and $q \equiv 1 \pmod 3$; this spread is locally hermitian at some line. As every spread of the hexagon $\mathsf{H}(q)$ is a 1-system of $Q(6,q)$, this spread yields another example of a 1-system of $Q(6,q)$ with an associated i-flock. This example will be studied in detail later on.

3. An algebraic description of flocks

Let L be the line $X_0 = X_1 = X_2 = 0$ of $\mathsf{PG}(4,q)$ and let $Q(2,q)$ be the conic $X_3 = X_4 = X_0X_2 - X_1^2 = 0$. Then the cone $LQ(2,q)$ has equation $X_0X_2 - X_1^2 = 0$. A flock $\mathcal{F}$ of $LQ(2,q)$ consists of q^2 disjoint conics $\mathcal{C}_i$ in planes π_i, disjoint from L. In general, a plane disjoint from L has equations of the form

$$\pi_i \equiv \begin{cases} a_iX_0 + b_iX_1 + c_iX_2 + X_3 = 0 \\ d_iX_0 + e_iX_1 + f_iX_2 + X_4 = 0, \end{cases}$$

with $a_i, b_i, c_i, d_i, e_i, f_i \in \mathsf{GF}(q)$. Let

$$\varphi_{ij} := (a_{ij}f_{ij} - c_{ij}d_{ij})^2 - (a_{ij}e_{ij} - b_{ij}d_{ij})(b_{ij}f_{ij} - c_{ij}e_{ij}),$$

where a_{ij} is a short notation for $a_i - a_j$ and similarly for $b_{ij}, \ldots, f_{ij}$. It is an easy exercise to show that $\langle \pi_i, \pi_j \rangle \neq \mathsf{PG}(4,q)$ implies $\varphi_{ij} = 0$, $i \neq j$.

Theorem 3.1 (i) *With the above notation, the conics $\mathcal{C}_i = LQ(2,q) \cap \pi_i$, $i = 1, 2, \ldots, q^2$, form a flock of $LQ(2,q)$ if and only if, for all $i \neq j$,*

$$\varphi_{ij} \neq 0.$$

(ii) *For q odd, the conics $\mathcal{C}_1, \mathcal{C}_2, \ldots, \mathcal{C}_{q^2}$ form an* i-*flock or an* e-*flock of $LQ(2,q)$ if and only if, for all $i \neq j$,*

$$\varphi_{ij} \textit{ is a non-square or a non-zero square in } \mathsf{GF}(q).$$

(iii) *For q even, the conics $\mathcal{C}_1, \mathcal{C}_2, \ldots, \mathcal{C}_{q^2}$ form a general flock of $LQ(2,q)$ if and only if, for all $i \neq j$,*

$$\varphi_{ij} \neq 0 \textit{ and } \mathrm{rank} \begin{bmatrix} 1 & b_i & e_i \\ 1 & b_j & e_j \end{bmatrix} = 2.$$

Proof For $i \neq j$, the intersection of the planes π_i and π_j is determined by

$$\begin{aligned} &a_i X_0 + b_i X_1 + c_i X_2 + X_3 = 0, \\ &d_i X_0 + e_i X_1 + f_i X_2 + X_4 = 0, \\ &(a_i - a_j)X_0 + (b_i - b_j)X_1 + (c_i - c_j)X_2 = 0, \\ &(d_i - d_j)X_0 + (e_i - e_j)X_1 + (f_i - f_j)X_2 = 0. \end{aligned}$$

For $\langle \pi_i, \pi_j \rangle = \mathsf{PG}(4,q)$, the point $\langle \pi_i \cap \pi_j, L \rangle \cap \langle Q(2,q) \rangle = \{p_{ij}\}$ is the intersection of the lines

$$\begin{aligned} &a_{ij} X_0 + b_{ij} X_1 + c_{ij} X_2 = 0, \\ &d_{ij} X_0 + e_{ij} X_1 + f_{ij} X_2 = 0, \\ &X_3 = X_4 = 0, \end{aligned}$$

in the plane of $Q(2,q)$. Hence $p_{ij} = (r_{ij}, s_{ij}, t_{ij}, 0, 0)$, with

$$\begin{aligned} r_{ij} &= b_{ij} f_{ij} - c_{ij} e_{ij}, \\ s_{ij} &= c_{ij} d_{ij} - a_{ij} f_{ij}, \\ t_{ij} &= a_{ij} e_{ij} - b_{ij} d_{ij}. \end{aligned}$$

Clearly $\{\mathcal{C}_1, \mathcal{C}_2, \ldots, \mathcal{C}_{q^2}\} = \mathcal{F}$ is a flock of $LQ(2,q)$ if and only if $p_{ij} \notin Q(2,q)$ for all $i \neq j$, that is, if and only if $\varphi_{ij} \neq 0$ for all $i \neq j$.

The set $\mathcal{F}$ is an i-flock of $LQ(2,q)$, with q odd, if and only if p_{ij} is an internal point of $LQ(2,q)$ for all $i \neq j$, that is, if and only if φ_{ij} is a non-square for all $i \neq j$. Similarly, $\mathcal{F}$ is an e-flock of $LQ(2,q)$, with q odd, if and only if φ_{ij} is a non-zero square for all $i \neq j$.

Finally, let q be even. For $i = 1, 2, \ldots, q^2$, the kernel of the conic $\mathcal{C}_i$ is the point $(0, 1, 0, b_i, e_i)$. These nuclei are all distinct if and only if

$$\mathrm{rank} \begin{bmatrix} 1 & b_i & e_i \\ 1 & b_j & e_j \end{bmatrix} = 2$$

for all $i \neq j$. □

4. The i-flock related to the classical spread of $Q^-(5,q)$, q odd

Unless stated otherwise, we will always assume that q is odd and work on the parabolic quadric $Q(6,q)$ in $\mathsf{PG}(6,q)$ with equation

$$X_0 X_4 + X_1 X_5 + X_2 X_6 - X_3^2 = 0.$$

The classical spread of $Q^-(5,q)$, embedded in $Q(6,q)$, yields a 1-system of $Q(6,q)$ which is locally hermitian at any line. As mentioned above, the conic planes of the associated flock all share one point p such that $p^\perp = \langle Q^-(5,q)\rangle$. The line L is chosen to be the line $X_0 = X_1 = X_2 = X_3 = X_4 = 0$; so $L^\perp$ is the subspace $X_1 = X_2 = 0$ and the cone $L^\perp \cap Q(6,q)$ has equations $X_1 = X_2 = X_0X_4 - X_3^2 = 0$. It is easily seen that an algebraic description of this flock is given by, for all $k_i, b_i \in \mathsf{GF}(q)$,

$$\pi(k_i,b_i) \equiv \begin{cases} X_1 = X_2 = 0 \\ \gamma k_i X_0 + 2b_i X_3 + k_i X_4 + X_5 = 0 \\ \gamma b_i X_0 + 2\gamma k_i X_3 + b_i X_4 + X_6 = 0, \end{cases}$$

where γ denotes a non-square in $\mathsf{GF}(q)$. Indeed, the planes $\pi(k_i,b_i)$ have the point $p = (-1,0,0,0,\gamma,0,0)$ in common and the polar space of p with respect to $Q(6,q)$ intersects $Q(6,q)$ in the elliptic quadric

$$Q^-(5,q) \equiv \begin{cases} \gamma X_0 - X_4 = 0 \\ \gamma X_0^2 + X_1X_5 + X_2X_6 - X_3^2 = 0. \end{cases}$$

For completeness, we check the condition on φ_{ij} from Theorem 3.1:

$$\varphi_{ij} = 4\gamma[\gamma(k_i-k_j)^2 - (b_i-b_j)^2]^2,$$

which is a non-square in $\mathsf{GF}(q)$ since it cannot be equal to zero as long as $(k_i,b_i) \neq (k_j,b_j)$.

5. The i-flock associated with the semi-classical non-hermitian spread of $\mathsf{H}(q)$, q odd and $q \equiv 1 \pmod 3$

5.1. Algebraic description

In [2], Bloemen, Thas and Van Maldeghem show that the classical hexagon $\mathsf{H}(q)$, q odd and $q \equiv 1 \pmod 3$, admits a unique semi-classical non-hermitian spread, which they call $\mathcal{S}_{[9]}$. They describe this spread explicitly by using the coordinatization of $\mathsf{H}(q)$ introduced in [3], as follows:

$$\mathcal{S}_{[9]} = \{[\infty]\} \cup \{[9\gamma b', -\gamma k'', k', b', k''] \mid k', b', k'' \in \mathsf{GF}(q)\}.$$

Here, γ again denotes a fixed non-square in $\mathsf{GF}(q)$. Translated to the parabolic quadric $Q(6,q)$, this gives a 1-system consisting of the lines

$$L' = \langle (1,0,0,0,0,0,0), (0,0,0,0,0,0,1) \rangle,$$

and, for $k', b', k'' \in \mathsf{GF}(q)$,

$$\langle (k' - \gamma k'' b', 9\gamma b', 1, -\gamma k'', 0, b', \gamma^2 k''^2 - 9\gamma b'^2), \\ (b'^2 - \gamma k''^2, \gamma k'', 0, -b', 1, k'', -7\gamma b' k'' - k') \rangle.$$

The spread $\mathcal{S}_{[9]}$ of $\mathsf{H}(q)$ is locally hermitian at the line $[\infty]$ of $\mathsf{H}(q)$, corresponding to the line $L' = \langle (1,0,0,0,0,0,0), (0,0,0,0,0,0,1) \rangle$ of $Q(6,q)$. We determine the i-flock associated with $\mathcal{S}_{[9]}$, considered as a 1-system of $Q(6,q)$. To that end we first determine the 3-spaces containing the reguli through L' which constitute $\mathcal{S}_{[9]}$; one can calculate that these 3-spaces have the following equations, for all $k'', b' \in \mathsf{GF}(q)$:

$$R(k'', b') \equiv \begin{cases} X_1 = 9\gamma b' X_2 + \gamma k'' X_4 \\ X_3 = -b' X_4 - \gamma k'' X_2 \\ X_5 = b' X_2 + k'' X_4. \end{cases}$$

The polar space of $R(k'', b')$ with respect to $Q(6,q)$ is given by, for all $k'', b' \in \mathsf{GF}(q)$,

$$\pi(k'', b') \equiv \begin{cases} X_2 = X_4 = 0 \\ b' X_1 + 2\gamma k'' X_3 + 9\gamma b' X_5 + X_6 = 0 \\ X_0 + k'' X_1 + 2b' X_3 + \gamma k'' X_5 = 0; \end{cases}$$

the planes $\pi(k'', b')$ are contained in $L'^{\perp}$, which has equations $X_2 = X_4 = 0$. The cone $L'^{\perp} \cap Q(6,q)$ has equations $X_2 = X_4 = X_1 X_5 - X_3^2 = 0$. To have more similarities to the algebraic setting in Section 4, we apply a permutation of the coordinates: $X_0 \leftrightarrow X_5$, $X_1 \leftrightarrow X_4$, $X_i \leftrightarrow X_i$ for $i = 2, 3, 6$. This permutation obviously stabilizes the quadric $Q(6,q)$, it maps the line L' onto the line L with equations $X_0 = X_1 = X_2 = X_3 = X_4 = 0$ and the plane $\pi(k'', b')$ onto the following plane, where we now write k_i and b_i instead of k'' and b':

$$\pi(k_i, b_i) \equiv \begin{cases} X_1 = X_2 = 0 \\ \gamma k_i X_0 + 2b_i X_3 + k_i X_4 + X_5 = 0 \\ 9\gamma b_i X_0 + 2\gamma k_i X_3 + b_i X_4 + X_6 = 0, \end{cases} \tag{5.1}$$

for all $k_i, b_i \in \mathsf{GF}(q)$. For completeness, we again check the condition on φ_{ij} (see Theorem 3.1):

$$\varphi_{ij} = 4\gamma[3(b_i - b_j)^2 + \gamma(k_i - k_j)^2]^2.$$

As γ is a non-square in $\mathsf{GF}(q)$, so φ_{ij} is a non-square provided it is not equal to zero. This condition is satisfied for all $(k_i, b_i) \neq (k_j, b_j)$, because -3 is a square in $\mathsf{GF}(q)$ with $q \equiv 1 \pmod 3$.

We will denote this i-flock by $\mathcal{F}_{[9]}$.

5.2. A geometric description of $\mathcal{F}_{[9]}$

Here we use the algebraic setting to describe the i-flock $\mathcal{F}_{[9]}$, and hence also the spread $\mathcal{S}_{[9]}$ of $\mathsf{H}(q)$, geometrically. We will study the properties of $\mathcal{F}_{[9]}$ step-by-step in a series of lemmas.

Lemma 5.1 *The intersection points of a fixed conic plane of the flock $\mathcal{F}_{[9]}$ with all other conic planes are exactly the $q+1$ points of an irreducible conic lying in this plane.*

Proof Without loss of generality, by [2], we may assume that the given conic plane is the plane $\pi(0,0)$; the other conic planes are given by (5.1). One can easily check that the intersection point of $\pi(0,0)$ with a plane $\pi(k_i,b_i)$, for $(k_i,b_i) \in \mathsf{GF}(q)^2 \setminus \{(0,0)\}$, is given by

$$(b_i^2 - \gamma k_i^2, 0, 0, 4\gamma k_i b_i, \gamma^2 k_i^2 - 9\gamma b_i^2, 0, 0).$$

Now suppose that two different planes $\pi(k_i,b_i)$ and $\pi(k_j,b_j)$, such that $(0,0) \neq (k_i,b_i) \neq (k_j,b_j) \neq (0,0)$, intersect $\pi(0,0)$ in the same point. Then there exists a non-zero constant $\mu \in \mathsf{GF}(q)$ such that

$$\begin{aligned} &b_j^2 - \gamma k_j^2 = \mu(b_i^2 - \gamma k_i^2),\\ &4\gamma k_j b_j = \mu(4\gamma k_i b_i),\\ &\gamma^2 k_j^2 - 9\gamma b_j^2 = \mu(\gamma^2 k_i^2 - 9\gamma b_i^2), \end{aligned}$$

which after some computation implies that μ is a square and $k_j = \sqrt{\mu} k_i$ as well as $b_j = \sqrt{\mu} b_i$. So two planes $\pi(k_i,b_i)$ and $\pi(k_j,b_j)$ share the same point with $\pi(0,0)$ if and only if there exists a constant $\lambda \in \mathsf{GF}(q)^*$ such that $(k_j,b_j) = (\lambda k_i, \lambda b_i)$. Hence every intersection point of $\pi(0,0)$ with some other conic plane is contained in q different conic planes, $\pi(0,0)$ included. This implies that the plane $\pi(0,0)$ contains exactly $\frac{q^2-1}{q-1} = q+1$ intersection points. We determine these $q+1$ points explicitly. If we put $b_i = 0$, we find $q-1$ planes $\pi(k_i,0)$, $k_i \in \mathsf{GF}(q)^*$, intersecting $\pi(0,0)$ in the same point $(-1,0,0,0,\gamma,0,0)$. For $b_i = 1$, there are q planes $\pi(k_i,1)$, $k_i \in \mathsf{GF}(q)$, which intersect $\pi(0,0)$ in q different points, namely the points, for all $k_i \in \mathsf{GF}(q)$,

$$(1 - \gamma k_i^2, 0, 0, 4\gamma k_i, \gamma^2 k_i^2 - 9\gamma, 0, 0).$$

It is straightforward to see that the $q+1$ intersection points we have found are indeed the points of the irreducible conic

$$\mathcal{C}' \equiv \begin{cases} X_1 = X_2 = X_5 = X_6 = 0\\ 9\gamma^2 X_0^2 - 4\gamma X_3^2 + X_4^2 + 10\gamma X_0 X_4 = 0, \end{cases}$$

in the plane $\pi(0,0)$ with equations $X_1 = X_2 = X_5 = X_6 = 0$. This proves the lemma. $\square$

Henceforth, we will often denote the plane $\pi(0,0)$ by π.

The previous lemma implies that there are $\frac{q^2(q+1)}{q} = q(q+1)$ intersection points of conic planes, lying on an irreducible conic in each of these conic planes. In the following lemma we have a closer look at the configuration of these intersection points.

Lemma 5.2 *For any point* $x \in \mathcal{C}'$ *the plane* $\langle x, L\rangle$ *has* q *points of some line in common with* $\bigcup_{k_i,b_i \in \mathsf{GF}(q)} \pi(k_i, b_i)$. *These* $q+1$ *lines are disjoint.*

Proof If p_∞ denotes the point $(-1,0,0,0,\gamma,0,0)$, then

$$\langle p_\infty, L\rangle \equiv \begin{cases} X_1 = X_2 = X_3 = 0 \\ \gamma X_0 + X_4 = 0. \end{cases}$$

It is easy to see that the plane $\pi(k_i, b_i)$, $k_i, b_i \in \mathsf{GF}(q)$, intersects $\langle p_\infty, L\rangle$ in the point $(1,0,0,0,-\gamma,0,-8\gamma b_i)$ and these q points lie on the line

$$N_\infty \equiv \begin{cases} X_1 = X_2 = X_3 = X_5 = 0 \\ \gamma X_0 + X_4 = 0, \end{cases}$$

which shares the point $(0,0,0,0,0,0,1)$ with the line L.

For $t \in \mathsf{GF}(q)$ we denote the point $(1-\gamma t^2, 0, 0, 4\gamma t, \gamma^2 t^2 - 9\gamma, 0, 0)$ of $\mathcal{C}'$ by p_t; thus

$$\langle p_t, L\rangle \equiv \begin{cases} X_1 = X_2 = 0 \\ 4\gamma t X_0 - (1-\gamma t^2) X_3 = 0 \\ (\gamma^2 t^2 - 9\gamma) X_0 - (1-\gamma t^2) X_4 = 0. \end{cases}$$

Now, the plane $\pi(k_i, b_i)$, with $k_i, b_i \in \mathsf{GF}(q)$, has the point

$$(1-\gamma t^2, 0, 0, 4\gamma t, \gamma^2 t^2 - 9\gamma, 8\gamma(k_i - t b_i), 8\gamma^2 t(t b_i - k_i))$$

in common with $\langle p_t, L\rangle$. As a consequence, for every $t \in \mathsf{GF}(q)$, the plane $\langle p_t, L\rangle$ shares q points of the line

$$N_t \equiv \begin{cases} X_1 = X_2 = 0 \\ 4\gamma t X_0 - (1-\gamma t^2) X_3 = 0 \\ (\gamma^2 t^2 - 9\gamma) X_0 - (1-\gamma t^2) X_4 = 0 \\ \gamma t X_5 + X_6 = 0 \end{cases}$$

with $\bigcup_{k_i,b_i \in \mathsf{GF}(q)} \pi(k_i, b_i)$, and obviously N_t intersects L in the point

$$(0,0,0,0,0,1,-\gamma t).$$

We conclude that the lines $N_\infty, N_1, N_2, \ldots, N_q$ meet L in different points and since two planes $\langle x, L\rangle$ and $\langle y, L\rangle$, $x, y \in \mathcal{C}'$, only share the line L, these $q+1$ lines are disjoint. □

So far, we know that every two conic planes intersect in exactly one point and the intersection points of a conic plane with the other ones form an irreducible conic. Moreover, every intersection point is contained in q such conics and exactly one line which meets L. This yields $q+1$ disjoint lines, one line on every point of L, projecting a conic. In the following we identify the configuration of the intersection points as the projection of the Veronese surface $\mathcal{V}_2^4$ from a point on $\mathcal{V}_2^4$ onto a 4-dimensional space.

Lemma 5.3 *The $q(q+1)$ intersection points of conic planes together with the line L, form a rational normal cubic scroll $\mathcal{R}^3$ having L as directrix line.*

Proof The $q(q+1)$ intersection points of conic planes together with the line L form a set which is the union of the $q+1$ lines joining the points $(0,0,0,0,0,1,-\gamma t)$ of L to the respective points $(1-\gamma t^2, 0, 0, 4\gamma t, \gamma^2 t^2 - 9\gamma, 0, 0)$ of $\mathcal{C}'$, with $t \in \mathsf{GF}(q) \cup \{\infty\}$. Then, by VII-3.22 of [9], this set is a rational normal cubic scroll $\mathcal{R}^3$ having L as directrix line. □

Note that the variety $\mathcal{R}^3$ is the projection of a Veronese surface $\mathcal{V}_2^4$ from a point $u \in \mathcal{V}_2^4$ onto a hyperplane $\mathsf{PG}(4,q)$ of $\mathsf{PG}(5,q)$, with $\mathcal{V}_2^4 \subset \mathsf{PG}(5,q)$; see VII-3.22 of [9]. The line L is the intersection of $\mathsf{PG}(4,q)$ with the tangent plane of $\mathcal{V}_2^4$ at u.

6. A new construction of 1-systems of $Q(6,q)$ for q odd

In this section, we investigate whether the above construction can be reversed. This means the following. If one starts from a rational normal cubic scroll $\mathcal{R}^3$ with directrix line L in $L^\perp$, is it always possible to construct a 1-system of $Q(6,q)$ in the same way as above? As we shall see now, the construction turns out to be reversible as long as we start from a conic all points of which are internal points of the conic

$$\mathcal{C} \equiv \begin{cases} X_1 = X_2 = X_5 = X_6 = 0 \\ X_0 X_4 = X_3^2, \end{cases}$$

which is the intersection of π with $Q(6,q)$.

Lemma 6.1 *Consider a cone $L\mathcal{C}$ with vertex line L in $\mathsf{PG}(4,q)$, q odd, where $\mathcal{C}$ is a conic contained in some plane π skew to L. Let $\mathcal{C}'$ be* *another conic of* π, *disjoint from* $\mathcal{C}$ *and such that every point of* $\mathcal{C}'$ *is an internal point of* $\mathcal{C}$. *Then for every plane* $\alpha \neq \pi$ *which is also skew to* L, *the points of the conic* $L\mathcal{C}' \cap \alpha := \mathcal{C}'_\alpha$ *are internal points of the conic* $L\mathcal{C} \cap \alpha := \mathcal{C}_\alpha$, *and hence internal points of* $L\mathcal{C}$.

Proof This is clear. □

The following theorem describes a way to construct i-flocks and consequently also locally hermitian 1-systems of $Q(6,q)$.

Theorem 6.2 *Let the cone* $LQ(2,q)$ *be embedded in* $Q(6,q)$ *and let* π *be a plane of* $L^\perp$ *skew to* L, *with* $Q(6,q) \cap \pi := \mathcal{C}$. *Assume* $\mathcal{C}' \subseteq \pi$ *is a conic with the property that all of its points are internal points of* $\mathcal{C}$. *Then any rational normal cubic scroll* $\mathcal{R}^3$ *with directrix line* L *and* $\mathcal{C}' \subseteq \mathcal{R}^3$, *uniquely defines an* i-*flock of* $L\mathcal{C}$, *and consequently also a locally hermitian* 1-*system of* $Q(6,q)$.

Proof It is known that the scroll $\mathcal{R}^3$ consists of q^2 conics $\mathcal{C}'_i$, defining q^2 planes π_i, for $i = 1, 2, \ldots, q^2$, such that each two planes π_i and π_j, $i \neq j$, intersect in a unique point that belongs to both $\mathcal{C}'_i$ and $\mathcal{C}'_j$. Hence the planes π_i define a partition of $L\mathcal{C} \setminus L$ into q^2 disjoint conics, as $L\mathcal{C}$ and $L\mathcal{C}'$ only have the line L in common. By the choice of $\mathcal{C}'$ as a conic entirely consisting of internal points of $\mathcal{C}$, Lemma 6.1 yields that any two distinct planes π_i and π_j intersect in an internal point of $L\mathcal{C}$. Consequently, the conics $\pi_i \cap Q(6,q)$ form an i-flock of $L\mathcal{C}$, and hence define a locally hermitian 1-system of $Q(6,q)$. □

It is straightforward to check that the conics $\mathcal{C}$ and $\mathcal{C}'$ introduced in Section 5 are doubly tangent at two conjugate points, namely

$$p_1(1,0,0,\sqrt{-3\gamma},-3\gamma,0,0),\ p_2(1,0,0,-\sqrt{-3\gamma},-3\gamma,0,0).$$

Let $\mathcal{B}$ be the pencil of conics defined by $\mathcal{C}$ and $\mathcal{C}'$. The singular elements of $\mathcal{B}$ are

$$\mathcal{C}_1 : (3\gamma X_0 + X_4)^2 = 0,$$
$$\mathcal{C}_2 : 9\gamma^2 X_0^2 + 12\gamma X_3^2 + X_4^2 - 6\gamma X_0 X_4 = 0.$$

Hence the cross-ratio $\{\mathcal{C}_1, \mathcal{C}_2; \mathcal{C}, \mathcal{C}'\}$ is equal to 4.

Lemma 6.3 *Let* $\mathcal{C}$ *be any nonsingular conic of the plane* $\mathsf{PG}(2,q)$ *and let* r_1, r_2 *be distinct points on* $\mathcal{C}$ *which are conjugate with respect to the extension* $\mathsf{GF}(q^2)$ *of* $\mathsf{GF}(q)$. *Further, let* G *be either the subgroup of* $\mathsf{PGL}(3,q)$ *which fixes* $\mathcal{C}$ *and the line* r_1r_2, *or the subgroup which fixes* $\mathcal{C}$

and also the points r_1 *and* r_2*. If* y *is any point of* $\mathsf{PG}(2,q)$*, not on* $\mathcal{C}$*, not on the line* r_1r_2*, and distinct from the pole* k *of the line* r_1r_2 *with respect to* $\mathcal{C}$ *for* q *odd and distinct from the nucleus* k *of* $\mathcal{C}$ *for* q *even, then the orbit* G_y *is a nonsingular conic doubly tangent to* $\mathcal{C}$ *at* r_1 *and* r_2*.*

Proof Let $\mathcal{B}$ be the pencil of conics in $\mathsf{PG}(2,q)$ defined by the line r_1r_2 counted twice and the nonsingular conic $\mathcal{C}$. Then $\mathcal{B}$ also contains the singular conic consisting of the conjugate lines r_1k and r_2k. Any element σ of G fixes $\mathcal{C}$ and the two singular conics in $\mathcal{B}$. Hence σ fixes each element of $\mathcal{B}$.

Now, let σ be an element of $\mathsf{PGL}(3,q)$ which fixes $\mathcal{C}$ and also the points r_1, r_2. If $y \in \mathsf{PG}(2,q)$ is not on $\mathcal{C}$, not on r_1r_2 and distinct from k, then $y^\sigma = y$ implies that $(r_1y \cap \mathcal{C}) \setminus \{r_1\}$ is fixed for σ, so implies that $\mathcal{C}$ is pointwise fixed, and consequently $\sigma = 1$. It follows that the group G, consisting of all elements of $\mathsf{PGL}(3,q)$ fixing $\mathcal{C}$ and the points r_1, r_2, acts regularly on any nonsingular conic of $\mathcal{B}$. □

Corollary 6.4 *Let* $\mathcal{C}$ *be a nonsingular conic in the plane* $\mathsf{PG}(2,q)$*,* q *odd, and let* r_1*,* r_2 *be distinct points on* $\mathcal{C}$ *which are conjugate with respect to the extension* $\mathsf{GF}(q^2)$ *of* $\mathsf{GF}(q)$*. If* $\mathcal{B}$ *is the pencil of conics defined by* $\mathcal{C}$ *and the line* r_1r_2 *counted twice, then* $\mathcal{B}$ *contains* $\frac{q-3}{2}$ *nonsingular conics consisting entirely of internal points of* $\mathcal{C}$ *and* $\frac{q-1}{2}$ *nonsingular conics consisting entirely of external points of* $\mathcal{C}$*.*

Proof As any element of $\mathsf{PGL}(3,q)$ fixing $\mathcal{C}$ also fixes both the set of all internal points and the set of all external points of $\mathcal{C}$, this follows immediately from Lemma 6.3. □

Theorem 6.5 *Let* $\mathcal{C}$ *be a nonsingular conic in the plane* $\mathsf{PG}(2,q)$*,* q *odd. If* $\mathcal{C}'$ *is a nonsingular conic consisting entirely of internal or entirely of external points of* $\mathcal{C}$*, then* $\mathcal{C}'$ *is doubly tangent to* $\mathcal{C}$ *at two distinct points* r_1*,* r_2 *which are conjugate with respect to the quadratic extension* $\mathsf{GF}(q^2)$ *of* $\mathsf{GF}(q)$*.*

Proof Clearly $\mathcal{C} \cap \mathcal{C}' = \varnothing$ in $\mathsf{PG}(2,q)$. Let $r_1, r_2 \in \mathcal{C} \cap \mathcal{C}'$ in $\mathsf{PG}(2,q^2)$, with r_1 and r_2 conjugate with respect to the quadratic extension $\mathsf{GF}(q^2)$ of $\mathsf{GF}(q)$. Further, let b be the pole of r_1r_2 with respect to $\mathcal{C}$. Then b is a point of $\mathsf{PG}(2,q)$ which is internal for $\mathcal{C}$. The pencil of conics containing $\mathcal{C}$ and the line r_1r_2 counted twice is denoted by $\mathcal{B}$. We must prove that $\mathcal{C}' \in \mathcal{B}$.

Assume by way of contradiction that $\mathcal{C}' \notin \mathcal{B}$. Consider all nonsingular conics of $\mathcal{B}$ having at least one point in common with $\mathcal{C}'$. At most two of these conics are tangent to $\mathcal{C}'$ and possibly $b \in \mathcal{C}'$. In any case there are at least $\frac{q+1}{2}$ such conics. If $\mathcal{C}'$ consists entirely of internal points of $\mathcal{C}$, then, by Lemma 6.3, all these conics of $\mathcal{B}$ consist entirely of internal points of $\mathcal{C}$, contradicting Corollary 6.4 telling that exactly $\frac{q-3}{2}$ nonsingular conics of $\mathcal{B}$ consist entirely of internal points of $\mathcal{C}$.

A similar argument holds if $\mathcal{C}'$ consists entirely of external points of $\mathcal{C}$. So we conclude that $\mathcal{C}' \in \mathcal{B}$. □

According to Corollary 6.4, there are different choices possible for a conic $\mathcal{C}'$ consisting entirely of internal points of $\mathcal{C}$, which, by Theorem 6.2, all give rise to a 1-flock of a quadratic cone LC with line vertex. Also, to every such 1-flock corresponds a locally hermitian 1-system of the parabolic quadric $Q(6,q)$. We have further seen that one specific choice for $\mathcal{C}'$ yields the 1-system of $Q(6,q)$ which is at the same time the (unique) semi-classical non-hermitian spread $\mathcal{S}_{[9]}$ of the hexagon $\mathsf{H}(q)$. The obvious question is whether or not the 1-systems arising from different choices for $\mathcal{C}'$ can be isomorphic.

7. The isomorphism question

First of all we observe the following. In [2], Bloemen, Thas and Van Maldeghem show that every spread of the hexagon $\mathsf{H}(q)$, which is locally hermitian for at least two distinct lines, is a hermitian spread. In fact, the proof they give does not use any specific properties of $\mathsf{H}(q)$; on the contrary, it explicitly uses the fact that a spread of $\mathsf{H}(q)$ is a 1-system of $Q(6,q)$. Consequently it remains valid for 1-systems of $Q(6,q)$ as well.

Theorem 7.1 (Bloemen, Thas, Van Maldeghem [2]) *If a 1-system of the quadric $Q(6,q)$ is locally hermitian for at least two distinct of its elements, then it is a hermitian 1-system.*

Let $\mathcal{M}_i$ be a non-hermitian 1-system of $Q(6,q)$, which is locally hermitian at the line L_i, $i=1,2$, and arising as described in the previous section from a cubic scroll. Choose a regulus R_i in $\mathcal{M}_i$ containing the line L_i, $i=1,2$. Then there is an element $\sigma \in \mathsf{PGL}(7,q)$ fixing $Q(6,q)$, for which $L_1^\sigma = L_2$ and $R_1^\sigma = R_2$. So, investigating the isomorphism question, we may restrict ourselves to non-hermitian 1-systems $\mathcal{M}_1$ and $\mathcal{M}_2$ of $Q(6,q)$, which are locally hermitian at a common line L and for which the corresponding flocks $\mathcal{F}_1$ and $\mathcal{F}_2$ share a conic $\mathcal{C}$. If $\mathcal{R}_i$ is the cubic scroll corresponding to $\mathcal{F}_i$, then let $\mathcal{C}_i' = \mathcal{R}_i \cap \pi$ with π the plane of $\mathcal{C}$, $i=1,2$. Without loss of generality we may assume that $\mathcal{C}_1' \cap \mathcal{C} = \mathcal{C}_2' \cap \mathcal{C}$. Now let β be an element of $\mathsf{PGL}(7,q)$ fixing $Q(6,q)$, for which $\mathcal{M}_1^\beta = \mathcal{M}_2$. By Theorem 7.1 we have $L^\beta = L$. Clearly $\mathcal{F}_1^\beta = \mathcal{F}_2$ and $\mathcal{R}_1^\beta = \mathcal{R}_2$. Let $\mathcal{C}_1'^\beta = \mathcal{C}_2''$, where $\mathcal{C}_2'' = \mathcal{R}_2 \cap \pi'$ for some conic plane

$\pi' \neq \pi$ of $\mathcal{F}_2$. There is an element $\theta \in \mathsf{PGL}(7,q)$ fixing $Q(6,q)$, $\mathcal{R}_2$ and mapping $\mathcal{C}_2''$ onto $\mathcal{C}_2'$. Hence $\beta\theta$ fixes the plane π, the conic $\mathcal{C}$, and maps $\mathcal{C}_1'$ onto $\mathcal{C}_2'$. Then from Lemma 6.3 it follows that $\mathcal{C}_1' = \mathcal{C}_2'$. Also, for $\mathcal{C}_1' = \mathcal{C}_2'$, there is always an element $\alpha \in \mathsf{PGL}(7,q)$ fixing $Q(6,q)$, L, $\mathcal{C}$ and mapping $\mathcal{R}_1$ onto $\mathcal{R}_2$. Consequently, under the action of the subgroup of $\mathsf{PGL}(7,q)$ fixing $Q(6,q)$, there arise $\frac{q-3}{2}$ orbits in the set of all non-hermitian 1-systems of $Q(6,q)$, constructed as described in Section 6.

Let $\mathcal{C}$ be a nonsingular conic of $\mathsf{PG}(2,q)$ and let r_1, r_2 be a pair of conjugate points of $\mathcal{C}$. Let $\mathcal{B}$ be the pencil of conics defined by $\mathcal{C}$ and the line r_1r_2 counted twice, and let $\mathcal{D}$ be the subset of size $\frac{q-3}{2}$ of $\mathcal{B}$ formed by the nonsingular conics of $\mathcal{B}$ consisting entirely of internal points of $\mathcal{C}$. Now let H be the subgroup of $\mathsf{P\Gamma L}(3,q)$ formed by the elements fixing $\mathcal{B}$ and $\mathcal{C}$; then H fixes $\mathcal{D}$. We have $H \cong \mathrm{Aut}(\mathsf{GF}(q))$. Further, let d be the number of orbits of H in $\mathcal{D}$. Relying on the previous paragraph we have the following theorem.

Theorem 7.2 *Under the action of the subgroup of* $\mathsf{P\Gamma L}(7,q)$ *that fixes* $Q(6,q)$ *there arise* d *orbits in the set of all non-hermitian 1-systems of* $Q(6,q)$*, constructed as described in Section 6.*

Remarks 1. It is readily checked that d is equal to the number of orbits of $\mathrm{Aut}(\mathsf{GF}(q))$ in the set of squares of $\mathsf{GF}(q) \setminus \{0,1\}$ (consider the action of H on the parameter $t \in \mathsf{GF}(q) \cup \{\infty\}$ of $\mathcal{B}$).

2. The orbit corresponding to the spread $\mathcal{S}_{[9]}$ of $\mathsf{H}(q)$, q odd and $q \equiv 1 \pmod 3$, has size one. This follows immediately from the remark on the cross-ratio $\{\mathcal{C}_1, \mathcal{C}_2; \mathcal{C}, \mathcal{C}'\}$ preceding Lemma 6.3; this cross-ratio is 4, and so is fixed by $\mathrm{Aut}(\mathsf{GF}(q))$.

8. Characterization and determination of all semi-classical non-hermitian 1-systems of the quadric $Q(6,q)$, q odd

Let $\mathcal{M}$ be a 1-system of $Q(6,q)$, q odd, which is locally hermitian at $L \in \mathcal{M}$. For $x \in L$ arbitrary we consider a 4-dimensional space $\gamma \subseteq x^\perp \setminus \{x\}$, so $\gamma \cap Q(6,q) = Q(4,q)$. Denote the q^2 reguli through L which constitute $\mathcal{M}$ by $R_1, R_2, \ldots, R_{q^2}$; the opposite regulus of R_i will be denoted by $\overline{R}_i$. Each R_i, $i = 1,2,\ldots,q^2$, has a unique transversal $N_i \in \overline{R}_i$ through x, which intersects γ in a point n_i. As $\langle R_i, R_j\rangle$, with $i \neq j$, intersects $Q(6,q)$ in a nonsingular elliptic quadric, the points n_i and n_j are not collinear on $Q(4,q)$. If l denotes the point $L \cap \gamma$, the set $\{l, n_1, n_2, \ldots, n_{q^2}\}$ clearly is an ovoid of $Q(4,q)$. This ovoid will be called the *projection of* $\mathcal{M}$ *from* x *along reguli.* The 1-system $\mathcal{M}$ is called *semi-classical* if for each $x \in L$ the projection of $\mathcal{M}$ from x along reguli is the classical ovoid of $Q(4,q)$. In the sequel, we will denote the projection of $\mathcal{M}$ from x along reguli by $\mathcal{O}_x$.

Consider two reguli R_i, R_j of $\mathcal{M}$ through L. Then $\langle R_i, R_j\rangle \cap Q(6,q)$ is an elliptic quadric $Q^-_{ij}(5,q)$. The quadric $Q^-_{ij}(5,q)$ defines a generalized quadrangle which is the dual of the generalized quadrangle $H(3,q^2)$ arising from a nonsingular hermitian variety in $\mathsf{PG}(3,q^2)$. With R_i and R_j correspond point sets W_i and W_j on $H(3,q^2)$, each consisting of collinear points in $\mathsf{PG}(3,q^2)$ and having exactly one point of $H(3,q^2)$ in common. So W_i and W_j are contained in exactly one common plane which intersects $H(3,q^2)$ in a classical ovoid $\mathcal{O}$ of $H(3,q^2)$. Hence the reguli R_i and R_j uniquely define a hermitian spread $\mathcal{S}$ of $Q^-_{ij}(5,q)$. In $\mathsf{PG}(5,q) = \langle Q^-_{ij}(5,q)\rangle$, consider a $\mathsf{PG}(3,q)$ skew to L. If $\mathcal{S} = \{L, M_1, M_2, \dots, M_{q^3}\}$ and τ is the tangent space of $Q^-_{ij}(5,q)$ at L, then the lines $\langle L, M_i\rangle \cap \mathsf{PG}(3,q)$, $i = 1,2,\dots,q^3$, together with $\tau \cap \mathsf{PG}(3,q) := L'$ form a regular spread S of $\mathsf{PG}(3,q)$; see Bloemen, Thas and Van Maldeghem [2]. Let $x \in L$ and consider the projection $\mathcal{O}'_x$ of $\mathcal{S}$ from x along reguli. Then $\mathcal{O}'_x$ is an elliptic quadric $Q^-(3,q)$, as $\mathcal{S}$ is hermitian. The planes $\langle L, y\rangle$ with $y \in \mathcal{O}'_x$ and $y \notin L$ intersect $\mathsf{PG}(3,q)$ in q^2 points which, together with L', form a plane π_x. Each conic of $\mathcal{O}'_x$ through $l := L \cap \mathcal{O}'_x$ corresponds to a line of π_x, different from L'; this line belongs to the opposite regulus of a regulus of S through L'. Hence each regulus of S containing L' defines a conic of $\mathcal{O}'_x$ through l, and conversely. The set of q reguli of $\mathcal{S}$ defined by a regulus of S containing L' will be called an *$\mathcal{R}$-conic* of $Q(6,q)$. If $x' \in L\setminus\{x\}$, then the plane $\pi_{x'}$ defines the same $\mathcal{R}$-conics as $x \in L$. Hence the projection of an $\mathcal{R}$-conic from $x \in L$ along reguli is a conic of $\mathcal{O}'_x$ for each point $x \in L$.

Now assume that $\mathcal{M}$ is semi-classical. Then, by Bloemen, Thas and Van Maldeghem [2], the unique $\mathcal{R}$-conic containing the reguli R_i and R_j belongs completely to $\mathcal{M}$.

Theorem 8.1 *If q is odd, a locally hermitian, non-hermitian* 1*-system of the quadric $Q(6,q)$ is semi-classical if and only if its associated* i*-flock arises from a rational normal cubic scroll $\mathcal{R}^3$ as described in Section 6.*

Proof Suppose that $\mathcal{M}$ is a non-hermitian 1-system of $Q(6,q)$, q odd, which is locally hermitian at L, and denote the associated i-flock by $\mathcal{F}$.

First assume that $\mathcal{F}$ arises from a rational normal cubic scroll $\mathcal{R}^3$ with directrix line L as described in Section 6. For $x \in L$ arbitrary, we have to show that the projection $\mathcal{O}_x$ of $\mathcal{M}$ from x along reguli onto a 4-dimensional space $\gamma \subseteq x^\perp \setminus \{x\}$, is an elliptic quadric $Q^-(3,q)$. On $\mathcal{R}^3$, there exists a unique generator N through x. Denote the points of

$N \setminus L$ by $s_1, s_2, \ldots, s_q$ and assume that $s_1 \in \gamma$. Then $N^\perp \cap Q(6,q) = xQ^-(3,q)$, where $Q^-(3,q)$ is the intersection with $Q(4,q) = \gamma \cap Q(6,q)$ of the polar space of s_1 with respect to $Q(4,q)$. As $s_i \in N$, it holds that $N^\perp \cap Q(6,q) = xQ^-(3,q)$ is contained in $s_i^\perp \cap Q(6,q) := Q_i^-(5,q)$, $i = 1, 2, \ldots, q$. Each of the quadrics $Q_i^-(5,q)$, $i = 1, 2, \ldots, q$, is the intersection of $\langle R, R' \rangle$ with $Q(6,q)$ for some reguli R, R' of $\mathcal{M}$ through L. Suppose that there exists a regulus $\tilde{R} \in \mathcal{M}$ containing L, the transversal of which through x meets γ in some point $\tilde{n} \notin Q^-(3,q)$. This regulus $\tilde{R}$ defines a conic plane π of $\mathcal{F}$, which intersects each generator of $\mathcal{R}^3$ in a unique point. Hence $s_j \in \pi$ for some $j \in \{1, 2, \ldots, q\}$, so $\tilde{R} \subseteq Q_j^-(5,q)$. This is a contradiction as $xQ^-(3,q) \subseteq Q_j^-(5,q)$ and $x\tilde{n} \subseteq Q_j^-(5,q)$ cannot hold at the same time if $\tilde{n} \notin Q^-(3,q)$. We conclude that the projection of $\mathcal{M}$ from x along reguli is completely contained in $Q^-(3,q)$, so $Q^-(3,q)$ must coincide with $\mathcal{O}_x$. As $x \in L$ was chosen arbitrarily, it follows that $\mathcal{M}$ is semi-classical.

Now suppose that $\mathcal{M}$ is semi-classical and locally hermitian at L, but not hermitian. We will show in several steps that the i-flock associated with $\mathcal{M}$ arises from a rational normal cubic scroll $\mathcal{R}^3$ with directrix line L.

First we prove that each $Q_{ij}^-(5,q) := \langle R_i, R_j \rangle \cap Q(6,q)$, with R_i, R_j reguli of $\mathcal{M}$ through L, contains exactly q reguli of $\mathcal{M}$ and that these reguli form an $\mathcal{R}$-conic. For $x \in L$ arbitrary, the projection $\mathcal{O}_x$ of $\mathcal{M}$ from x along reguli is an elliptic quadric $Q^-(3,q)$. There are exactly q elliptic quadrics $Q_{xi}^-(5,q)$, $i = 1, 2, \ldots, q$, on $Q(6,q)$ through $x\mathcal{O}_x$. Also, if R_j is an arbitrary regulus of $\mathcal{M}$ through L, then $\langle x\mathcal{O}_x, R_j \rangle \cap Q(6,q) = Q_{xk}^-(5,q)$ for some $k \in \{1, 2, \ldots, q\}$. As $\mathcal{M}$ consists of q^2 reguli through L, it is clear that the average number of reguli of $\mathcal{M}$ through L contained in a $Q_{xi}^-(5,q)$ equals $\frac{q^2 \cdot 1}{q} = q$. Assume that there exists a $Q_{xl}^-(5,q)$ which contains at least $q+1$ reguli of $\mathcal{M}$ through L. Then, for $y \in L$, the projection $\mathcal{O}_y$ of $\mathcal{M}$ from y along reguli meets $Q_{xl}^-(5,q)$ in at least $q+1$ points different from $l' := L \cap \mathcal{O}_y$. As l' together with these $q+1$ points necessarily spans $\langle \mathcal{O}_y \rangle$, it follows that the cone $y\mathcal{O}_y$ is also contained in $Q_{xl}^-(5,q)$. So $Q_{xl}^-(5,q)$ contains each of the cones $y\mathcal{O}_y$, $y \in L$. In particular, all transversals of all reguli of $\mathcal{M}$ through L lie on $Q_{xl}^-(5,q)$ and consequently $\mathcal{M} \subseteq Q_{xl}^-(5,q)$. This is a contradiction as $\mathcal{M}$ is assumed not to be hermitian. Hence each $Q_{xi}^-(5,q)$ contains at most q reguli of $\mathcal{M}$ through L, that is, at most the average number of reguli of $\mathcal{M}$ through L in a $Q_{xi}^-(5,q)$, and so they must all contain exactly q reguli of $\mathcal{M}$ through L. We still have to show that every $Q_{ij}^-(5,q) = \langle R_i, R_j \rangle \cap Q(6,q)$, with R_i and R_j distinct reguli of $\mathcal{M}$ through L, is obtained as $Q_{xi}^-(5,q)$ for some $x \in L$. Fix a regulus R_i of

$\mathcal{M}$ through L. Then for each $x \in L$, R_i uniquely defines a $Q^-_{xi}(5,q)$ and every $Q^-_{xi}(5,q)$ contains $q-1$ reguli of $\mathcal{M}\setminus\{R_i\}$ through L. Hence there are at most $(q+1)(q-1) = q^2-1$ reguli of $\mathcal{M}\setminus\{R_i\}$ through L which span a $Q^-_{xi}(5,q)$, $x \in L$, together with R_i. If two of these reguli would coincide, then $Q^-_{xi}(5,q) = Q^-_{yi}(5,q)$ for some $x \neq y$, and so $\mathcal{M}$ belongs to $Q^-_{xi}(5,q)$, a contradiction. Hence there arise exactly q^2-1 distinct reguli of $\mathcal{M}\setminus\{R_i\}$ through L. So every $Q^-_{ij}(5,q)$ is indeed obtained as $Q^-_{xi}(5,q)$ for some $x \in L$. Finally it remains to show that the q reguli in a $Q^-_{ij}(5,q)$ form an $\mathcal{R}$-conic, but this follows immediately from a similar argument as was used to show that $Q^-_{xi}(5,q)$ contains at most q reguli of $\mathcal{M}$ through L.

We summarise the properties of the structure of the conic planes of the i-flock $\mathcal{F}$, associated with $\mathcal{M}$, and the intersection points of these conic planes. Clearly, there are q^2 conic planes, and it follows from the previous part that each intersection point is contained in q conic planes; consequently there are $\frac{q^2(q^2-1)}{q(q-1)} = q(q+1)$ intersection points and each conic plane contains $\frac{q^2-1}{q-1} = q+1$ intersection points.

Next, we show that there exist $q+1$ lines $N_1, N_2, \ldots, N_{q+1} \in L^\perp$ which are disjoint and such that the q points on $N_i \setminus L$ are intersection points of conic planes, for $i = 1, 2, \ldots, q+1$. From the first part of the proof, we know that for each $x \in L$, $x\mathcal{O}_x$ is contained in exactly q elliptic quadrics $Q^-_{xi}(5,q)$, $i = 1, 2, \ldots, q$, which all contain an $\mathcal{R}$-conic of reguli of $\mathcal{M}$. Now $\langle x\mathcal{O}_x\rangle^\perp$ is a line $R_x \subseteq L^\perp$ and tangent to $Q(6,q)$ at x. As $x\mathcal{O}_x \subseteq Q^-_{xi}(5,q)$ for $i = 1, 2, \ldots, q$, the q intersection points of conic planes which correspond to the quadrics $Q^-_{xi}(5,q)$, $i = 1, 2, \ldots, q$ (that is, the points $\langle Q^-_{xi}(5,q)\rangle^\perp$) are exactly the q points of $R_x \setminus L$. Consequently $\{R_x \mid x \in L\}$ is the desired set of lines $\{N_1, N_2, \ldots, N_{q+1}\}$.

Consider a fixed conic plane π of $\mathcal{F}$ and the set of $q+1$ intersection points contained in π. We show that no three of these intersection points are collinear, hence they form an irreducible conic in π. Suppose there exist three collinear intersection points $s_1, s_2, s_3 \in \pi$, with $s_i^\perp \cap Q(6,q) = Q^-_i(5,q)$, $i = 1, 2, 3$. Remark that $Q^-_i(5,q)$, $i = 1, 2, 3$, contains the regulus corresponding to the conic plane π. Then $Q^-_1(5,q) \cap Q^-_2(5,q) = Q^-_2(5,q) \cap Q^-_3(5,q) = Q^-_3(5,q) \cap Q^-_1(5,q) = \langle s_1, s_2, s_3\rangle^\perp \cap Q(6,q)$. As s_1 is an internal point of $\pi \cap Q(6,q)$, the line s_1s_2 is not tangent to $Q(6,q)$ and consequently $\langle s_1, s_2, s_3\rangle^\perp \cap Q(6,q)$ is not singular, so $\langle s_1, s_2, s_3\rangle^\perp \cap Q(6,q) = Q(4,q)$. There exists a point x on L such that $x\mathcal{O}_x \subseteq Q^-_1(5,q)$; clearly $x\mathcal{O}_x \not\subseteq Q^-_i(5,q)$, so $x\mathcal{O}_x \cap Q^-_i(5,q) = x\mathcal{C}_i$, with $\mathcal{C}_i \subseteq \mathcal{O}_x$, for $i = 2, 3$. The conic $\mathcal{C}_i$ defines an $\mathcal{R}$-conic on $Q^-_i(5,q)$, for $i = 2, 3$. It follows that $x\mathcal{C}_i \subseteq Q^-_1(5,q) \cap Q^-_i(5,q)$ and consequently $x\mathcal{C}_i \subseteq Q(4,q)$

for $i = 2, 3$. But $\mathcal{C}_2 \neq \mathcal{C}_3$, as otherwise $Q_2^-(5,q) = Q_3^-(5,q)$, and a contradiction follows. We conclude that π contains an irreducible conic of intersection points of conic planes of $\mathcal{F}$.

Finally we prove that the intersection points of conic planes, together with the line L, form a rational normal cubic scroll $\mathcal{R}^3$ having L as directrix line. Let π be an arbitrary conic plane and p an arbitrary point of the conic $\mathcal{C}$ of intersection points in π. Each of the $q+1$ intersection points in π lies on exactly one line N_i, $i \in \{1, 2, \ldots, q+1\}$; say $p \in N_{q+1}$. Consider a 3-dimensional space $\mathsf{PG}(3,q) \subseteq L^\perp$, skew to p, and project from p onto $\mathsf{PG}(3,q)$. Clearly, the q conic planes through p are projected onto q disjoint lines $L_1', L_2', \ldots, L_q'$ and L is projected onto a line L' disjoint from $L_1', L_2', \ldots, L_q'$. The lines $N_1, N_2, \ldots, N_q$ are projected onto q lines $N_1', N_2', \ldots, N_q'$ of $\mathsf{PG}(3,q)$ and as every conic plane has a unique point in common with each line N_i, every line N_i' meets $L_1', L_2', \ldots, L_q'$ in exactly one point. It is clear that each N_i' also meets L' in a unique point, hence the lines $L', L_1', L_2', \ldots, L_q'$ form a regulus in $\mathsf{PG}(3,q)$. The tangent lines at p of the conics of intersection points lying in the conic planes through p are projected onto q points on respectively $L_1', L_2', \ldots, L_q'$, lying on a line N_{q+1}' belonging to the opposite regulus of $\{L', L_1', L_2', \ldots, L_q'\}$. For all $i, j, k, l \in \{1, 2, \ldots, q+1\}$, the cross-ratio $\{\mathcal{C} \cap N_i, \mathcal{C} \cap N_j; \mathcal{C} \cap N_k, \mathcal{C} \cap N_l\}$ equals the cross-ratio $\{L_1' \cap N_i', L_1' \cap N_j'; L_1' \cap N_k', L_1' \cap N_l'\}$, where we assume that L_1' is the projection of π onto $\mathsf{PG}(3,q)$. But $\{L_1' \cap N_i', L_1' \cap N_j'; L_1' \cap N_k', L_1' \cap N_l'\} = \{N_i', N_j'; N_k', N_l'\} = \{L' \cap N_i', L' \cap N_j'; L' \cap N_k', L' \cap N_l'\} = \{L \cap N_i, L \cap N_j; L \cap N_k, L \cap N_l\}$, so there exists a linear isomorphism which maps the points $N_i \cap L$ onto the points $N_i \cap \mathcal{C}$. It follows (see Semple and Roth [9], p. 133) that the intersection points of the conic planes of $\mathcal{F}$, together with the line L, indeed form a rational normal cubic scroll $\mathcal{R}^3$ with directrix line L. □

Theorem 8.1 characterizes, for q odd, all locally hermitian, semi-classical 1-systems of $Q(6,q)$: either such a 1-system is hermitian or it arises from a rational normal cubic scroll $\mathcal{R}^3$ as described in Section 6. Moreover, Theorem 6.5 determines the cubic scrolls that give rise to a non-hermitian locally hermitian 1-system of $Q(6,q)$. This solves the problem of the characterization and determination of all locally hermitian, semi-classical 1-systems of $Q(6,q)$, q odd, completely.

Acknowledgements

The first author is Research Assistant of the Fund for Scientific Research – Flanders (Belgium) (F.W.O.).

References

[1] L. Bader, G. Lunardon and J.A. Thas, Derivation of flocks of quadratic cones, *Forum Math.* **2** (1990), 163–174.

[2] I. Bloemen, J.A. Thas and H. Van Maldeghem, Translation ovoids of generalized quadrangles and hexagons, *Geom. Dedicata* **72** (1998), 19–62.

[3] V. De Smet and H. Van Maldeghem, The finite Moufang hexagons coordinatized, *Beiträge Algebra Geom.* **34** (1993), 217–232.

[4] J.C. Fisher and J.A. Thas, Flocks in $\mathsf{PG}(3, q)$, *Math. Z.* **169** (1979), 1–11.

[5] H. Gevaert, N.L. Johnson and J.A. Thas, Spreads covered by reguli, *Simon Stevin* **62** (1988), 51–62.

[6] J.W.P. Hirschfeld, *Projective Geometries over Finite Fields*, Second edition, Oxford University Press, Oxford, 1998.

[7] J.W.P. Hirschfeld and J.A. Thas, *General Galois Geometries*, Oxford University Press, Oxford, 1991.

[8] S.E. Payne and J.A. Thas, Conical flocks, partial flocks, derivation and generalized quadrangles, *Geom. Dedicata* **38** (1991), 229–243.

[9] J.G. Semple and L. Roth, *Introduction to Algebraic Geometry*, Oxford University Press, Oxford, 1949.

[10] E.E. Shult and J.A. Thas, m-systems of polar spaces, *J. Combin. Theory Ser. A* **68** (1994), 184–204.

[11] J.A. Thas, Generalized quadrangles and flocks of cones, *European J. Combin.* **8** (1987), 441–452.

DIAGRAMS FOR EMBEDDINGS OF POLYGONS

B. Mühlherr
Fachbereich Mathematik
Universität Dortmund
D-44221 Dortmund
Germany
bernhard.muehlherr@mathematik.uni-dortmund.de

H. Van Maldeghem
Department of Pure Mathematics and Computer Algebra
Ghent University
B 9000 Gent
Belgium
hvm@cage.rug.ac.be

Abstract

We introduce the concept of a "convex embedding" for generalized polygons. This concept emerges from a study of convex subcomplexes of buildings. We review some results on embeddings of generalized polygons in this perspective. We also relate it to some (known) characterization theorems.

1. Introduction

The main examples of buildings, in particular of spherical buildings, are obtained from a restricted class of buildings (the so called *split buildings*) by considering fixed point sets of automorphisms. This way, one obtains *Tits diagrams* from Coxeter diagrams. Geometrically, this fixed point structure is a *convex subcomplex* of the building and one might ask whether there are such complexes not related to automorphisms. Examples are given by buildings related to groups of mixed type in small characteristics. In this paper, we describe this phenomenon in a geometric way. This enables us to recognize other examples, and to relate this to the theory of embeddings of polygons in projective spaces. For

A. Blokhuis et al. (eds.), Finite Geometries, 277–293.

instance, the classical hexagon defined on a quadric in projective 6-space can be seen as such a convex subcomplex (called a folding below), but it does not arise from an automorphism of that quadric. Let us also mention by passing that, even if there is an automorphism around, it might still be hard to recognize the subcomplex. A striking example is given by the (relatively new) Moufang quadrangles discovered by Richard Weiss in 1997, which are convex subcomplexes of certain buildings of type F_4, and which escaped Tits' conjecture about Moufang quadrangles for about 20 years.

More exactly, we will define below *foldings* and *twistings*, which are both convex subcomplexes, but the former uses all types of the original building, while the latter does not. The results for foldings are easier and more convenient to state, but the proofs are basically the same, and therefore, we have chosen to concentrate on foldings for the general theory. We give examples, though, of both foldings and twistings. Our main result is in fact that foldings and twistings in spherical buildings are Moufang buildings. Relating this to the theory of embeddings, this firstly explains why only Moufang polygons turn up in classification results of embeddings, and, secondly, explains the conditions that are frequently used to characterize and classify certain embeddings. This will be explained in detail in Section 4.

The beauty of the theory exposed in this paper lies partly in the following observations. When twisting a building $\mathcal{B}$, the type of the new building $\widetilde{\mathcal{B}}$ can be determined by a calculation in the Weyl group, hence the standard apartment of $\mathcal{B}$ is responsible for the type. Now, if one wants to go back from $\widetilde{\mathcal{B}}$ to $\mathcal{B}$, then one needs to reconstruct the missing elements of $\mathcal{B}$. The existence of these missing elements puts some structure on $\widetilde{\mathcal{B}}$, hence some geometric conditions under which the reconstruction is possible, if starting from a general building of the same type as $\widetilde{\mathcal{B}}$. This produces a characterization of $\widetilde{\mathcal{B}}$. The main observation is now that these conditions in general can already be deduced from the (standard) apartments of $\widetilde{\mathcal{B}}$ and $\mathcal{B}$. Very good examples are provided by the symplectic quadrangles, the split Cayley hexagons and Ree-Tits octagons. We discuss these in Sections 4 and 5. The procedure to convert the standard apartment of $\widetilde{\mathcal{B}}$ into the standard apartment of $\mathcal{B}$ will be called *unfolding*.

Let us finally remark that the folded and twisted diagrams related to the folded and twisted buildings are a generalization of the above mentioned Tits diagram. In the lecture of the conference, the second author called these generalizations *Mühlherr diagrams*, which, for obvious reasons due to the first author, will not be done in the present paper.

2. Folding buildings

2.1. Foldings of Coxeter diagrams

Let I be a set. A *Coxeter diagram over* I is a symmetric matrix $M = (m_{ij})_{i,j \in I}$ such that $m_{ij} \in \{2, 3, \dots\} \cup \{\infty\}$ for all $i \neq j \in I$ and such that all entries on the diagonal are equal to 1. Given a Coxeter diagram over I and a subset J of I then M_J denotes the restriction of M to J.

Let M be a Coxeter diagram over I. A Coxeter system of type M is a pair consisting of a group W and a set $S = \{s_i \mid i \in I\}$ of generators of W such that $s_i s_j$ has order m_{ij} for all $i, j \in I$ and such that the set S together with these relations form a presentation for W. Given a Coxeter system (W, S) of type M and a subset J of I, then we put $S_J = \{s_j \mid j \in J\}$ and $W_J = \langle S_J \rangle$. If (W, S) is a Coxeter system, the we have a natural length function from W into the set of natural numbers, which assigns to each element of W the length of a shortest representation as a product of elements of S; the length of $w \in W$ will be denoted by $l(w)$.

For a given Coxeter diagram M, there exists up to isomorphism only one Coxeter system (W, S) of type M. The diagram is called *spherical* if the corresponding Coxeter group W is finite. In that case there exists a unique element $r \in W$ such that $l(r) > l(w)$ for all $w \in W \setminus \{r\}$; moreover r is an involution.

Let M be a Coxeter diagram over I. A subset J of I is called spherical if M_J is a spherical diagram. In that case we let r_J denote the longest element in W_J. A partition $\tilde{I}$ of I is called a *spherical partition of* M if all elements of $\tilde{I}$ are spherical subsets of I. Given a spherical partition $\tilde{I}$ of M, then we define $\tilde{W} := \langle r_{\tilde{i}} \mid \tilde{i} \in \tilde{I} \rangle$. The spherical partition $\tilde{I}$ is called a *folding of* M if for all $\tilde{w} \in \tilde{W}$ and all $\tilde{i} \in \tilde{I}$ one has $l(\tilde{w} r_{\tilde{i}}) = l(\tilde{w}) + \epsilon l(r_{\tilde{i}})$ for an $\epsilon \in \{1, -1\}$. Given a folding, then we put $R = \{r_{\tilde{i}} \mid \tilde{i} \in \tilde{I}\}$. It follows from [4] that $(\tilde{W}, R)$ is a Coxeter system of type $(o(r_{\tilde{i}} r_{\tilde{j}}))_{\tilde{i}, \tilde{j} \in \tilde{I}}$. This matrix is denoted by $\tilde{M}$.

Coxeter diagrams will also be considered as graphs in the usual way; in particular we will talk about the *nodes* of a diagram. An *irreducible* diagram is then a connected one.

2.2. Foldings of rank 2

Let M be a Coxeter diagram over I and let $\tilde{I}$ be folding of M. The rank of the folding is defined to be the cardinality of $\tilde{I}$. Later on we will be interested in rank 2 foldings of irreducible spherical diagrams. These can be described as follows.

Proposition 2.1 *Let M be an irreducible spherical Coxeter diagram over I and let $\tilde{I} = \{\alpha, \beta\}$ be a rank 2 folding of M. Then either $\tilde{I}$ corresponds to the bipartite partition of the diagram and $\tilde{m}_{\alpha\beta}$ is equal to the Coxeter number of M or one of the following holds, with notation from* [1]:

Type I $M = A_{2n}, \alpha = \{\dots n-4, n-2, n, n+1, n+3, n+5, \dots\}, \beta = \{\dots, n-5, n-3, n-1, n+2, n+4, n+6, \dots\}$ *and* $\tilde{m}_{\alpha\beta} = 2n$;

Type II $M = E_6, \alpha = \{1, 2, 6\}, \beta = \{3, 4, 5\}$ *and* $\tilde{m}_{\alpha\beta} = 8$;

Type III $M = F_4, \alpha = \{1, 4\}, \beta = \{2, 3\}$ *and* $\tilde{m}_{\alpha\beta} = 8$.

The previous proposition is proved in [5] for the diagrams A_n, C_n and D_n; there are only finitely many cases remaining, which can easily be handled.

2.3. Buildings

Let I be a set, let M be a Coxeter matrix over I and let (W, S) be the Coxeter system of type M. A *building* of type M is a pair $\mathcal{B} = (\mathcal{C}, \delta)$ where $\mathcal{C}$ is a set and where $\delta : \mathcal{C} \times \mathcal{C} \to W$ is a *distance function* satisfying the following axioms where $x, y \in \mathcal{C}$ and $w = \delta(x, y)$:

(Bu 1) $w = 1$ if and only if $x = y$;

(Bu 2) if $z \in \mathcal{C}$ is such that $\delta(y, z) = s \in S$, then $\delta(x, z) = w$ or ws, and if, furthermore, $l(ws) = l(w) + 1$, then $\delta(x, z) = ws$;

(Bu 3) if $s \in S$, there exists $z \in \mathcal{C}$ such that $\delta(y, z) = s$ and $\delta(x, z) = ws$.

Given a building $\mathcal{B} = (\mathcal{C}, \delta)$, then the elements of $\mathcal{C}$ are called *chambers*. Given a set $J \subseteq I$ and $x \in \mathcal{C}$, the *J-residue* of x is the set $R_J(x) = \{y \in \mathcal{C} | \delta(x, y) \in W_J\}$. Each J-residue is a building of type M_J with the distance function induced by δ (cf [8];p.30). Note that a set consisting of one element $c \in \mathcal{C}$ is the $\varnothing$-residue of c. The rank of a J-residue is the cardinality of J. The building $\mathcal{B}$ is called *thick* if each rank 1 residue of $\mathcal{B}$ contains at least 3 chambers. The numerical distance between two chambers $c, d \in \mathcal{C}$ is defined by $l(c, d) := l(\delta(c, d))$.

Let $\tilde{I}$ be a folding of M. A subset $\tilde{\mathcal{C}}$ of $\mathcal{C}$ is called folding of type $\tilde{I}$ if $\delta(\tilde{c}, \tilde{d}) \in \tilde{W}$ for all $\tilde{c}, \tilde{d} \in \tilde{\mathcal{C}}$ and if $\tilde{\mathcal{B}} := (\tilde{\mathcal{C}}, \tilde{\delta})$ is a building of type $\tilde{M}$, where $\tilde{\delta}$ denotes the restriction of δ to $\tilde{\mathcal{C}}$. The folding $\tilde{\mathcal{C}}$ is called thick if the building $\tilde{\mathcal{B}}$ is a thick building.

2.4. Projections and opposition

Throughout this subsection let M be a Coxeter diagram over a set I and let $\mathcal{B} = (\mathcal{C}, \delta)$ be a building of type M. Let $R \subseteq \mathcal{C}$ be a residue in $\mathcal{B}$ and let c be a chamber of $\mathcal{B}$. Then there exists a unique chamber $d \in R$

such that $l(c,x) = l(c,d) + l(d,x)$ for all $x \in R$. This chamber is called the *projection of* c *onto* R and it will be denoted by $\mathrm{proj}_R c$.

Suppose now that M is spherical and let r denote the longest element in the corresponding Coxeter group. Two chambers in a building of type M are called *opposite* if their distance is equal to r.

The following observations follow easily from the definitions:

Lemma 2.2 *Let $\tilde{\mathcal{C}}$ be a folding of the building $\mathcal{B} = (\mathcal{C}, \delta)$ of type $\tilde{I}$. Let $\tilde{J}$ be a subset of $\tilde{I}$ and put $J = \cup_{\tilde{j} \in \tilde{J}} \tilde{j}$. Let $\tilde{c}, \tilde{d} \in \tilde{\mathcal{C}}$ and let R (respectively $\tilde{R}$) denote the J-residue (respectively the $\tilde{J}$-residue) of $\tilde{d}$. Then the projection of $\tilde{c}$ on R in $\mathcal{B}$ and the projection of $\tilde{c}$ on $\tilde{R}$ in $\tilde{\mathcal{B}}$ coincide. In particular,* $\mathrm{proj}_R \tilde{c} \in \tilde{\mathcal{C}}$.

Lemma 2.3 *Let $\tilde{I}$ be a folding of the Coxeter diagram M over I and let $\tilde{J}$ be a spherical subset of $\tilde{I}$ with respect to the Coxeter diagram $\tilde{M}$. Put $J = \cup_{\tilde{j} \in \tilde{J}} \tilde{j}$. Then M_J is spherical and the longest element of $\tilde{W}_{\tilde{J}}$ coincides with the longest element in W_J. In particular, if $\tilde{\mathcal{C}}$ is a folding of a building $\mathcal{B}$ of type M and if $\tilde{c}, \tilde{d}$ are contained in a $\tilde{J}$-residue of $\tilde{\mathcal{B}}$. Then they are opposite in that residue if and only if they are opposite in the corresponding J-residue of $\mathcal{B}$.*

Using the previous two lemmas and similar ideas as in the proof of Tits' Rigidity-Theorem for spherical buildings (Theorem 4.1.1 in [18]) one can easily prove the following proposition.

Proposition 2.4 *Let $\tilde{\mathcal{C}}$ be a thick folding of the spherical building $\mathcal{B}$ of type $\tilde{I}$, let $\tilde{c}, \tilde{d}$ be two opposite chambers in $\tilde{\mathcal{B}}$. Given a folding $\tilde{\mathcal{C}}'$ of $\mathcal{B}$ of type $\tilde{I}$ such that $\tilde{c}, \tilde{d} \in \tilde{\mathcal{C}}'$ and such that $R_{\tilde{i}}(\tilde{c}) \cap \tilde{\mathcal{C}} = R_{\tilde{i}}(\tilde{c}) \cap \tilde{\mathcal{C}}$ for all $\tilde{i} \in \tilde{I}$, then $\tilde{\mathcal{C}} = \tilde{\mathcal{C}}'$.*

2.5. The Moufang property of spherical buildings

Let $\mathcal{B}$ be a building. An automorphism of $\mathcal{B}$ is called *unipotent* if it fixes at least one chamber and if it fixes all chambers in each panel, in which it fixes at least 2 chambers. The following lemma is immediate.

Lemma 2.5 *Suppose that a unipotent automorphism of the building $\mathcal{B}$ fixes two opposite chambers in a spherical residue R of $\mathcal{B}$. Then it fixes each chamber in R.*

Suppose now that $\mathcal{B} = (\mathcal{C}, \delta)$ is a spherical building. Given a chamber $c \in \mathcal{C}$ then c^{op} denotes the set of all chambers of $\mathcal{B}$ which are opposite to c. Moreover, given a chamber $d \in c^{op}$, then c_d^{op} denotes the set of all chambers x in c^{op} such that $\mathrm{proj}_P x = \mathrm{proj}_P d$ for all panels P containing c.

The equivalence of (i) and (ii) of the following condition is proved in [5]; the equivalence of (i) and (iii) can be established in a similar way.

Proposition 2.6 *Let $\mathcal{B}$ be a thick and irreducible building of spherical type. Then the following conditions are equivalent:*

(i) *$\mathcal{B}$ is Moufang in the sense of* [18];

(ii) *for each chamber c of $\mathcal{B}$ there is a transitive group of unipotent automorphisms of $\mathcal{B}$ fixing c and acting transitively on c^{op};*

(iii) *for any pair c, d of opposite chambers in $\mathcal{B}$ there is a group of unipotent automorphisms of $\mathcal{B}$ fixing c and acting transitively on c^{op}_d.*

2.6. Foldings of spherical Moufang buildings

Throughout this subsection we assume that M is an irreducible spherical Coxeter diagram over a set I, that $\tilde{I}$ is a folding and that $\tilde{M}$ is the Coxeter diagram over $\tilde{I}$ which is obtained by the folding $\tilde{I}$. Note that $\tilde{M}$ is also an irreducible spherical Coxeter diagram. We assume furthermore that $\mathcal{B} = (\mathcal{C}, \delta)$ is a Moufang building of type M and that $\tilde{\mathcal{C}}$ is a thick folding of $\mathcal{B}$ of type $\tilde{I}$. We obtain hence in this way a building $\tilde{\mathcal{B}} = (\tilde{\mathcal{C}}, \tilde{\delta})$.

The following observation is an easy consequence of Lemmas 2.3 and 2.5.

Lemma 2.7 *Let u be a unipotent automorphism of $\mathcal{B}$ which stabilizes an element of $\tilde{\mathcal{C}}$ and which normalizes $\tilde{\mathcal{C}}$. Then the restriction of u onto $\tilde{\mathcal{C}}$ is a unipotent automorphism of $\tilde{\mathcal{B}}$.*

Let $\tilde{c}$ be a chamber in $\tilde{\mathcal{C}}$, let $\tilde{d}, \tilde{d}' \in \tilde{\mathcal{C}}$ be two chambers opposite to $\tilde{c}$ and suppose that $\mathrm{proj}_{R_{\tilde{i}}(\tilde{c})}\tilde{d} = \mathrm{proj}_{R_{\tilde{i}}(\tilde{c})}\tilde{d}'$ for each $\tilde{i} \in \tilde{I}$. The unique unipotent automorphism u of $\mathcal{B}$ which sends $\tilde{d}$ to $\tilde{d}'$ stabilizes $x_{\tilde{i}} := \mathrm{proj}_{R_{\tilde{i}}(\tilde{c})}\tilde{d}$ for each $\tilde{i}$ and as $x_{\tilde{i}}$ is opposite to $\tilde{c}$ in $R_{\tilde{i}}(\tilde{c})$ it follows from Lemma 2.5 that u induces the identity on this residue. Let $\mathcal{C}'$ denote the image of $\mathcal{C}$ under u. It follows that $\mathcal{C}'$ is also a folding of $\mathcal{B}$ of type $\tilde{I}$. Moreover, $\tilde{c}, \tilde{d}'$ are opposite chambers contained in both foldings and by the considerations above we have $R_{\tilde{i}}(\tilde{c}) \cap \tilde{\mathcal{C}} = R_{\tilde{i}}(\tilde{c}) \cap \tilde{\mathcal{C}}'$. We conclude by Proposition 2.4 that the foldings $\tilde{\mathcal{C}}$ and $\tilde{\mathcal{C}}'$ coincide and hence that u stabilizes $\tilde{\mathcal{C}}$. Now, by Lemma 2.7, the restriction of u to $\tilde{\mathcal{C}}$ is a unipotent automorphism of $\tilde{\mathcal{B}}$, which sends $\tilde{d}$ onto $\tilde{d}'$. Finally we obtain by the third characterization of irreducible spherical Moufang buildings in Proposition 2.6 the following theorem.

Theorem 2.8 *A thick folding of rank at least 2 of an irreducible spherical Moufang building is Moufang.*

3. Twisting buildings

3.1. Twisting Coxeter diagrams

Let M be a Coxeter diagram over I. A *generalized spherical partition of M* is a pair $(K, \tilde{J})$ consisting of a subset K of I and a partition $\tilde{J}$ of $J := I \setminus K$ such that $K \cup \tilde{j}$ is a spherical subset of I and such that $r_{K\cup\tilde{j}}$ centralizes $r_{\tilde{j}}$ for each $\tilde{j} \in \tilde{J}$. Given a generalized spherical partition of M, then we put $\tilde{W} = \langle r_{K\cup\tilde{j}} \mid \tilde{j} \in \tilde{J}\rangle$. We call the generalized partition $(K, \tilde{J})$ a twisting of M if for each $\tilde{j} \in \tilde{J}$ and each $\tilde{w} \in \tilde{W}$ the following holds: $l(\tilde{w}r_{K\cup\tilde{j}}) = l(\tilde{w}) + \epsilon l(r_{K\cup\tilde{j}})$ for an $\epsilon \in \{1, -1\}$. Given a twisting $(K, \tilde{J})$ then $(\tilde{W}, R)$ is a Coxeter system of a certain type $\tilde{M}$ over $\tilde{J}$, where $R = \{r_{K\cup\tilde{j}} \mid \tilde{j} \in \tilde{J}\}$.

3.2. Twisting spherical buildings

In this section we consider buildings as simplicial complexes and we freely make use of the definitions of [18]. Let M be a Coxeter diagram over I. A building Δ of type M is hence a simplicial complex $(\mathbf{S}, \subseteq)$ which is endowed with an apartment system. The elements of $\mathbf{S}$ are the simplices of Δ and we have natural type function $\tau : \mathbf{S} \to 2^I$. The buildings considered in this section will always be assumed to be of spherical type. A twisting of a spherical building $\Delta = (\mathbf{S}, \subseteq)$ is a subset $\tilde{\mathbf{S}}$ such that $\mathrm{proj}_{\tilde{A}}\tilde{B}$ is contained in $\tilde{\mathbf{S}}$ for all $\tilde{A}, \tilde{B} \in \mathbf{S}$, such that for each $\tilde{A} \in \tilde{\mathbf{S}}$ there exists $\tilde{B} \in \tilde{\mathbf{S}}$ which is opposite to $\tilde{A}$ in Δ and such that each comaximal element in $\tilde{\mathbf{S}}$ is contained in at least 3 maximal ones.

Given a twisting $\tilde{\mathbf{S}}$ of a spherical building $\Delta = (\mathbf{S}, \subseteq)$ then the partial ordered set $\tilde{\Delta} = (\tilde{\mathbf{S}}, \subseteq)$ is a thick spherical building and the restriction of the type function τ onto $\tilde{\mathbf{S}}$ is a type function on $\tilde{\Delta}$. In particular, all chambers of $\tilde{\Delta}$ have the same type J, which is a subset of I and the types of the vertices in $\tilde{\Delta}$ yield a partition $\tilde{J}$ of J. The data consisting of $I \setminus J$ and the partition $\tilde{J}$ of J is called the type of the twisting $\tilde{\mathbf{S}}$. It is a twisting of the Coxeter diagram M in the sense of the previous subsection.

Just as in the case of a folding, one can show in a similar fashion, the following result.

Theorem 3.1 *A thick twisting of rank at least 2 of an irreducible spherical Moufang building is Moufang.*

4. Embeddings of polygons

4.1. Convex embeddings

In the rank 2 case, twisted and folded buildings very often produce embeddings of generalized polygons in projective spaces in the classical sense. Let us first define what we mean by an *embedding.*

Let $\mathcal{P}$ be a generalized polygon, that is, a spherical rank 2 building; we view $\mathcal{P}$ as a simplicial complex of dimension 2, where we call the vertices of one type "points" and the ones of the other type "lines". Also, a simplex of dimension 1 is called a "flag" and the vertices of such a simplex are called "incident". Let $\mathrm{PG}(d,k)$ be the d-dimensional projective space defined over the skew field k; for the definitions below, this projective space may be infinite dimensional, but for the purpose of the present paper, we may assume that d is a natural number. An *embedding* of $\mathcal{P}$ in $\mathrm{PG}(d,k)$ is an injective identification of the point set and the line set of $\mathcal{P}$ with a point set and a line set, respectively, of $\mathrm{PG}(d,k)$ in such a way that incident elements in $\mathcal{P}$ are also incident in $\mathrm{PG}(d,k)$ (but the converse is not required), and with the additional condition that the point set of $\mathcal{P}$ generates $\mathrm{PG}(d,k)$.

Suppose that we are given a generalized polygon $\mathcal{P}$ as a folding of a certain spherical building $\mathcal{B}$ of type M, where M is a Coxeter diagram over I, and let $\widetilde{I}$ be its type. Suppose also that the building $\mathcal{B}$ is a folding of an A_n-building $\mathcal{S}$, that is, a projective space, and putting $J=\{1,2,\dots,n\}$, let $\widetilde{J}$ be its type. Suppose these foldings satisfy the following two properties.

(i) Denote by a the node in J corresponding to the point set of $\mathcal{S}$. Let $\widetilde{a}$ be the class of $\widetilde{J}$ containing a. Let b be the node of I corresponding to the class $\widetilde{a}$ of $\widetilde{J}$, and let $\widetilde{b}$ be the class of $\widetilde{I}$ containing b. Then $\widetilde{b}$ corresponds to the point set of $\mathcal{P}$.

(ii) Denote by a' the node in J corresponding to the line set of $\mathcal{S}$. Let $\widetilde{a}'$ be the class of $\widetilde{J}$ containing a'. Let b' be the node of I corresponding to the class $\widetilde{a}'$ of $\widetilde{J}$, and let $\widetilde{b}'$ be the class of $\widetilde{I}$ containing b'. Then $\widetilde{b}'$ corresponds to the line set of $\mathcal{P}$.

We then obtain a folding of $\mathcal{S}$ into $\mathcal{P}$ such that the simplices of $\mathcal{S}$ that correspond to the points of $\mathcal{P}$ contain points of $\mathcal{S}$, and the simplices of $\mathcal{S}$ that correspond to the lines of $\mathcal{P}$ contain lines of $\mathcal{S}$. If we now consider the points and lines of $\mathcal{S}$ that are precisely contained in a simplex belonging to the folding, then we obviously obtain an embedding of $\mathcal{P}$ into the projective subspace of $\mathcal{S}$ generated by the points and lines of the embedded polygon.

Obviously, one can generalize the previous observations to twistings.

Let us call an embedding obtained from a folding (or, more generally, a twisting) as just described a *convex embedding*. Two main questions immediately arise.

(i) Can one classify all convex embeddings of generalized polygons?

(ii) Under which conditions is a particular class of embeddings convex?

Both questions have indirectly received much attention in the literature. Indeed, these questions are very closely related to the classification and characterization of (lax) embeddings of polygons in the sense of [14]. It is well known that it is for the moment an impossible task to classify all embeddings of polygons, even in the finite case. Hence, one considers extra conditions. These additional conditions are very often just part of the geometric interpretation of the appropriate twisted A_n-diagram; see the examples below. Also, these conditions are chosen in the best economical way, and very often the first steps in the classification is to prove the remaining part of the geometrical interpretation. In this way, one singles out the convex embeddings, which are usually called the "standard embeddings" or the "natural embeddings", obtaining characterizations for these. This is a job worthwhile doing because there are a lot of non-convex embeddings, and even some classes of those can be classified under some mild conditions; see examples below.

5. Examples

In this section, we review some results on embeddings and put these in the perspective of the present paper.

5.1. Generalized quadrangles

In all examples of convex embeddings of generalized quadrangles, the corresponding twisted diagram has the property that the nodes belonging to points and hyperplanes are in the same class. For now, think of the "natural embeddings" of classical quadrangles as quadrics or Hermitian varieties in projective space. Hence, for each such embedding of a quadrangle $\mathcal{Q}$, to every point p of $\mathcal{Q}$ corresponds a unique hyperplane H of the projective space. This is precisely the hyperplane spanned by the points of $\mathcal{Q}$ collinear with p, that is, incident with a common line with p. This motivates the following definition, which is phrased in general.

Let the generalized polygon $\mathcal{P}$ be embedded in the projective space $\mathrm{PG}(d,k)$, for some field k. The embedding is *polarized*, if for each point p of $\mathcal{P}$, the set of points of $\mathcal{P}$ not opposite p does not span $\mathrm{PG}(d,k)$.

Note that the condition involved in a polarized embedding is a little weaker than just hypothesizing a hyperplane for each point. But in most cases, one shows that for a polarized embedding, the set of points not opposite a given point p spans a hyperplane that does not contain any point opposite p. Note also that the definition of polarized embedding implies that $\mathcal{P}$ has even diameter.

Now we introduce another condition for embeddings.

Let the generalized polygon $\mathcal{P}$ be embedded in the projective space $\mathrm{PG}(d,k)$, for some field k, then we call the embedding *full* if for each line L of $\mathcal{P}$ the set of points of $\mathcal{P}$ incident with L in $\mathcal{P}$ coincides with the set of points of $\mathrm{PG}(d,k)$ incident with L in $\mathrm{PG}(d,k)$.

Applied to the case of generalized quadrangles, we see that the notion of polarized embedding is a geometric translation of one single feature of the twisted diagram behind the convex embedding. But it appears to be enough to be able to classify them. By inspection of the list obtained in [10, 11], together with the results in [3] and [13], the following result may be stated.

Theorem 5.1 *Every full embedding of a generalized quadrangle into projective space is a polarized embedding. Also, all polarized embeddings can be explicitly enumerated. They are all convex, except if the quadrangle is isomorphic to the unique quadrangle $W(2)$ with three points per line and three lines per point, and the projective space has characteristic different from 2 . In the finite case, every convex embedding is obtained from a full embedding by just extending the field of the ambient projective space.*

Note that $W(2)$ has a unique exceptional non-convex embedding in $\mathrm{PG}(4,k)$ for every field k of characteristic $\neq 2$.

The results in [15] show that in may cases the geometric conditions can be weakened; sometimes, just a bound on the dimension of the ambient projective space is enough. Also, all projections of the convex embeddings in lower-dimensional spaces are characterized in [15]: these are examples of classifications of non-convex embeddings under very mild conditions.

5.2. Generalized hexagons

Here are some remarkable cases. Let us first determine the twisted diagrams of some Moufang hexagons. We consider the hexagons obtained from a triality, as discovered and explained by Tits in [16]. First, let $\mathcal{H}$

be the (so-called split Cayley) generalized hexagon corresponding to the group $G_2(k)$, for some field k, where we choose the line set in such a way that it corresponds to a line set of the triality quadric. It is known, see [16] again, that $\mathcal{H}$ lives on a non-degenerate quadric $Q(6,k)$ of maximal Witt index in projective 6-space, and that the points of $\mathcal{H}$ collinear to any point p of $\mathcal{H}$ are exactly the points of a plane $p^\perp$ of $Q(6,k)$; the lines of $\mathcal{H}$ are some lines of $Q(6,k)$. Hence we may regard the points of $\mathcal{H}$ really as point-plane flags and we obtain a convex subcomplex of $Q(6,k)$. Hence the embedding is convex and the folded diagram is the bipartite C_3-diagram. Combined with the folded diagram of $Q(6,k)$, we obtain a folded A_6-diagram of type I for $\mathcal{H}$; see Proposition 2.1. Geometrically, we consider the flags of type $\{p, p^\perp, (p^\perp)^\theta, p^\theta\}$, where p is any point of $\mathcal{H}$, and where θ is the (possibly degenerate) polarity associated with $Q(6,k)$. These flags form the vertices of one type. The vertices of the other type are obtained by considering a line L of $\mathcal{H}$ and defining the flag $\{L, L^\theta\}$. These two sets of flags form a folding, as defined in Section 2.

If the characteristic of k is equal to 2, then we can project the whole situation from the kernel of $Q(6,k)$ into a hyperplane of the ambient projective space and we obtain as folded diagram the bipartite A_5-diagram. For later reference, we call this full embedding a *symplectic* embedding, since it lives inside a symplectic space; see [23, 2.4.14] for more details.

Now consider any Moufang hexagon $\mathcal{T}$ obtained from a triality τ, and choose as line set the elements of $\mathcal{T}$ that are lines of the corresponding triality quadric $Q^+(7,k)$, a quadric of maximal Witt index in $\mathrm{PG}(7,k)$. Let θ be the polarity (possibly degenerate) associated with $Q^+(7,k)$. To each point p of $\mathcal{T}$, we associate the flag $F_p = \{p, p^\tau \cap p^{\tau^2}, (p^\tau \cap p^{\tau^2})^\theta, p^\theta\}$. To each line L of $\mathcal{T}$ we associate the flag $F_L = \{L, L^\theta\}$. Then the set of all flags F_p, F_L, with p a point of $\mathcal{T}$ and L a line of $\mathcal{T}$, forms a convex subcomplex. The diagram is not a folded one, but a twisted A_7, where one partitions the subset of types $\{1,2,3,5,6,7\}$ into $\{1,3,5,7\}$ and $\{2,6\}$, where type i means as usual "$(i-1)$-dimensional subspaces". The corresponding embeddings are hence convex. If the field automorphism involved in τ is trivial, then the embedding actually is contained in a hyperplane, and it is the previous example: this is an example for which the embedding derived from the twisting does not generate the whole projective space in which the twisting occurs.

Now, in all these examples of twisted diagrams, the class of types containing the node corresponding to the points of the projective space also contains the nodes corresponding to planes and hyperplanes. So a partial geometric interpretation of these diagrams is to say that the corresponding embedding is polarized and also *flat*, if flat means the following.

Let the generalized polygon $\mathcal{P}$ be embedded in the projective space $\mathrm{PG}(d,k)$, for some field k, then we call the embedding *flat* if for each point p of $\mathcal{P}$ the set of points of $\mathcal{P}$ collinear to p is contained in a plane of $\mathrm{PG}(d,k)$.

It turns out that, at least in the finite case, the flat polarized embeddings of hexagons can be classified, and they are all convex. In fact, we have the following theorem; see [14].

Theorem 5.2 *Let $\mathcal{H}$ be a generalized hexagon embedded in the projective space $\mathrm{PG}(d,k)$. If the embedding is flat and polarized, then $\mathcal{H}$ is a Moufang hexagon and every unipotent automorphism of $\mathcal{H}$ is induced by a linear transformation of $\mathrm{PG}(d,k)$. If moreover k is finite, then the embedding is convex, $\mathcal{H}$ arises from a triality and the embedding arises by field extension from either the full embedding corresponding to that triality, or from a symplectic embedding and, in this case, $\mathcal{H}$ is the split Cayley hexagon over a field of characteristic* 2.

The classification in the infinite case is still open, but this should not be too hard a problem in view of the explicit list of Moufang hexagons obtained by Tits and Weiss [20].

Also, it is an open question whether the dual of the triality hexagons admit convex embeddings.

Again, [14] shows that weaker conditions can be used to characterize the convex embeddings. Let us remark that there are no characterizations known of non-convex embeddings.

5.3. Generalized octagons

There is not much literature about embeddings of generalized octagons. In fact, the Main Result in [12] says that there can be no flat embedding of any (thick) octagon, hence, in view of the classification of the rank 2 foldings of A_n-diagrams (see Proposition 2.1), we deduce that there can be no convex embedding arising from a folding, for any generalized octagon (finite or infinite).

The Moufang octagons arise naturally from a polarity in a building of type F_4. Hence these octagons admit convex embeddings in these rank 4 buildings, and the folded diagram is of type III, in the notation of Proposition 2.1. One could ask for a characterization of this embedding in terms of geometric conditions. Our conjecture is that these conditions are again the translation of the folded diagram. Work in progress shows that this should indeed be the case. We leave it to the reader to formulate the geometric properties corresponding to the folded diagram of type III.

6. Unfolding apartments

In this section, we look at a problem at the same time related and opposite to the the main ideas of the previous sections. Indeed, we constructed buildings *inside* other buildings, and used twisted diagrams to determine the type of the smaller building. Now we are interested in the following question: given that smaller building, can we reconstruct the bigger ambient building? The fact that this is possible should induce some specific properties in the smaller building that distinguishes it from other buildings of the same type. In this paragraph, we want to show that in many cases the apartments already tell us the relevant conditions, in contrast to the twisting, where the diagrams hold the geometric information and hence we talk about unfolding the apartments. We do so with three very characteristic examples.

6.1. The symplectic quadrangle

Consider the symplectic quadrangle $W(k)$, naturally embedded in the projective 3-space $\mathrm{PG}(3,k)$ and related to a symplectic polarity. The embedding is convex and has a bipartite A_3-diagram. The apartment of $W(k)$ is an ordinary quadrangle Q. The apartment of $\mathrm{PG}(3,k)$ is a tetrahedron T, that is, a complete graph on 4 points. Adding the diagonals of Q to Q, one obtains T. This observation suggests that, in order to reconstruct $\mathrm{PG}(3,k)$ from $W(k)$, we should take all points and lines of $W(k)$, and define some new lines by joining opposite points. In the apartment Q, this addition creates triangles. Hence in the thick case, we should create generalized triangles, that is, projective planes. We translate the situation in the apartment in a straightforward way as follows: the points of the new lines are collinear with two opposite points, and the new lines that are "collinear" with a certain point p exactly form, together with the old lines through p, a generalized triangle. Indeed, we have the following theorem, due to Schroth [9], where $p^\perp$ denotes the set of points collinear with p.

Theorem 6.1 *Let $\mathcal{Q}$ be any generalized quadrangle. Let p be any point of $\mathcal{Q}$ and define the geometry $\mathcal{Q}_p$ as follows. The points of $\mathcal{Q}_p$ are the points in $\mathcal{Q}$ collinear with p; the lines are the lines of $\mathcal{Q}$ through p and the intersections $p^\perp \cap p'^\perp$ for p' opposite p; incidence is the natural one. If $\mathcal{Q}_p$ is a projective plane for all points p of $\mathcal{Q}$, then $\mathcal{Q}$ is isomorphic to $W(k)$ for some field k.*

In view of the remarks preceding the theorem, the proof is straightforward: one defines $\mathrm{PG}(3,k)$ is the obvious way; the symplectic polarity follows readily. As was the case with the diagrams, we can sometimes consider weaker conditions, or alternative ones. For instance, in the finite case one just requires that $\mathcal{Q}_p$ is a linear space and that the number of points on every line is equal to the number of lines through every point, see [6]. In [23], the following general weaker theorem is proved.

Theorem 6.2 *Let $\mathcal{Q}$ be any generalized quadrangle. If for every point p of $\mathcal{Q}$ the geometry $\mathcal{Q}_p$ (defined in the previous theorem) is a partial linear space, and if for some point x this geometry $\mathcal{Q}_x$ is a projective plane, then $\mathcal{Q}$ is isomorphic to $W(k)$ for some field k.*

6.2. The split Cayley hexagon

We can try to do a similar thing with other generalized n-gons. Let us first try to find the geometric property that is symbolized by joining two points p, p' of an apartment A, an ordinary n-gon, which are joined to a common point x. Let v be the element of A opposite x; here, v can be a line or a point. The points p, p' are characterized by the property that they are all points of A at distance 2 from x and $n-2$ from v, where distances are taken in the incidence graph of A. So in an arbitrary generalized n-gon $\mathcal{P}$ containing A, we can define the geometry $\mathcal{P}_x$ as follows. The points of $\mathcal{P}_x$ are the points of $\mathcal{P}$ collinear with x; the lines are the lines of $\mathcal{P}$ through x, together with the sets of points collinear with x and at distance $n-2$ from some element v opposite x, and such a set is denoted by x^v. Incidence is the natural one. In the thin case, that of A, this geometry is a triangle; so $\mathcal{P}_x$ should be a projective plane. If this is the case, then we call x a *projective point.* The interesting thing is that now we can formulate an equivalent condition which will enable us to generalize this approach in a more elegant way to points p, p' at distance $2j > 4$.

Let P be a set of points of the generalized polygon Γ. Let us denote by $P^{\perp\!\!\perp}$ the set of points in Γ not opposite any element of P. For two non-opposite elements v, w, let us call the *projection of v onto v'* the element incident with v' closest to v. Then we have the following result by Proposition 2.8 of [2], where the *codistance* in a generalized n-gon equals n minus the distance in the incidence graph.

Proposition 6.3 *A point p of a generalized polygon is projective if and only if for every two non-collinear points x, y in $p^\perp$, the projection of the set $\{x,y\}^{\perp\!\!\perp\perp\!\!\perp}$ onto any element at codistance 1 of every point in $\{x,y\}^{\perp\!\!\perp\perp\!\!\perp}$ is either constant or surjective.*

Applied to a generalized pentagon, we obtain a condition under which such a geometry would admit an embedding in a projective 4-space such that all points of the space are covered and such that the points collinear with a fixed point all lie in a plane. Unfortunately, no generalized pentagon can have two collinear projective points by [21]. This explains partly why there are no natural "beautiful" generalized pentagons.

Let us have a look at the generalized hexagons. Joining the points at distance 4 of an ordinary hexagon gives us the apartment of a building of type C_3. Indeed, one can easily deduce from [7] the following theorem; see also 6.3.1 of [23].

Theorem 6.4 *If every point of a generalized hexagon $\mathcal{H}$ is projective, then $\mathcal{H}$ is isomorphic to the split Cayley hexagon associated to Dickson's group of type G_2 naturally embedded in the quadric $Q(6,k)$ of maximal Witt index in projective 6-space. The points and lines of $Q(6,k)$ are the points and the lines of $\mathcal{H}$ together with the sets $\{x,y\}^{\perp\!\!\perp\perp\!\!\perp}$, for all non-collinear non-opposite points x,y.*

In fact it again suffices that every $\mathcal{H}_p$ is a linear space and at least one $\mathcal{H}_p$ is a projective plane.

Now if we moreover join opposite points of an ordinary hexagon, then we obtain a complete graph on 6 vertices, hence the apartment of projective 5-space. Translating this to the thick case, and in view of Proposition 6.3, we require, for a generalized hexagon $\mathcal{H}$, that for every two opposite points x,y, the projection of the set $\{x,y\}^{\perp\!\!\perp\perp\!\!\perp}$ onto any element at codistance 1 of every point in $\{x,y\}^{\perp\!\!\perp\perp\!\!\perp}$ is either constant or surjective. This explains Theorem 5.1. of [2]. We rephrase that result as follows.

Theorem 6.5 *If, in a generalized hexagon $\mathcal{H}$, for every two non-collinear points x,y, the projection of the set $\{x,y\}^{\perp\!\!\perp\perp\!\!\perp}$ onto any element at codistance 1 of every point in $\{x,y\}^{\perp\!\!\perp\perp\!\!\perp}$ is either constant or surjective, then the points of $\mathcal{H}$ and the lines of $\mathcal{H}$ together with all sets $\{x,y\}^{\perp\!\!\perp\perp\!\!\perp}$, for x,y non-collinear, are the points and the lines, respectively, of a projective 5-space. In this case, $\mathcal{H}$ is isomorphic to the split Cayley hexagon over a perfect field of characteristic 2.*

In fact, it turns out that we only need to require the given condition for *opposite* points x,y; see again [2].

Since not all points of a generalized 7-, 8- or 9-gons can be projective, no unfolding of the apartments of these polygons will extend to the thick case. And it readily follows from [21] that the same is true for all generalized n-gons, $n \geq 10$. This is in conformity with the fact that Moufang n-gons exist only for $n = 3,4,6,8$, although also the Moufang octagons do not behave nicely in that respect.

6.3. Ree–Tits octagons

All Moufang octagons are isomorphic to Ree–Tits octagons, arising from the Ree groups of type ${}^2F_4$; these have been generalized by Tits [19] to non-perfect fields. A geometric characterization of the *perfect* Ree–Tits octagons, that is, where the underlying field is perfect, is presented in [22]. There, the method is precisely to reconstruct the ambient building of type F_4, from which the Ree–Tits octagon arises as a folding. Also, in this case, restriction to the apartment shows how to obtain the 24-cell, the apartment of a building of type F_4, from an ordinary octagon. In fact, the position of the missing points in the apartment helped to prove the thick case. So in this case it is not just an *a posteriori* explanation; the role of the apartments were crucial for the result.

It would take too much space to reproduce and explain here the conditions of the characterization in [22]. However, let us simply write down how one constructs the 24-cell from an ordinary octagon, and then comment shortly on the necessary geometric interpretations.

Let b_i, $i \in \mathbf{Z}$ mod 8, be a natural cyclic ordering of the vertices of an ordinary octagon. Then we define new vertices a_j, $j \in \mathbf{Z} + \frac{1}{2}$ mod 8, and join a_j to b_i with $i \in \{j - \frac{3}{2}, j - \frac{1}{2}, j + \frac{1}{2}, j + \frac{3}{2}\}$. Also, we define new vertices c_ℓ, $\ell \in \mathbf{Z}$ mod 8 and join c_ℓ to $b_{\ell-1}$, $b_{\ell+1}$, $c_{\ell-3}$ and $c_{\ell+3}$, and also to a_j for $j \in \{\ell - \frac{5}{2}, \ell - \frac{1}{2}, \ell + \frac{1}{2}, \ell + \frac{5}{2}\}$.

In the thick case, the vertices b_i are the points of the Ree–Tits octagon $\mathcal{O}$, the vertices a_j are the points of the building $\mathcal{B}$ of type F_4 collinear to collinear points of the octagon; they are recognized inside $\mathcal{O}$ as sets of lines satisfying certain geometric conditions. The vertices c_ℓ are the vertices of $\mathcal{B}$ collinear to all points of a trace in $\mathcal{O}$, and they are recognized inside $\mathcal{O}$ as certain classes of points related to the corresponding trace, but one trace gives rise to different vertices in $\mathcal{B}$; this is an extra complication. For more details, see [22].

References

[1] N. Bourbaki, *Groupes et Algèbres de Lie, Chap. IV-VI*, Hermann, Paris, 1968.

[2] J. van Bon, H. Cuypers and H. Van Maldeghem, Hyperbolic lines in generalized polygons, *Forum Math.* **8** (1994), 343–362.

[3] F. Buekenhout, C. Lefèvre-Percsy, Generalized quadrangles in projective spaces, *Arch. Math.* **25** (1974), 540–552.

[4] B. Mühlherr, Coxeter groups in Coxeter groups, *Finite Geometry and Combinatorics* (eds. F. De Clerck et al.), London Math. Soc. Lecture Note Series **191**, Cambridge University Press, Cambridge, 1993.

[5] B. Mühlherr, Some contributions to the theory of buildings based on the gate property, Doctoral Dissertation, University of Tübingen, 1994.

[6] S.E. Payne and J.A. Thas *Finite Generalized Quadrangles*, Pitman Research Notes **110**, Longman, London, 1984.

[7] M.A. Ronan, A geometric characterization of Moufang generalized hexagons, *Invent. Math.* **57** (1980), 227–262.

[8] M.A. Ronan, *Lectures on Buildings*, Academic Press, San Diego, 1989.

[9] A.E. Schroth, Characterizing symplectic quadrangles by their derivations, *Arch. Math.* **58** (1992), 98–104.

[10] A. Steinbach and H. Van Maldeghem, Generalized quadrangles weakly embedded of degree > 2 in projective space, *Forum Math.* **11** (1999), 139–176.

[11] A. Steinbach and H. Van Maldeghem, Generalized quadrangles weakly embedded of degree 2 in projective space, *Pacific J. Math.* **193** (2000), 227–248.

[12] J.A. Thas and H. Van Maldeghem, Embedded thick finite generalized hexagons in projective space, *J. London Math. Soc.* **54** (1996), 566–580.

[13] J.A. Thas and H. Van Maldeghem, Generalized quadrangles weakly embedded in finite projective space *J. Statist. Plann. Inference* **73** (1998), 353–361.

[14] J.A. Thas and H. Van Maldeghem, Flat lax and weak lax embeddings of finite generalized hexagons, *European J. Combin.* **19** (1998), 733–751.

[15] J.A. Thas and H. Van Maldeghem, Lax embeddings of finite generalized quadrangles, *Proc. London Math. Soc.*, to appear.

[16] J. Tits, Sur la trialité et certains groupes qui s'en déduisent, *Inst. Hautes Études Sci. Publ. Math.* **2** (1959), 14–60.

[17] J. Tits, Géometries polyédriques et groupes simples, *Deuxième réunion du groupement de mathématiciens d'expression latine*, Florence (1961), 66–88.

[18] J. Tits, *Buildings of Spherical Type and Finite BN-Pairs*, Lecture Notes in Math. **386**, Springer, Berlin, 1974.

[19] J. Tits, Moufang octagons and the Ree groups of type ${}^2F_4$, *Amer. J. Math.* **105** (1983), 539–594.

[20] J. Tits and R. Weiss, *The Classification of Moufang Polygons*, to appear.

[21] H. Van Maldeghem, The non-existence of certain regular generalized polygons, *Arch. Math.* **64** (1995), 86–96.

[22] H. Van Maldeghem, A geometric characterization of the perfect Ree–Tits generalized octagons, *Proc. London Math. Soc.* **76** (1998), 203–256.

[23] H. Van Maldeghem, *Generalized Polygons*, Monographs in Math. **93**, Birkhäuser, Basel, 1998.

THE LAW–PENTTILA q-CLAN GEOMETRIES

Stanley E. Payne
Department of Mathematics
Campus Box 170
University of Colorado at Denver
P.O. Box 173364
Denver, Colorado 80217-3364
United States of America
spayne@carbon.cudenver.edu

1. Introduction

For a prime power q, a q-clan is a set of q 2×2 matrices over $GF(q)$ whose pairwise differences give anisotropic quadratic forms. These objects are used to construct translation planes of order q^2, flocks of quadratic cones in $PG(3,q)$, and generalized quadrangles with parameters (q^2, q).

Law and Penttila [5] have constructed a q-clan for each q with $q = 3^e$, $e \geq 1$. This family generalizes an example they found and investigated by computer for $q = 27$. That example was shown to be new, but they left open the question of whether or not all their examples for $q > 27$ are new. We show that their examples for $q > 27$ are indeed new by computing the collineation group of the associated generalized quadrangles. For each q only one generalized quadrangle arises, but as q increases the number of pairwise inequivalent flocks and planes grows quite rapidly. Moreover, when e is a power of an odd prime, many of the flocks that arise are rigid, that is, they have trivial automorphism group. It seems that these examples provide the first infinite family of rigid flocks.

2. The Law–Penttila construction

For any prime power $q = p^e$ let $F = GF(q)$ be the Galois field of order q. A q-clan is a set $\mathcal{C} = \{A_t = \begin{pmatrix} x_t & y_t \\ 0 & z_t \end{pmatrix} : t \in F\}$ of pairwise distinct 2×2 matrices over F whose pairwise differences $A_t - A_s$ (for $s \neq t$) are

A. Blokhuis et al. (eds.), Finite Geometries, 295–303.

anisotropic. Associated with a q-clan is a flock of a quadratic cone, a line-spread of $PG(3,q)$ and its associated translation plane, and a generalized quadrangle $GQ(\mathcal{C})$ of order (q^2,q). A reader of this paper is presumed to be familiar with these constructions so that most the technical details of these constructions will not be repeated here. The main result to be applied is the Fundamental Theorem of q-Clan Geometry given in Payne [6]. However, the survey by Johnson and Payne [3] contains all the necessary definitions and results to be quoted here along with many other related results and many references. It should be noted that the term "rigid" has been used differently by other authors; see [1].

For the remainder of this paper let $q = 3^e$ and let $n = \boxbslash \in F$, that is, n is a fixed nonsquare of F. For $t \in F = GF(q)$ put

$$A_t = \begin{pmatrix} t & t^4 + nt^2 \\ t^4 + nt^2 & -\frac{1}{n}t^9 + t^7 + n^2t^3 - n^3t \end{pmatrix}$$

$$\equiv \begin{pmatrix} t & -t^4 - nt^2 \\ 0 & -\frac{t}{n}(t^2 - n)^4 \end{pmatrix} = \begin{pmatrix} x_t & y_t \\ 0 & z_t \end{pmatrix}.$$

Then according to [5], $\mathcal{C} = \{A_t : t \in F\}$ is a q-clan. The case $q = 27$ was first handled by computer and given in [4], where the stabilizer of the associated flock was also given. Hence in the present work we find it very convenient to assume that $q \geq 81$. Also, recall the standard notation: $K_t = A_t + A_t^T = 2A_t$.

We thank M. Law and T. Penttila for communicating to us their results in [5] long before the publication of that paper.

3. The stabilizer of the line $[A(\infty)]$

We want to determine all collineations of the generalized quadrangle $GQ(\mathcal{C})$ where $\mathcal{C}$ is the Law–Penttila q-clan given just above. Since $GQ(\mathcal{C})$ is not classical, the point (∞) is fixed by all collineations; see [7]. Hence it suffices to determine those collineations fixing the point $(\overline{0}, 0, \overline{0})$. In this section we start by determining those collineations that fix the line $[A(\infty)]$ and the point $(\overline{0}, 0, \overline{0})$. By the Fundamental Theorem [6] this is equivalent to determining all permutations $\pi : t \mapsto \bar{t}$ of the elements of F, $0 \neq \lambda \in F$, $\begin{pmatrix} a & b \\ c & d \end{pmatrix} \in GL(2,q)$, $\sigma \in Aut(F)$, for which

$$\begin{bmatrix} x_{\bar{t}} \\ y_{\bar{t}} \\ z_{\bar{t}} \\ 1 \end{bmatrix} = \begin{pmatrix} \lambda a^2 & \lambda ab & \lambda b^2 & x_{\bar{0}} \\ 2\lambda ac & \lambda(ad+bc) & 2\lambda bd & y_{\bar{0}} \\ \lambda c^2 & \lambda cd & \lambda d^2 & z_{\bar{0}} \\ 0 & 0 & 0 & 1 \end{pmatrix} \begin{bmatrix} x_t^\sigma \\ y_t^\sigma \\ z_t^\sigma \\ 1 \end{bmatrix}. \tag{3.1}$$

Here $x_t = t$, $y_t = -t^4 - nt^2$, $z_t = -\frac{1}{n}t^9 + t^7 + n^2t^3 - n^3t$.

From Equation (3.1) we get three equations:

$$\bar{t} = \lambda a^2 t^\sigma + \lambda ab(-t^4 - nt^2)^\sigma + \lambda b^2(-\frac{1}{n}t^9 + t^7 + n^2t^3 - n^3t)^\sigma + \bar{0}. \tag{3.2}$$

$$-\bar{t}^4 - n\bar{t}^2 = 2\lambda act^\sigma - \lambda(ad+bc)[t^{4\sigma} + n^\sigma t^{2\sigma}] + 2\lambda bd[-\frac{1}{n^\sigma}t^{9\sigma} + t^{7\sigma} + n^{2\sigma}t^{3\sigma} - n^{3\sigma}t^\sigma] - \bar{0}^4 - n\bar{0}^2. \tag{3.3}$$

$$-\frac{1}{n}\bar{t}^9 + \bar{t}^7 + n^2\bar{t}^3 - n^3\bar{t} = \lambda c^2 t^\sigma - \lambda cd[t^{4\sigma} + n^\sigma t^{2\sigma}] + \lambda d^2[-\frac{1}{n^\sigma}t^{9\sigma} + t^{7\sigma} + n^{2\sigma}t^{3\sigma} - n^{3\sigma}t^\sigma] - \frac{1}{n}\bar{0}^9 + \bar{0}^7 + n^2\bar{0}^3 - n^3\bar{0}. \tag{3.4}$$

Now use Equation (3.2) to write $-\bar{t}^4 - n\bar{t}^2$ as a polynomial in t^σ, in fact as an *even* polynomial, and equate it to the right hand side of Equation (3.3). Since the degree of the resulting polynomial that t^σ must satisfy (for all $t^\sigma \in F$) is less than 81, all the coefficients must be zero. In particular, the odd degree part on the right hand side of Equation (3.3) must be zero; that is, for all $t^\sigma \in F$ we must have that

$$0 = 2\lambda act^\sigma + 2\lambda bd[-\frac{1}{n^\sigma}t^{9\sigma} + t^{7\sigma} + n^{2\sigma}t^{3\sigma} - n^{3\sigma}t^\sigma]. \tag{3.5}$$

The coefficient of $(t^\sigma)^9$ must be 0; so $bd = 0$. The coefficient of t^σ must be 0; so $ac = 0$. This implies that

$$a = d = 0 \text{ and } bc \neq 0. \tag{3.6}$$

or

$$b = c = 0 \text{ and } ad \neq 0. \tag{3.7}$$

In particular, from Equation (3.2) we now have

$$\bar{t} = \frac{-\lambda b^2}{n^\sigma}(t^\sigma)^9 + \lambda b^2(t^\sigma)^7 + \lambda b^2 n^{2\sigma}(t^\sigma)^3 + \lambda(a^2 - b^2 n^{3\sigma})t^\sigma + \bar{0}. \tag{3.8}$$

From Equation (3.3) we have

$$-\bar{t}^4 - n\bar{t}^2 = -\lambda(ad+bc)[(t^\sigma)^4 + n^\sigma(t^\sigma)^2] - \bar{0}^4 - n\bar{0}^2. \tag{3.9}$$

When we use Equation (3.7) in Equation (3.9) we get a polynomial equation in t^σ whose highest degree term is $-\left(\frac{-\lambda b^2}{n^\sigma}\right)^4 (t^\sigma)^{36}$. It follows that $b = 0$. So $b = c = 0$ and $ad \neq 0$.

This simplifies matters considerably. We now have the following:

$$\bar{t} = \lambda a^2 t^\sigma + \bar{0}; \tag{3.10}$$

$$-\bar{t}^4 - n\bar{t}^2 = -\lambda ad[(t^\sigma)^4 + n^\sigma (t^\sigma)^2] - \bar{0}^4 - n\bar{0}^2. \tag{3.11}$$

After some computation we find that

$$\begin{aligned} -\bar{t}^4 - n\bar{t}^2 &= -\lambda^4 a^8 (t^\sigma)^4 - \bar{0}\lambda^3 a^6 (t^\sigma)^3 - n\lambda^2 a^4 (t^\sigma)^2 \\ &\quad + \lambda a^2 (\bar{0}n - \bar{0}^3) t^\sigma - \bar{0}^4 - n\bar{0}^2 \\ &= -\lambda ad(t^\sigma)^4 - \lambda adn^\sigma (t^\sigma)^2 - \bar{0}^4 - n\bar{0}^2. \end{aligned} \tag{3.12}$$

The coefficient on $(t^\sigma)^3$ forces $\bar{0} = 0$. The remaining conditions are now:

(i) $\lambda ad = \lambda^4 a^8$, implying $d = \lambda^3 a^7$;

(ii) $n\lambda^2 a^4 = \lambda adn^\sigma$, implying $d = n^{1-\sigma}\lambda a^3$.

From the above it follows quickly that the collineations of $GQ(\mathcal{C})$ fixing the line $[A(\infty)]$ and the point $(\bar{0}, 0, \bar{0})$ are completely determined as follows. Choose nonzero $a \in F$ in $q - 1$ ways. Recall that $n = \not\square$ is fixed. Let $\sigma \in Aut(F)$. Put $\lambda = \pm n^{\frac{1-\sigma}{2}} a^{-2}$ and $d = \lambda^3 a^7 = \pm n^{\frac{3(1-\sigma)}{2}} a$. Then it can be checked that Equations (3.2), (3.3) and (3.4) all hold. So the group $\mathcal{G}_0$ fixing $[A(\infty)]$ and $(\bar{0}, 0, \bar{0})$ has order $2(q-1)e$, where $q = 3^e$. If we replace a with a^{-1}, the corresponding collineation of $GQ(\mathcal{C})$, in the usual notation of [6, p.192], has the form

$$(\alpha, c, \beta) \mapsto$$
$$\left(a\alpha^\sigma \begin{pmatrix} 1 & 0 \\ 0 & \pm n^{\frac{3(\sigma-1)}{2}} \end{pmatrix}, \pm n^{\frac{1-\sigma}{2}} a^2 c^\sigma, a\beta^\sigma \begin{pmatrix} \pm n^{\frac{1-\sigma}{2}} & 0 \\ 0 & n^{2(1-\sigma)} \end{pmatrix} \right). \tag{3.13}$$

Putting $\pm = +$ and $\sigma = id$ gives the $q - 1$ elements of the q-clan kernel. This kernel is the kernel of the action of $\mathcal{G}_0$ on the lines through (∞), so the action on those lines is given by the group of order $2e$ given by

$$t \mapsto \bar{t} = \pm n^{\frac{1-\sigma}{2}} t^\sigma.$$

The kernel mentioned above is also the kernel of a homomorphism $\theta : T \to T_\theta$ mapping $\mathcal{G}_0$ onto the group of collineations of $PG(3, q)$ leaving invariant the corresponding flock of a quadratic cone. Hence the group leaving the flock invariant has order $2e$. This group is the direct product of a cyclic group of order e with a group of order 2.

4. The orbit of $[A(\infty)]$

In [6] it is shown that with each line $[A(s)]$ through the point (∞) in $GQ(\mathcal{C})$ we may associate a q-clan $\mathcal{C}^{i_s}$ in such a way that two such lines are in the same orbit of $\mathcal{G}_0$ if and only if their associated q-clans are equivalent. Here the original q-clan $\mathcal{C}$ is associated with the line $[A(\infty)]$. Then for $s \in F$, the q-clan associated with $[A(s)]$ may be described as

$$\mathcal{C}^{i_s} = \mathcal{C}' = \{A'_t = -\left(\det(K_{t+s} - K_s)\right)^{-1}(A_t - A_s) : t \in F\}.$$

Then $\mathcal{C}'$ is normalized, that is, $A'_0 = \begin{pmatrix} 0 & 0 \\ 0 & 0 \end{pmatrix}$, and since here q is odd, $A'_t = \begin{pmatrix} x_t & y_t/2 \\ y_t/2 & z_t \end{pmatrix}$ is symmetric.

The important point here is that $GQ(\mathcal{C})$ has a collineation moving $[A(\infty)]$ to $[A(s)]$ if and only if $\mathcal{C}$ and $\mathcal{C}'$ are equivalent, which is if and only if there exist nonzero $\lambda \in F$, $B = \begin{pmatrix} a & b \\ c & d \end{pmatrix} \in GL(2,q)$, $\sigma \in Aut(F)$, $\pi \in Sym_F$ such that for all $t \in F$:

$$\begin{bmatrix} x_{t^\pi} \\ y_{t^\pi} \\ z_{t^\pi} \\ 1 \end{bmatrix} = \begin{pmatrix} \lambda a^2 & \lambda ab & \lambda b^2 & x_{0^\pi} \\ 2\lambda ac & \lambda(ad+bc) & 2\lambda bd & y_{0^\pi} \\ \lambda c^2 & \lambda cd & \lambda d^2 & z_{0^\pi} \\ 0 & 0 & 0 & 1 \end{pmatrix} \begin{bmatrix} (x'_t)^\sigma \\ (y'_t)^\sigma \\ (z'_t)^\sigma \\ 1 \end{bmatrix}. \tag{4.1}$$

Using $q = 3^e$ so that $A_t = \begin{pmatrix} x_t & -y_t \\ -y_t & z_t \end{pmatrix}$, it is easy to check that

$$\det(K_{t+s} - K_s) = \det(A_{t+s} - A_s).$$

As a polynomial in t,

$$\Delta_t = \det(A_{t+s} - A_s) = -\frac{1}{n}t^{10} + \cdots t^2(-n^3 - (s^3 - ns)^2) + 0 \cdot t + 0.$$

The point here is that the coefficient on t and the constant term in Δ_t are both zero.

Calculating the matrices $A'_t = -\frac{1}{\Delta_t}(A_{t+s} - A_s) = \begin{pmatrix} x'_t & -y'_t \\ -y'_t & z'_t \end{pmatrix}$, we find

$$\begin{aligned} x'_t &= \frac{-t}{\Delta_t}; \\ -y'_t &= \frac{-(t^4 + t^3 sts^3 + n(t^2 - st))}{\Delta_t}; \\ z'_t &= \frac{\frac{1}{n}t^9 - t^7 - t^6 s + t^4 s^3 + t^3 s^4 - ts^6 - n^2 t^3 + n^3 t}{\Delta_t}. \end{aligned} \tag{4.2}$$

Using Equations (4.1) (top row) and (4.2) and the fact that $x_{t^\pi} = t^\pi$, we obtain an expression for t^π:

$$\begin{aligned} t^\pi &= \frac{1}{\Delta_t^\sigma}\Big(0^\pi \Delta_t^\sigma + \frac{1}{n^\sigma}(t^\sigma)^9 + \cdots \\ &\qquad -\lambda t^\sigma(a^2 + ab(n^\sigma - s^{2\sigma})s^\sigma + b^2(s^{6\sigma} - n^{3\sigma}))\Big). \end{aligned}$$

Hence

$$t^\pi = \frac{1}{\Delta_t^\sigma}(0^\pi \cdot \Delta_t^\sigma + f_9(t^\sigma)), \tag{4.3}$$

where $f_9(t^\sigma)$ is a polynomial in t^σ of degree 9 with zero constant term and whose coefficient on the term of degree 1 is clearly visible just above. Keep in mind that Δ_t^σ is a polynomial in t^σ with degree 10 and divisible by $(t^\sigma)^2$.

Now we can use the second row of Equations (4.1) and Equation (4.3) to find two formulas for $-(t^\pi)^4 - (t^\pi)^2$ and equate them. Then multiply both sides by $(\Delta_t^\sigma)^4$ to obtain:

$$\begin{gathered} -(0^\pi \Delta_t^\sigma + f_9(t^\sigma))^4 - (\Delta_t^\sigma)^2(0^\pi \cdot \Delta_t^\sigma + f_9(t^\sigma))^2 = \\ (\Delta_t^\sigma)^3 \left\{ (t^\sigma)^9 \left(\frac{-\lambda bd}{n^\sigma}\right) + \right. \\ \cdots + (t^\sigma)[\lambda ac + \lambda(ad+bc)(s^{3\sigma} - n^\sigma s^\sigma) - \lambda bd(n^{3\sigma} - s^{6\sigma})] + \\ \left. \Delta_t^\sigma(-(0^\pi)^4 - (0^\pi)^2) \right\}. \end{gathered}$$

Compute the coefficient of $(t^\sigma)^4$ on both sides of the above equation. On the right it is 0 and on the left it is

$$[-\lambda(a^2 + ab(n^\sigma - s^{2\sigma})s^\sigma + b^2(s^{6\sigma} - n^{3\sigma}))]^4,$$

which then must be zero. Since $\lambda \neq 0$, this is equivalent to having

$$(a,b)\begin{pmatrix} 1 & (s^{2\sigma} - n^\sigma)s^\sigma \\ (s^{2\sigma} - n^\sigma)s^\sigma & s^{6\sigma} - n^{3\sigma} \end{pmatrix}\begin{pmatrix} a \\ b \end{pmatrix} = 0.$$

But, if Δ is the determinant of the 2×2 matrix in this equation, an easy computation shows that $-\Delta = n^\sigma(s^{2\sigma} - n^\sigma)^2 = \square$, since $n^\sigma = \square$ and $0 \neq (s^2 - n)^{2\sigma}$ is a square. Hence the inner matrix is anisotropic, forcing $(a,b) = (0,0)$. But this is impossible since $\begin{pmatrix} a & b \\ c & d \end{pmatrix} \in GL(2,q)$.

5. Rigid flocks

For each choice of $\sigma \in Aut(F)$ and either choice of $\pm$, we have a collineation of $GQ(\mathcal{C})$ that maps $[A(t)]$ to $[A(\bar{t})]$, where

$$\bar{t} = \pm n^{\frac{1-\sigma}{2}} t^{\sigma}.$$

So $[A(\infty)]$ and $[A(0)]$ each form an orbit of size 1. For $0 \neq t \in F$, if the orbit of $[A(t)]$ has size h, then the stabilizer of $[A(t)]$ has order $2e/h$, so the stabilizer of the corresponding flock also has order $2e/h$. So the flocks corresponding to $[A(\infty)]$ and $[A(0)]$ both have stabilizers of order $2e$. We claim that there are many cases where there are orbits of size $2e$, so that the corresponding flocks have trivial stabilizer, that is, they are *rigid.*

For the remainder of this article, let $q = 3^e$ where e is odd. In this case we may take $n = -1$. Then for $\epsilon = \pm 1$, $\sigma : x \mapsto x^{3^i}$ (for $1 \leq i \leq e$), define $\pi_{i,\epsilon}$ to be the permutation of the elements of $F^* = F \setminus \{0\}$ given by

$$\pi_{i,\epsilon} : t \mapsto \epsilon \cdot (-1)^{\frac{1-3^i}{2}} t^{3^i} = \epsilon \cdot (-1)^i t^{3^i}.$$

We want to use the usual orbit-counting lemma to count the orbits of $\mathcal{H} = \{\pi_{i,\epsilon} : 1 \leq i \leq e;\ \epsilon = \pm 1\}$. Clearly $\pi_{i,\epsilon} : t \mapsto t$ if and only if $t^{3^i-1} = \epsilon(-1)^i$. Hence the number N of orbits of $\mathcal{H}$ acting on F^* is given by

$$N = \frac{1}{2e} \sum_{i,\epsilon} |\{t : t^{3^i-1} = \epsilon(-1)^i\}|.$$

We first note that $t^{3^i-1} = -1$ has no solution, since -1 is a nonsquare in F, and $t^{3^i-1} = +1$ has $3^d - 1$ solutions where $d = gcd(i,e)$. From this it is straightforward to check the following:

$t^{3^i-1} = \epsilon(-1)^i$ has *no* solution when
  $\epsilon = +1$ and i is odd, or $\epsilon = -1$ and i is even;
$t^{3^i-1} = \epsilon(-1)^i$ has $3^d - 1$ solutions when
  $\epsilon = +1$ and i is even, or $\epsilon = -1$ and i is odd; here $d = \gcd(i,e)$.

If $d|e$, so d is odd, then $d = (i,e)$ if and only if $i = d \cdot i'$ where $1 \leq i' \leq h/e$ and $\gcd(i', h/e) = 1$. Moreover, $i \equiv i' \pmod 2$. Half the i' are even; half are odd. It now follows that

$$N = \sum_{d|e} (3^d - 1) \cdot \phi(\frac{e}{d}).$$

A special case that is amusing to consider is where $e = p^h$ where p is an odd prime, $1 \leq h$. Here the number N of orbits is given by

$$\begin{aligned}
N &= \frac{1}{2p^h}\sum_{i=0}^{h}(3^{p^i}-1)\phi(p^{h-i}) \\
&= \frac{1}{2p^h}\left\{\left[\sum_{i=0}^{h-1}(3^{p^i}-1)(p^{h-i}-p^{h-i-1})\right]+(3^{p^h}-1)\cdot 1\right\} \\
&= \frac{1}{2p^h}\left[p^h(3-1)+\sum_{i=1}^{h-1}p^i(3^{p^{h-1}}-1-3^{p^{h-i-1}}+1)\right. \\
&\qquad\left. +3^{p^h}-1-3^{p^{h-1}}+1\right] \\
&= 1+\sum_{j=1}^{h}\frac{3^{p^j}-3^{p^{j-1}}}{2p^j}.
\end{aligned}$$

A finite induction shows that the number of orbits on $GF(3^{p^j})\backslash GF(3^{p^{j-1}})$ is $\frac{3^{p^j}-3^{p^{j-1}}}{2p^j}$. Moreover, for $j \geq 1$ these are all orbits of length $2p^j$. So each $t \in GF(3^{p^j}) \setminus GF(3^{p^{j-1}})$ corresponds to a flock whose stabilizer has order $\frac{2p^h}{2p^j} = p^{h-j}$. Hence the distinct $3^{p^{h-1}}\left(\frac{3^{\phi(p^h)}-1}{2p^h}\right)$ orbits in $GF(3^{p^h}) \setminus GF(3^{p^{h-1}})$ correspond to rigid flocks.

The translation plane associated with a given flock is derivable to give several "Bose-Barlotti Δ-planes." Johnson [2] has shown that their isomorphism types are in bijective correspondence with the orbits of the generators of the cone $x_0x_2 = x_1^2$ under the group stabilizing the flock. We can show that this action is given (on the points of the generators lying in the plane $x_3 = 0$) by

$$(0,0,1,0) \mapsto (0,0,1,0)$$

$$(1,d,d^2,0) \mapsto (1,\pm n^{\frac{3(\sigma-1)}{2}}d^\sigma, n^{3(\sigma-1)}d^{2\sigma},0)$$

With h odd and $n = -1$, this is essentially the same action as that studied above. This is only for the q-clan $\mathcal{C}$ and flock $\mathcal{F}(\mathcal{C})$. If we start with a rigid flock, we get $3^h + 1$ planes. Hence in general for $q = 3^e$ we have huge numbers of distinct translation planes of order q^2 and having dimension 2 over their kernel.

References

[1] V. Jha and N. L. Johnson, Rigidity in conical flocks, *J. Combin. Theory Ser. A* **73** (1996), 60–76.

[2] N. L. Johnson, The derivation of dual translation planes, *J. Geom.* **36** (1989), 63–90.

[3] N. L. Johnson and S. E. Payne, Flocks of Laguerre planes and associated geometries, *Mostly Finite Geometries* (ed. N. L. Johnson), Dekker, New York, 1997, 51–122.

[4] M. Law and T. Penttila, BLT-sets over small fields II, preprint, (2000).

[5] M. Law and T. Penttila, Some flocks in characteristic 3, preprint, 1999.

[6] S. E. Payne, The fundamental theorem of q-clan geometry, *Des. Codes Cryptogr.* **8** (1996), 181–202.

[7] S. E. Payne and J. A. Thas, Conical flocks, partial flocks, derivation and generalized quadrangles, *Geom. Dedicata* **38** (1991), 229–243.

IMPLICATIONS OF LINE-TRANSITIVITY FOR DESIGNS

Cheryl E. Praeger
Department of Mathematics and Statistics
University of Western Australia
35 Stirling Highway
Crawley
WA 6009
Australia
praeger@maths.uwa.edu.au

Abstract For a t-design $\mathcal{D}$ ($t \geq 2$) transitivity of a group G of automorphisms on the blocks (or lines) of $\mathcal{D}$ implies transitivity on points. Delandtsheer and Doyen proved in 1989 that if in addition the group G leaves invariant a non-trivial partition of the v points, that is, G is point-imprimitive, then v is bounded above by $(\binom{k}{2} - 1)^2$, where k is the block size. Such designs may be 2-designs, or 3-designs, and we discuss a recent proof by Mann and Tuan of a conjecture of Cameron and the author that for a block-transitive, point-imprimitive 3-design, the number v of points is bounded above by $\binom{k}{2} + 1$. The subclass of line-transitive point-imprimitive 2-designs that are linear spaces have also received much attention: apart from finite projective planes, there are exactly 469 known examples and they all have one of two sets of parameters: $(v, k) = (91, 6)$ or $(729, 8)$. Efforts to obtain restrictions on the parameters and the structure of the group G may lead to new constructions of line-transitive, point-imprimitive linear spaces.

1. Block-transitive t-designs with $t \geq 2$

For positive integers t, v, k, λ satisfying $k > t$ and $v > k + t$, a *t-design* $\mathcal{D} = (\mathcal{P}, \mathcal{L})$ with parameters t, v, k, λ, sometimes called a t-(v, k, λ) design, consists of a set $\mathcal{P}$ of v points and a set $\mathcal{L}$ of k-subsets of $\mathcal{P}$, called the lines or blocks of $\mathcal{D}$, such that every t-subset of $\mathcal{P}$ is contained in exactly λ blocks. Note that, in this definition, we assume that there are 'no repeated blocks', so that a block can be identified with a k-subset of points. The *trivial t-(v, k, λ) design* $\mathcal{D}$ is the design in which every

A. Blokhuis et al. (eds.), Finite Geometries, 305–317.

k-subset of points is a block. We shall always assume that the t-design $\mathcal{D}$ is not the trivial design, and we shall almost always assume that $t \geq 2$. An *automorphism* of $\mathcal{D}$ is a permutation of $\mathcal{P}$ which leaves the block set $\mathcal{L}$ invariant. The set of all automorphisms forms a subgroup $\mathrm{Aut}(\mathcal{D})$ of $\mathrm{Sym}(\mathcal{P})$, called the *automorphism group* of $\mathcal{D}$. In this paper we are concerned with block-transitive (or line-transitive) t-designs, that is, those where $\mathrm{Aut}(\mathcal{D})$ is transitive on the block set $\mathcal{L}$.

By a result of Richard Block [1], transitivity of a group G of automorphisms on the blocks of $\mathcal{D}$ implies transitivity on points. In [4] Cameron and the author observed that Block's result, together with a result of Ray-Chaudhuri and Wilson [21], imply that a group G acting block-transitively on a t-design $\mathcal{D}$ must also act transitively on the $[t/2]$-subsets of points of $\mathcal{D}$, where $[t/2]$ denotes the greatest integer $n \leq t/2$. They went on to prove [4, Theorem 1.1] that, for a non-trivial block-transitive t-design, the parameter t is at most 7, and they made the following conjecture.

Conjecture 1.1 *For a non-trivial block-transitive t-design, the parameter t is at most 5.*

The proofs of [4, Theorem 1.1] relied on the finite simple group classification, and showed [4, Section 2] that any block-transitive 6-design $\mathcal{D}$ must have automorphism group satisfying either

1. [Affine type] $\mathrm{Aut}(\mathcal{D}) = \mathrm{AGL}(d,2)$, with $\mathcal{P}$ the set of $v = 2^d \geq 8$ points of the affine geometry $\mathrm{AG}(d,2)$, or

2. [Projective type] $\mathrm{PSL}(2,q) \leq \mathrm{Aut}(\mathcal{D}) \leq \mathrm{P\Gamma L}(2,q)$, with $\mathcal{P}$ the set of $v = q+1 \geq 6$ points of the projective line $\mathrm{PG}(1,q)$, and $\mathrm{Aut}(\mathcal{D})$ transitive on 3-subsets of $\mathcal{P}$.

Conjecture 1.1 remains open, and is equivalent to the following.

Conjecture 1.2 *There exist no block-transitive* 6*-designs of affine type or projective type.*

We are interested in the possibility that a block-transitive group G of a t-design $\mathcal{D}$ leaves invariant a non-trivial partition $\mathcal{C}$ of the point set $\mathcal{P}$, that is, both $|\mathcal{C}| \geq 2$ and $|C| \geq 2$ for $C \in \mathcal{C}$, and for each $g \in G$ and each $C \in \mathcal{C}$, the image $C^g \in \mathcal{C}$. In this case the group G is said to be *point-imprimitive*, and otherwise it is called *point-primitive*. Every automorphism group G which acts transitively on the 2-subsets of $\mathcal{P}$ is primitive on $\mathcal{P}$ (see for example, [13, Section 2.1]), and so the discussion above shows that block-transitive, point-imprimitive automorphism groups can exist only for t-designs with $t \leq 3$. There are in fact many examples for both $t = 2$ and $t = 3$, and this topic is the main theme of the paper.

In Section 2 we discuss a seminal result of Delandtsheer and Doyen, the first result to give an upper bound for v in terms of the block size k for block-transitive, point-imprimitive t-designs, $t \geq 2$. In the course of examining some consequences of this result, Cameron and the author were led to study a class of t-designs, which we describe in Section 3. All designs in the class are 1-designs, some are 2-designs, and some are 3-designs. This investigation led to a classification of all 2-designs with block size $k \neq 3, 4, 5, 8$, for which the Delandtsheer–Doyen upper bound is achieved, see Section 4. Certain experimental observations about the parameters of the 3-designs studied in [3] led to a conjectured upper bound on v for block-transitive, point-imprimitive 3-designs which was considerably smaller than the Delandtsheer–Doyen bound. This conjecture was recently proved by Mann and Tuan [17] and will be discussed in Section 4.

We turn then to the special case of linear spaces which are block-transitive and point-imprimitive. These are the block-transitive, point-imprimitive 2-(v, k, λ) designs for which $\lambda = 1$, and for these designs the blocks are usually called lines. None of the designs in the classification in [3] (see Theorem 4.1) were linear spaces, and there was some speculation that perhaps some stronger bound than the Delandtsheer–Doyen bound might hold for them. However it turned out that there are examples of linear spaces attaining the Delandtsheer–Doyen upper bound precisely when the block size is 8, one of the few cases not covered by the Cameron–Praeger classification stated in Theorem 5.1. At the time of writing this paper, the only known examples of line-transitive point-imprimitive linear spaces are the 467 examples in Theorem 5.1 which are all 2-$(729, 8, 1)$ designs, certain examples which are projective planes, and two further examples which are 2-$(91, 6, 1)$ designs. The values of k and v for these examples give a few clues to where one might search for new examples. The final sections of the paper contain, first, an examination of some of the combinatorial and algebraic properties of the known examples, and second, a report on recent theoretical analyses which may provide a pathway to discovering parameter sets and group structures where we may find more line-transitive point-imprimitive linear spaces.

2. The Delandtsheer–Doyen upper bound

Suppose that $\mathcal{D} = (\mathcal{P}, \mathcal{L})$ is a t-(v, k, λ)-design, with $t = 2$ or 3, and that $G \leq \mathrm{Aut}(\mathcal{D})$ is such that G is block-transitive and point-imprimitive on $\mathcal{D}$. We shall call such a pair $(\mathcal{D}, G)$ an *imprimitive pair.* Suppose that G leaves invariant a non-trivial partition $\mathcal{C}$ of the point set $\mathcal{P}$, with $d = |\mathcal{C}| \geq 2$ and $c = |C| \geq 2$, for $C \in \mathcal{C}$. Delandtsheer and Doyen [11] proved that there exist positive integers x, y such that

$$c = \frac{\binom{k}{2} - x}{y} \text{ and } d = \frac{\binom{k}{2} - y}{x}. \tag{2.1}$$

In particular this implies that

$$v = cd = \frac{\binom{k}{2} - x}{y} \cdot \frac{\binom{k}{2} - y}{x} \leq \left(\binom{k}{2} - 1\right)^2 \tag{2.2}$$

and hence that, for a given block size k, there are only finitely many imprimitive pairs $(\mathcal{D}, G)$. The integer x is equal to the number of pairs of points $\{\alpha, \beta\}$ in a block $L \in \mathcal{L}$ such that α and β lie in the same class of $\mathcal{C}$. Such pairs of points are called *inner pairs* or *$\mathcal{C}$-inner pairs.* There are many examples of imprimitive pairs for both $t = 2$ and $t = 3$, and two general constructions for them are discussed in the next section.

3. Cameron–Praeger designs

Suppose that $(\mathcal{D}, G)$ is an imprimitive pair with $\mathcal{D}$ a t-(v, k, λ)-design ($t = 2$ or 3), and that $\mathcal{C}$ is a non-trivial G-invariant point partition with d classes of size c, as in Section 2. It was observed in [3, Corollary 1.2 and Section 2] that, if we replace G by the largest subgroup $\hat{G}$ of $\mathrm{Sym}(\mathcal{P})$ which leaves $\mathcal{C}$ invariant, and if we replace $\mathcal{L}$ by the set $\hat{\mathcal{L}}$ of all images of blocks of $\mathcal{L}$ under elements of $\hat{G}$, then the resulting design $\hat{\mathcal{D}} = (\mathcal{P}, \hat{\mathcal{L}})$ is also a t-$(v, k, \hat{\lambda})$-design, where $\hat{\lambda} \geq \lambda$, and $(\hat{\mathcal{D}}, \hat{G})$ is an imprimitive pair with the same invariant point-partition $\mathcal{C}$. The group $\hat{G}$ is the wreath product $\hat{G} = S_c \wr S_d = \left(\prod_{C \in \mathcal{C}} \mathrm{Sym}(C)\right) \cdot \mathrm{Sym}(\mathcal{C})$.

Any block $L \in \mathcal{L}$ is partitioned by its intersections with the classes $C_1, \ldots, C_d$ of $\mathcal{C}$. Suppose that the sizes of these intersections are $x_i = |L \cap C_i|$, for $1 \leq i \leq d$. We may relabel the classes so that $x_1 \geq x_2 \geq \cdots \geq x_d$. Set $\mathbf{x} = (x_1, x_2, \ldots, x_d)$. Then $\hat{\mathcal{L}}$ consists of every k-subset L' of $\mathcal{P}$ for which

$$(|L' \cap C_1|, |L' \cap C_2|, \ldots, |L' \cap C_d|) \tag{3.1}$$

is a rearrangement of the sequence $\mathbf{x}$. The integers c, d and the sequence $\mathbf{x}$ determine the design $\hat{\mathcal{D}}$ completely. Designs of this type were called Cameron–Praeger designs by Camina [5] and were denoted $\hat{\mathcal{D}} = \mathrm{CP}(c, d; \mathbf{x})$. We will use the same notation.

We can define the incidence structure $\mathrm{CP}(c, d; \mathbf{x}) = (\mathcal{P}, \hat{\mathcal{L}})$ for any sequence $\mathbf{x} = (x_1, x_2, \ldots, x_d)$ with $\sum_{i=1}^{d} x_i = k$; we take $\hat{\mathcal{L}}$ to consist of all k-subsets L' of $\mathcal{P}$ for which the sequence in (3.1) is a rearrangement of $\mathbf{x}$. This incidence structure is always a 1-design, but not always a t-design for $t = 2$ or 3. The group $\hat{G}$ has exactly two orbits on pairs of points, namely the $\mathcal{C}$-inner pairs and the pairs which lie in different $\mathcal{C}$-classes. We will call the latter pairs *outer pairs*. Thus each inner pair lies in a constant number λ_{inn} of blocks of $\mathrm{CP}(c,d;\mathbf{x})$, and similarly each outer pair lies in a constant number λ_{out} of blocks of $\mathrm{CP}(c,d;\mathbf{x})$. Moreover $\mathrm{CP}(c,d;\mathbf{x})$ is a 2-design if and only if $\lambda_{inn} = \lambda_{out}$. An easy counting argument shows that this condition is equivalent to

$$\sum_{i=1}^{d} x_i(x_i - 1) = \frac{k(k-1)(c-1)}{v-1} \tag{3.2}$$

where $v = cd$. A similar argument shows that $\mathrm{CP}(c,d;\mathbf{x})$ is a 3-design if and only if it is a 2-design and

$$\sum_{i=1}^{d} x_i(x_i - 1)(x_i - 2) = \frac{k(k-1)(k-2)(c-1)(c-2)}{(v-1)(v-2)} \tag{3.3}$$

(see [3, Proposition 2.2]). Thus determining the existence of a block-transitive, point-imprimitive t-(v,k,λ) design, for given t, v, k is reduced to checking whether equation (3.2) and, if $t = 3$, also equation (3.3) hold for various factorisations $v = cd$. There are certainly some 3-designs in the family. In particular, if either $c = 2$ or $d = 2$ then $\mathrm{CP}(c,d;\mathbf{x})$ is a 3-design if and only if it is a 2-design, and in each case there are infinitely many 3-designs (see [3, Propositions 3.1 and 3.3]). Also, for each k, if $c = \binom{k}{2} - 1$ and $\mathbf{x} = (2,1,\ldots,1,0,\ldots,0)$ with entries adding to k, then $\mathrm{CP}(c,c;\mathbf{x})$ is a 2-design for which the upper bound of (2.2) is attained.

There was a second general construction given in [3, Section 3] for block-transitive, point-imprimitive t-designs. Consider the group $S_c \times S_d$ acting on the set $C \times D$ where $C = \{1,\ldots,c\}$ and $D = \{1,\ldots,d\}$, and $(g,h) \in S_c \times S_d$ maps $(i,j) \in C \times D$ to (i^g, j^h). Now $C \times D$ may be identified with the set of edges of the complete bipartite graph $K_{c,d}$ with vertex set $C \cup D$ and C, D the parts of the bipartition. A k-subset of $C \times D$ may then be identified with (the edge-set of) a subgraph of $K_{c,d}$ with k edges. If Γ is such a graph, let $\mathcal{D}(c,d;\Gamma) = (C \times D, \mathcal{L}(\Gamma))$ denote the design with block set $\mathcal{L}(\Gamma)$ the collection of images of Γ under $S_c \times S_d$. Let $\mathbf{x}$ and $\mathbf{y}$ be the degree sequences of the vertices of Γ in D and C respectively; where the *degree* of a vertex of Γ is the number of edges of Γ with which it is incident.

Proposition 3.1 [3, Proposition 3.6] *Let Γ, $\mathcal{D}(c,d;\Gamma)$, $\mathbf{x}$ and $\mathbf{y}$ be as defined above. Then*

(a) $\mathcal{D}(c,d;\Gamma)$ *is a block-transitive, point-imprimitive* 1*-design with respect to the group* $S_c \times S_d$*;*

(b) $\mathcal{D}(c,d;\Gamma)$ *is a* 2*-design if and only if both* $\mathrm{CP}(c,d;\mathbf{x})$ *and* $\mathrm{CP}(d,c;\mathbf{y})$ *are* 2*-designs;*

(c) $\mathcal{D}(c,d;\Gamma)$ *is a* 3*-design if and only if both* $\mathrm{CP}(c,d;\mathbf{x})$ *and* $\mathrm{CP}(d,c;\mathbf{y})$ *are* 3*-designs and the number of paths of length* 3 *of* Γ *is*

$$\frac{k(k-1)(k-2)(c-1)(d-1)}{(cd-1)(cd-2)}.$$

4. Achieving upper bounds on v

As we have seen, for each value of k, there exists at least one block-transitive, point-imprimitive 2-design for which the upper bound of (2.2) is attained, namely $\mathrm{CP}(c,c;\mathbf{x})$ with $\mathbf{x}=(2,1,\ldots,1,0,\ldots,0)$, $c=\binom{k}{2}-1$. We may construct a possibly smaller 2-design with all these properties by replacing $G=S_c \wr S_c$ by a subgroup $H=S_c \wr H_0$ for some 2-transitive subgroup H_0 of S_c (in its usual representation on c points), and replacing the block set $\mathcal{L}$ of $\mathrm{CP}(c,c;\mathbf{x})$ by an H-orbit in $\mathcal{L}$. However, these are not the only kinds of block-transitive, point-imprimitive 2-designs attaining the upper bound of (2.2).

If a member $\mathcal{D}(c,d;\Gamma)$ of the second family of designs constructed in Section 3 is a t-design, then another t-design with smaller block set can sometimes be constructed as follows. The block set $\mathcal{L}(\Gamma)$ of $\mathcal{D}(c,d;\Gamma)$ is replaced by an $(H\times K)$-orbit in $\mathcal{L}(\Gamma)$, where H,K are t-transitive subgroups of S_c, S_d respectively.

For $\mathcal{D}(c,d;\Gamma)$ to be a 2-design attaining the upper bound of (2.2), we require $c=d=\binom{k}{2}-1$ and, by Proposition 3.1, the degree sequences $\mathbf{x}=\mathbf{y}=(2,1,\ldots,1,0,\ldots,0)$ with entries adding to k. There are up to isomorphism exactly two bipartite graphs Γ with such degree sequences; either (i) the connected components of Γ are a path of length 3, $k-2$ copies of the complete graph K_2, and $2(c-k)$ isolated vertices, or (ii) the connected components of Γ are two paths of length 2, $k-3$ copies of K_2, and $2(c-k)$ isolated vertices. It was proved in [3, Theorem 5.2] that all block-transitive, point-imprimitive 2-designs attaining the upper bound of (2.2), and having block size $k\neq 3,4,5,8$, can be obtained by this or the previous construction.

Theorem 4.1 [3, Theorem 5.2] *Let* $\mathcal{D}$ *be a block-transitive and point-imprimitive* 2-(v,k,λ) *design, with* $v=c^2$, $c=\binom{k}{2}-1$, $k>5$, $k\neq 8$. *Then, setting* $\mathbf{x}=(2,1,\ldots,1,0,\ldots,0)$ *with* $k-2$ *entries equal to* 1*, either*

(a) $\mathcal{D} = \mathrm{CP}(c, c; \mathbf{x})$ *with* $\mathrm{Aut}(\mathcal{D}) = S_c \wr S_c$, *or* Δ *is a subdesign with block set a* $(S_c \wr H)$*-orbit on the blocks of* $\mathrm{CP}(c, c; \mathbf{x})$, *where* H *is a 2-transitive subgroup of* S_c; *or*

(b) $\Delta(c, c; \Gamma)$ *with* $\mathrm{Aut}(\Delta) = S_c \times S_c$, *where the degree sequences of* Γ *for vertices in the bipartite halves are both equal to* $\mathbf{x}$, *or* $\mathcal{D}$ *is a subdesign with block set an* $(H \times K)$*-orbit on* $\mathcal{L}(\Gamma)$, *where* H, K *are 2-transitive subgroups of* S_c.

We have presented two general constructions for t-(v, k, λ) designs, each of which produces examples of block-transitive, point-imprimitive 3-designs. However it was observed in [3], that the number of points of the 3-designs obtained from these constructions seemed always to be at most $\binom{k}{2} + 1$, considerably less than the Delandtsheer–Doyen upper bound. Many of the examples attained this observed upper bound, and it was verified, for all possible examples with $k \leq 70$, that $v \leq \binom{k}{2} + 1$. Thus it was conjectured [3, Conjecture 3.5] that for all block-transitive, point-imprimitive 3-designs, the inequality $v \leq \binom{k}{2} + 1$ should hold. This conjecture has very recently been proved by Avinoam Mann and Ngo Dac Tuan. Their proof is elementary and elegant, exploiting the equations (3.2) and (3.3) and divisibility arguments.

Theorem 4.2 [17] *For* $3 < k < v - 3$, *let* $\mathcal{D}$ *be a block-transitive, point-imprimitive* 3-(v, k, λ) *design. Then* $v \leq \binom{k}{2} + 1$.

There remains still the problem of understanding the structure of typical examples of 3-designs meeting the bound established by Theorem 4.2.

Problem Classify the block-transitive, point-imprimitive 3-(v, k, λ) designs with $3 < k < v - 3$ and $v = \binom{k}{2} + 1$.

All the examples of block-transitive, point-imprimitive 2-designs in this section have rather large values for the parameter λ. It would be interesting to discover, given parameters c, d, k for which there exists a 2-design $\mathrm{CP}(c, d; \mathbf{x})$ with $\sum_i x_i = k$, the minimum value of λ for which there exists a block-transitive, point-imprimitive 2-(cd, k, λ) design with automorphism group preserving a point-partition having d classes of size equal to c.

5. Line-transitive linear spaces

The linear spaces to be discussed here have constant block size, and so may be defined as the 2-(v, k, λ) designs with $\lambda = 1$. Note that, in the case of linear spaces we frequently refer to blocks as lines. Many examples of line-transitive linear spaces exist, but most of them seem to be point-primitive. For example, linear spaces with automorphism group transitive on incident point-line pairs, that is, transitive on flags, are always point-primitive, and all flag-transitive linear spaces have been classified (see [2]; there remains an unresolved case where the automorphism group is a one-dimensional affine group). This classification provides numerous infinite families of examples of flag-transitive linear spaces.

Turning however to the case of special interest in this paper, namely the line-transitive, point-imprimitive case, it turns out that the collection of examples is rather small. We shall say that $(\mathcal{D}, G)$ is an *imprimitive pair* if $\mathcal{D}$ is a linear space, and $G \leq \mathrm{Aut}(\mathcal{D})$ with G acting line-transitively and point-imprimitively on $\mathcal{D}$. There is an infinite family of examples of imprimitive pairs $(\mathcal{D}, G)$ with $\mathcal{D}$ a Desarguesian projective plane of order q such that the number of points $v = 1 + q + q^2$ is not prime, and G is a cyclic group of order v; G is sometimes called a *Singer cycle*, and acts transitively on both the points and the lines of $\mathcal{D}$. Apart from finite projective planes however, there are up to isomorphism exactly 469 known examples of line-transitive point-imprimitive linear spaces, and they all have one of two sets of parameters: $(v, k) = (91, 6)$ (two linear spaces discovered by Mills [16] in 1975 and Colbourn and Colbourn [10] in 1980 or see [15]) or $(729, 8)$ (467 linear spaces discovered by Nickel, Niemeyer, O'Keefe, Penttila and the author [18] in 1992). (The line-transitive 2-$(91, 6, 1)$ design discovered by Colbourn and Colbourn is listed in their paper [10] with the name McCalla. Charlie Colbourn told me that he had been bemoaning the fact that his research did not seem to be commercially viable when his colleague McCalla offered to 'buy' one of his designs if he had a suitable one for sale. So Charlie 'sold' this design to his colleague, if I recall correctly, for 50 cents.)

At the time of the classification in [3] of block-transitive, point-imprimitive 2-designs (with $k \neq 3, 4, 5, 8$) attaining the Delandtsheer–Doyen upper bound on v, there were no known line-transitive, point-imprimitive linear spaces attaining that bound, for none of the examples classified in Theorem 4.1 are linear spaces. By Theorem 4.1, any linear space examples would need to have block size $k = 3, 4, 5$ or 8, and it turned out that there are many examples in the case where the block size is 8.

Theorem 5.1 [18, 19] *Let $\mathcal{D}$ be a line-transitive, point-imprimitive linear space, that is a 2-$(v, k, 1)$ design, with $v = \left(\binom{k}{2} - 1\right)^2$. Then $v = 729, k = 8$, and up to isomorphism there are 467 possibilities for $\mathcal{D}$.*

Table 1. Parameters for the known imprimitive pairs

Design	k	c	d	x	y	$\mathbf{x}$
Colbourn and Mills	6	7	13	1	2	$(2,1,1,1,1)$
designs [10, 16]	6	13	7	2	1	$(2,2,1,1)$
N^2OP^2 designs [18]	8	27	27	1	1	$(2,1,\ldots,1)$

This paucity of examples of line-transitive point-imprimitive linear spaces is both disappointing, and tantalising. Is it that we simply do not know where to look? When the search focused on the 'correct' parameters, as in the case of Theorem 5.1, there was spectacular success in finding new constructions. The known examples, however, provide little insight into what one might expect in a typical example. In the next sections we examine some of the properties of the known examples, and look at certain recent theoretical analyses which suggest possible parameter sets and group structures where we hope to find more line-transitive point-imprimitive linear spaces.

6. Examining the examples

Here we look at some features of the known examples of line-transitive point-imprimitive linear spaces. Let $(\mathcal{D}, G)$ be an imprimitive pair, where $\mathcal{D} = (\mathcal{P}, \mathcal{L})$ and $\mathcal{D}$ is not a projective plane, and let $\mathcal{C}$ be a G-invariant partition of $\mathcal{P}$ with $d = |\mathcal{C}| > 1$ and $c = |C| > 1$ $(C \in \mathcal{C})$. Let x, y be the Delandtsheer–Doyen parameters satisfying equations (2.1). Choose a line $L \in \mathcal{L}$, and set $x_i := |L \cap C_i|$ for $i = 1, \ldots, d$. Rearrange the C_i if necessary so that $x_1 \geq x_2 \geq \cdots \geq x_d$, and set $\mathbf{x} := (x_1, x_2, \ldots, x_d)$. We usually omit any zero values from $\mathbf{x}$. Then the values for $k = |L|$, c, d, and $\mathbf{x}$ for the known examples are recorded in Table 1.

In addition, in all cases the full automorphism group G of the linear space is soluble, and the stabiliser of a line has order 1 or 2. If, for a transitive permutation group G, the stabilisers are trivial then G is said to be *regular*. Line stabilisers of order 2 occur only in the case of one of the two 2-$(91, 6, 1)$ designs, and in this case G has a subgroup of index 2 which is regular on lines. In all cases G has a normal subgroup N which is regular on points, and another normal subgroup K, contained in N, such that the point-partition $\mathcal{C}$ is the set of K-orbits. A G-invariant partition which is the set of orbits of a normal subgroup of G is called a *G-normal partition*. For both of the 2-$(91, 6, 1)$ designs, and for some of the 2-$(729, 8, 1)$ designs, there are two such subgroups K, giving rise to two different G-normal partitions with the property that blocks of different partitions intersect in a unique point. In addition, in the case of the 2-$(91, 6, 1)$ design for which a line stabiliser has order 2, the kernel of the G-action on one of these partitions contains all the line-stabilisers. Two of the 467 line-transitive designs on 729 points have the exceptional property that there are 28 distinct G-invariant partitions of the point set, corresponding to the 28 parallel classes of lines in the Desarguesian affine plane of order 27.

7. Parameters of line-transitive point-imprimitive linear spaces

Throughout this section we assume that $(\mathcal{D}, G)$ is an imprimitive pair, such that $\mathcal{D} = (\mathcal{P}, \mathcal{L})$ is not a projective plane. Let $\mathcal{C}$ be a G-invariant partition of $\mathcal{P}$ with $d = |\mathcal{C}| > 1$ and $c = |C| > 1$ $(C \in \mathcal{C})$. Let x, y be the Delandtsheer—Doyen parameters as in equations (2.1), and for a chosen line $L \in \mathcal{L}$, let $x_i := |L \cap C_i|$, for $i = 1, \ldots, d$, and $\mathbf{x} = (x_1, x_2, \ldots, x_d)$, rearranged so that $x_1 \geq x_2 \geq \cdots \geq x_d$ (omitting any zero values from $\mathbf{x}$). A starting point for studying imprimitive pairs is the following result which follows implicitly from [14].

Proposition 7.1 [Higman and McLaughlin] *There are at least two distinct non-zero values for the x_i. Moreover, if the only values for the x_i are $0, 1, h$, then the restriction of $\mathcal{D}$ to a class C is a 2-$(c, h, 1)$ design.*

By Table 1, in the known examples the values of the x_i are $0, 1, 2$, and the restriction of $\mathcal{D}$ to a class C is the trivial 2-design in which every pair of points is a block. This proposition was used in [12] to obtain general restrictions on the parameters c, d. Some of the simply stated results are the following. Part (a) for the case $c = 3$ was also proved in [7, Lemma 8].

Lemma 7.2 [12, Proposition 2.3, Corollary 2.5, and Lemma 3.1]

(a) $\min\{c, d\} \geq 3$, *and in the case of equality $\mathcal{D}$ is a projective plane.*

(b) $x_1 < \sqrt{c} + 1/2$, *and* $x_1 < \sqrt{2x} + 1$.

To emphasise how limited our knowledge is about line-transitive linear spaces, we note that it is still not known, for example, whether an imprimitive pair exists for which c or d is equal to 4.

Question 7.3 *Does there exist an imprimitive pair such that* $\min\{c, d\} = 4$*?*

An imprimitive pair with $\min\{c, d\} = 4$ would have one of the Delandtsheer–Doyen parameters equal to 1, since c and d must satisfy certain inequalities involving x, y stated below.

Lemma 7.4 [20]

(a) $\binom{k}{2} \geq 2xy + x + y$, $c \geq 2x + 1$, *and* $d \geq 2y + 1$, *and if any of these three inequalities is an equality, then* $\mathcal{D}$ *is a projective plane.*

(b) *If* $\binom{k}{2} > (y - 2)xy + x + y$, *then* G *induces a primitive action on the partition* $\mathcal{C}$.

(c) *If* $\binom{k}{2} > (x - 2)xy + x + y$, *then the setwise stabiliser* G_C *of a class* $C \in \mathcal{C}$ *induces a primitive action on* C.

Thus if $c = 4$ then by Lemmas 7.2 (b) and 7.4 (a), $x_1 = 2$ and $x_i = 1$ for $2 \leq i \leq k - 2$. In this case it was shown in [7, Lemma 8] that $k = 8h + 2$ and $v = 4(24h^2 + 9h + 1)$ for some positive integer h, and the group G_C^C induced on C is A_4 or S_4. We do not know if such a linear space exists.

The inequalities involving x, y in (b) and (c) imply that, if k is large relative to x and y, then the group actions induced on $\mathcal{C}$ and on a class C are both primitive. In the known examples, these inequalities hold. This suggests that it might be fruitful to investigate the cases where the values of x, y are small, and in particular to study the group structure further in this case.

Problem Classify the imprimitive pairs $(\mathcal{D}, G)$ for which the Delandtsheer–Doyen parameters x, y satisfy $\max\{x, y\} \leq 2$.

In the known examples, we remarked that the partition $\mathcal{C}$ is G-normal, that is, $\mathcal{C}$ is the orbit set of a normal subgroup of G. If a transitive group G on $\mathcal{P}$ has no non-trivial G-invariant partitions which are G-normal, then G is said to be *quasiprimitive* on $\mathcal{P}$. It was proved by Camina and the author that, for an imprimitive pair $(\mathcal{D}, G)$ with G quasiprimitive on $\mathcal{P}$, the group G is an *almost simple group*, that is, $T \leq G \leq \mathrm{Aut}(T)$ for some non-abelian simple group T.

Theorem 7.5 [8] *If* $(\mathcal{D}, G)$ *is an imprimitive pair with* G *quasiprimitive on* $\mathcal{P}$, *then* G *is an almost simple group.*

It would seem sensible, therefore, to investigate separately the imprimitive pairs $(\mathcal{D}, G)$ where G is quasiprimitive on $\mathcal{P}$, from those where there exists a non-trivial G-normal point-partition. Some progress has been made in the case where G is almost simple and the simple group is an alternating group [6], a sporadic simple group [9] or certain of the classical simple groups such as $\mathrm{PSL}(2, q)$. No imprimitive pairs other than small flag-transitive projective geometries occurred in these classifications.

Table 2. Possible parameters when $x = 1, y \leq 2$

x	y	c	d	k	Comments
1	1	27	27	8	N^2OP^2 designs
1	2	7	13	6	Colbourn and Mills designs
1	2	27	53	11	existence unknown

Question 7.6 *Does there exist an imprimitive pair $(\mathcal{D}, G)$ where $\mathcal{D}$ is not a projective plane, and G is an almost simple group?*

We will assume from now on that $\mathcal{C}$ is G-normal, say $\mathcal{C}$ is the set of orbits in $\mathcal{P}$ of a normal subgroup K of G, and moreover we will assume that K is equal to the kernel of the action of G on $\mathcal{C}$ (that is, the subgroup of G fixing each class of $\mathcal{C}$ setwise). A permutation group is said to be *semiregular* if the only element fixing a point is the identity element. If k is large enough relative to x, then K must be semiregular.

Theorem 7.7 [20] *If $(\mathcal{D}, G)$ is an imprimitive pair such that $\mathcal{C}$ is G-normal relative to a normal subgroup K, and if $k > 2x + \frac{3}{2} + \sqrt{4x - \frac{7}{4}}$, then*

(a) *K is semiregular;*

(b) *c is odd, and at least one of the x_i is 1;*

(c) *each minimal normal subgroup of G contained in K is elementary abelian of odd order.*

There are further results in both [12] and [20] concerning the case where $\max x_i = 2$ (which holds for all the known examples). These results in the case when x and y are small lead to parameters k, c, d and group structures which may enable the discovery of new imprimitive pairs $(\mathcal{D}, G)$. We conclude by stating one of these results. It produces exactly one set of parameters where existence of an imprimitive pair is not yet settled. A computer search for examples in this case is under way.

Theorem 7.8 [12] *If $(\mathcal{D}, G)$ is an imprimitive pair such that $\mathcal{C}$ is G-normal, and if the Delandtsheer–Doyen parameters satisfy $x = 1$ and $y \leq 2$, then one of the lines of Table 2 holds.*

Acknowledgements

This work was partially supported by an Australian Research Council large grant A69941071.

References

[1] R.E. Block, On the orbits of collineation groups, *Math. Z.* **96** (1967), 33–49.

[2] F. Buekenhout, A. Delandtsheer, J. Doyen, P.B. Kleidman, M.W. Liebeck and J. Saxl, Linear spaces with flag-transitive automorphism groups, *Geom. Dedicata* **36** (1990), 89–94.

[3] P.J. Cameron and C.E. Praeger, Block-transitive t-designs. I. Point-imprimitive designs, *Discrete Math.* **118** (1993), 33–43.

[4] P.J. Cameron and C.E. Praeger, Block-transitive t-designs. II. Large t, *Finite Geometry and Combinatorics* (eds. F. De Clerck et al.), Cambridge University Press, Cambridge, 1993, 103–119.

[5] A.R. Camina, A survey of the automorphism groups of block designs, *J. Combin. Des.* **2** (1994), 79–100.

[6] A.R. Camina, P.M. Neumann, and C.E. Praeger, Finite alternating groups as line-transitive automorphism groups of linear spaces, *in preparation.*

[7] A.R. Camina and S. Mischke, Line-transitive automorphism groups of linear spaces, *Electronic J. Combin.* **3**, 1996.

[8] A.R. Camina and C.E. Praeger, Line-transitive, point-quasiprimitive automorphism groups of finite linear spaces are affine or almost simple, *Aequaetiones Math.*, to appear.

[9] A. R. Camina and F. Spiezia, Sporadic groups and automorphisms of linear spaces, *J. Combin. Designs* **8** (2000), 353–362.

[10] C.J. Colbourn and M.J. Colbourn, Cyclic Steiner systems having multiplier automorphisms, *Utilitas Math.* **17** (1980), 127–149.

[11] A. Delandtsheer and J. Doyen, Most block transitive t-designs are point primitive, *Geom. Dedicata* **29** (1989), 307–310.

[12] A. Delandtsheer, A.C. Niemeyer, and C.E. Praeger, Finite line-transitive linear spaces: parameters and normal point-partitions, preprint, 2000.

[13] J.D. Dixon and B. Mortimer, *Permutation Groups*, Springer, New York, 1996.

[14] D.G. Higman and J.E. McLaughlin, Geometric ABA-groups, *Illinois J. Math.* **5** (1961), 382–397.

[15] Z. Janko and V.D. Tonchev, Cyclic 2-$(91, 6, 1)$ designs with multiplier automorphisms, *Discrete Math.* **97** (1991), 265–268.

[16] W.H. Mills, Two new block designs, *Utilitas Math.* **7** (1975), 73–75.

[17] A. Mann and Ngo Dac Tuan, Block-transitive point-imprimitive t-designs, in preparation.

[18] W. Nickel, A.C. Niemeyer, C.M. O'Keefe, T. Penttila, and C.E. Praeger, The block-transitive, point-primitive 2-$(729, 8, 1)$ designs, *Appl. Algebra Engrg. Comm. Comput.* **3** (1992), 47–61.

[19] C.M. O'Keefe, T. Penttila, and C.E. Praeger, Block-transitive, point-imprimitive designs with $\lambda = 1$, *Discrete Math.* **115** (1993), 231–244.

[20] C.E. Praeger and Ngo Dac Tuan, Inequalities involving the Delandtsheer–Doyen parameters for finite line-transitive linear spaces, in preparation.

[21] D.K. Ray-Chaudhuri and R.M. Wilson, On t-designs, *Osaka J. Math.* **12** (1975), 737–744.

EXPONENT BOUNDS

Bernhard Schmidt
Institut für Mathematik
Universiät Augsburg
86135 Augsburg
Germany
schmidt@math.uni-augsburg.de

Abstract We survey the presently known exponent bounds for difference sets and relative difference sets. The cases where these bounds are sufficient for the existence of the objects in question are discussed in some detail.

1. Introduction

One of the central themes of the theory of difference sets is to search for necessary and sufficient conditions on the group structure for the existence of difference sets. Most known necessary conditions have the form of *exponent bounds.* Here the *exponent* of a group means the order of its largest cyclic subgroup. An exponent bound gives an upper bound on the exponent of groups containing difference sets. Turyn's exponent bound [49] from 1965 is the most prominent example. Since Turyn's work, exponent bounds have played an important role in the study of difference sets.

Turyn's exponent bound has two important features: it relies on a *self-conjugacy assumption* and it is obtained by considering a *single homomorphic image* of a putative difference set.

Turyn's bound has been refined in two ways. Firstly, under the self-conjugacy assumption, one can try to improve Turyn's result by considering several homomorphic images simultaneously. It turns out that this approach, though it needs sophisticated arguments to combine information from different homomorphic images, often leads to striking results. In particular, necessary and sufficient conditions for the existence of several infinite families of difference sets can be derived in this way. For difference sets, only five results of this category are known: For Hadamard difference sets in abelian 2-groups (Turyn, Davis, Kraemer), McFar-

A. Blokhuis et al. (eds.), Finite Geometries, 319–331.

land difference sets (McFarland, Ma/Schmidt, Davis/Jedwab), McFarland difference sets without self-conjugacy (Schmidt), Chen difference sets (Chen, Schmidt), and Davis–Jedwab difference sets (Davis–Jedwab, Schmidt). All these results will be treated in this paper.

The second, more important refinement of Turyn's method is to try to get rid of the very restrictive self-conjugacy assumption. Results in this direction have been obtained by McFarland [38], Chan [9] and Ma [32]. These results are remarkable, however, they only apply in quite special situations. Substantial progress has been obtained by the author [46, 47] by introducing the new method of *field descent*. These *field descent exponent bounds* are very general and do not rely on restrictive assumptions like self-conjugacy.

2. Preliminaries

In this section, we briefly mention the basic notions concerning difference sets. For thorough treatments see [7, Chapter VI] and [39].

Let G be a finite group of order nm, and let N be a subgroup of G of order n. A subset R of G is called an (m, n, k, λ) *difference set in G relative to N* if every $g \in G \setminus N$ has exactly λ representations $g = r_1 r_2^{-1}$ with $r_1, r_2 \in R$, and no non-identity element of N has such a representation. The subgroup N is called the *forbidden subgroup.*

In the case $n = 1$, that is, when the forbidden subgroup consists only of the identity element, we write v instead of m and speak of a (v, k, λ) *difference set* in G. The nonnegative integer $n = k - \lambda$ is called the *order* of the difference set. If $n \in \{0, 1\}$, the difference set is called *trivial.* By a "difference set" we will always mean a nontrivial difference set. Sometimes it is useful to attach n to the parameters of a difference set. Thus we also speak of (v, k, λ, n) *difference sets.*

3. Self-conjugacy exponent bounds

We first explain Turyn's classical exponent bound. Turyn's argument relies on the the so-called *self-conjugacy assumption.*

Definition 3.1 Let m be a positive integer. A prime p is called *self-conjugate* modulo m if $p^j \equiv -1 \pmod{m'}$ for some j where m' is the p-free part of m. A composite integer is called self-conjugate modulo m if all its prime divisors are self-conjugate modulo m.

The following is a straightforward generalization of Turyn's result [49, Cor. 1] using projections. The most important applications will be given later.

Theorem 3.2 (Turyn) *Assume that there is an (m,n,k,λ) difference set in a (not necessarily abelian) group G relative to a subgroup N. Let U be a normal subgroup of G such that G/U is cyclic and let l be a positive integer self-conjugate modulo $|G/U|$ such that*

(a) *N is not contained in U and l^2 divides k or*

(b) *N is contained in U, $k - \lambda n \neq 0$ and l^2 divides $k - \lambda n$.*

Then

$$\frac{|U|}{|U \cap N|} \geq \frac{l}{2^{r-1}}$$

where $r = \max(1, s)$, and s is the number of prime divisors of $\gcd(l, |G/U|)$.

3.1. Difference sets

In the case of difference sets, Turyn's exponent bound reads as follows.

Corollary 3.1 (Turyn) *Assume the existence of a (v,k,λ,n) difference set in a group G. Let U be a subgroup of G such that G/U is cyclic, and let l^2 be a divisor of n which is self-conjugate modulo $|G/U|$. Then*

$$|U| \geq \frac{l}{2^{r-1}}$$

where $r = \max(1, s)$, and s is the number of prime divisors of $\gcd(l, |G/U|)$.

A *Hadamard difference set* is a difference set with parameters

$$(v,k,\lambda,n) = (4u^2, 2u^2 - u, u^2 - u, u^2)$$

for some positive integer u. In the case $u = p^a$ for a prime p, the Turyn bound is quite strong since then p is always self-conjugate modulo $v/2$.

Corollary 3.2 *Assume that an abelian group G contains a Hadamard difference set with $u = p^a$ for some prime p. Then*

$$\exp G \leq \begin{cases} 2^{a+2} & \text{if } p = 2, \\ p^a & \text{if } p > 2. \end{cases}$$

It took almost 30 years until it was shown by Davis [13] and Kraemer [28] that Turyn's bound is also *sufficient* for the existence of Hadamard difference sets in abelian 2-groups:

Theorem 3.3 (Davis, Kraemer) *A Hadamard difference set in an abelian 2-group G of order 2^{2a+2} exists if and only if* $\exp G \leq 2^{a+2}$.

The case of *nonabelian* 2-groups is very different: Davis and Iiams [15] extended an example of Liebler and Smith [30] to an infinite family of difference sets in high exponent 2-groups.

Theorem 3.4 (Davis, Iiams) *For every positive integer t, there is a nonabelian group of order 2^{4t+2} and exponent 2^{3t+3} containing a Hadamard difference set.*

Iiams [23] proved that the exponent obtained in Theorem 3.4 is the highest possible under some conditions:

Theorem 3.5 (Iiams) *Let G be a 2-group of order 2^{4t+2^m}, $m \in \{1,2\}$, with a normal cyclic subgroup $\langle x \rangle$ of order $\exp G$ such that x and x^{-1} are not conjugate in G. If G contains a Hadamard difference set, then $\exp G \leq 2^{3t+m+1}$.*

The result of Davis and Kraemer is one of the few known necessary and sufficient conditions for the existence of an infinite family of difference sets. The following theorem on McFarland difference sets also belongs to this category. A *McFarland difference set* is a difference set with parameters

$$\begin{aligned} v &= q^{d+1}[1+(q^{d+1}-1)/(q-1)], \\ k &= q^d(q^{d+1}-1)/(q-1), \\ \lambda &= q^d(q^d-1)/(q-1), \\ n &= q^{2d}, \end{aligned}$$

where $q = p^f$ is a prime power and d is a positive integer. McFarland [37] constructed such difference sets in all abelian groups G of order $v = q^{d+1}[1+(q^{d+1}-1)/(q-1)]$ which contain an elementary abelian subgroup order q^{d+1}.

Theorem 3.6 (Ma, Schmidt [35]) *Assume that there is a McFarland difference set in an abelian group G of order $q^{d+1}[1+(q^{d+1}-1)/(q-1)]$ where $q = p^f$ and p is a prime self-conjugate modulo $\exp G$. Let P be the Sylow p-subgroup of G. Then the following hold:*

(a) *if p is odd, then P is elementary abelian;*

(b) *if $p = 2$ and $f \geq 2$, then $\exp P \leq 4$.*

In view of McFarland's result mentioned above, part a of Theorem 3.6 is *sufficient* for the existence of a McFarland difference set. Davis and Jedwab [16] showed that part b is also sufficient for infinitely many cases.

Theorem 3.7 (Ma/Schmidt, Davis/Jedwab) *A McFarland difference set in an abelian group G of order $2^{2d+3}(2^{2d+1}+1)/3$ exists if and only if the Sylow 2-subgroup of G has exponent at most 4.*

Quite recently, two new families of difference sets were discovered by Chen [11] and Davis, Jedwab [16]. Chen's difference sets have parameters:

$$\begin{aligned} v &= 4q^{2t}\frac{q^{2t}-1}{q^2-1}, \\ k &= q^{2t-1}[\frac{2(q^{2t}-1)}{q+1}+1], \\ \lambda &= q^{2t-1}(q-1)\frac{q^{2t-1}+1}{q+1}, \\ n &= q^{4t-2}, \end{aligned}$$

where $q = p^f$ is a power of 3 or a square of an odd prime power and t is a positive integer. For $t = 1$, such a difference set is a Hadamard difference set. For $t \geq 2$, *any* difference set with the above parameters, for *any* prime power q, will be called a *Chen difference set.* Chen's construction requires the underlying groups to have an elementary abelian Sylow p-subgroup.

The second recent series of difference sets was constructed by Davis and Jedwab [16] and has parameters

$$\begin{aligned} v &= 2^{2t+2}(2^{2t}-1)/3, \\ k &= 2^{2t-1}(2^{2t+1}+1)/3, \\ \lambda &= 2^{2t-1}(2^{2t-1}+1)/3, \\ n &= 2^{4t-2}, \end{aligned}$$

where $t \geq 2$ is an integer. *Any* difference set with such parameters will be called a *Davis–Jedwab difference set.* Note that Davis–Jedwab difference sets are also Chen difference sets (put $q = 2$). Davis and Jedwab [16] constructed Davis–Jedwab difference sets in all abelian groups of order $2^{2t+2}(2^{2t}-1)/3$ which have a Sylow 2-subgroup S_2 of exponent at most 4, with the single exception of $t = 2$ and $S_2 \cong \mathbb{Z}_4^3$. This exception was removed by Arasu and Chen [1] who constructed the necessary difference set in $\mathbb{Z}_4^3 \times \mathbb{Z}_5$. If we apply Turyn's bound 3.1 to Chen difference sets, we get the following.

Theorem 3.8 (Turyn) *Let $q = p^f$ be a prime power, and let G be an abelian group of order $4q^{2t}(q^{2t}-1)/(q^2-1)$ containing a Chen difference set. Assume that p is self-conjugate modulo $\exp G$. Denote the Sylow p-subgroup of G by S_p. Then the following hold:*

(a) *if p is odd, then $\exp S_p \leq q$;*

(b) *if $p = 2$, then $\exp S_2 \leq 4q$.*

Schmidt [45] improved Turyn's bound for Chen difference sets.

Theorem 3.9 (Schmidt) *Let $q = p^f$ be an odd prime power, and let t, f be integers ≥ 2. Let G be an abelian group of order $4q^{2t}(q^{2t} - 1)/(q^2 - 1)$ containing a Chen difference set. Assume that p is self-conjugate modulo* $\exp G$. *Then the Sylow p-subgroup of G has exponent at most p^{f-1}.*

Schmidt's result leads to the following necessary and sufficient conditions for the existence of Chen difference set for infinitely many cases.

Corollary 3.3 (Chen, Schmidt) *Let p be an odd prime. Let G be an abelian group of order $4p^8(p^4+1)$ whose Sylow 2-subgroup is elementary abelian. Then G contains a Chen difference set if and only if its Sylow p-subgroup is elementary abelian.*

For the formulation of the next result, we need a definition; see [7, Chapter VI] for the terminology.

Definition 3.10 Let D be a (v, k, λ, n) difference set in an abelian group. We say that D has the *character divisiblilty property* if $\chi(D)$ is divisible by $\sqrt{n}$ for all nontrivial characters χ of G.

Theorem 3.11 (Davis/Jedwab, Schmidt, Arasu/Chen) *Let G be an abelian group of order $2^{2t+2}(2^{2t} - 1)/3$ with $t \geq 2$. A Davis–Jedwab difference set in G with the character divisibility property exists if and only if the Sylow 2-subgroup of G has exponent at most 4.*

3.2. Relative difference sets

Turyn's Theorem 3.2 provides a useful exponent bound for relative difference sets. In the following, we concentrate on *semiregular* relative difference sets, that is, relative (m, n, k, λ) difference sets with $n > 1$ and $m = k$. For this family, there is a further exponent bound due to Pott [39, Thm. 4.1.1].

Theorem 3.12 (Pott) *If a relative (m, n, m, λ) difference set exists in an abelian group G with $|G| > 4$, then* $\exp G$ *divides m. In particular, there is no semiregular relative difference set in any cyclic group of order greater than 4.*

In the following, we focus on semiregular relative difference sets with *prime power parameters*, that is, relative difference sets with parameters of the form $(m, n, k, \lambda) = (p^a, p^b, p^a, p^{a-b})$, p prime. In this case, Turyn's exponent bound can be improved dramatically. The major result here is the following from [36] improving previous work of Ma/Pott [33] and Schmidt [43, 44].

Theorem 3.13 (Ma, Schmidt) *Let p be odd prime. If , in an abelian group G, a relative (p^a, p^b, p^a, p^{a-b}) difference set exists , then*

$$\exp G \leq p^{\lfloor a/2 \rfloor + 1}$$

where $\lfloor a/2 \rfloor$ denotes the largest integer not exceeding $a/2$.

The bound in Theorem 3.13 is sharp in the sense that it can be attained for all triples (p, a, b) with $b \leq \lfloor a/2 \rfloor$, see [39, Chapter 4].

For $p = 2$, the situation is quite different. We have the following results due to Ma/Pott [33] and Schmidt [43, 44].

Theorem 3.14 (Ma, Pott, Schmidt) (a) *If a $(2^{2a}, 2^b, 2^{2a}, 2^{2a-b})$ relative difference set exists in an abelian group G relative to N, then $\exp N \leq 2^a$ and $\exp G \leq 2^a \exp N$. Furthermore, if $b > a$, then N is not a direct factor of G.*

(b) *Assume the existence of a $(2^{2a+1}, 2^b, 2^{2a+1}, 2^{2a-b+1})$ relative difference set in an abelian group G relative to N. Then $\exp G \leq 2^{a+2}$ and $\exp N \leq 2^{a+1+\delta}$ where $\delta = 1$ if $\exp G = \exp N$ and $\delta = 0$ otherwise.*

Davis and Jedwab [16, Cor. 8.2] partially improved results of Davis [14] and Ma and Schmidt [34, 43] to show that the bounds in Theorems 3.13 and 3.14 are essentially *sufficient* in the case $b = 1$.

Theorem 3.15 (Ma/Schmidt, Davis/Jedwab) *With p a prime, apart from the possible exceptions $G \cong \mathbb{Z}_{p^{c+1}}^2$ or $\mathbb{Z}_{p^{c+1}} \times \mathbb{Z}_{p^c} \times \mathbb{Z}_p$, p odd, $c \geq 1$, an abelian p-group G contains a (p^a, p, p^a, p^{a-1}) relative difference set if and only if $\exp G \leq p^{\lfloor a/2 \rfloor + 1 + \delta}$ where $\delta = 1$ if $p = 2$ and $a \equiv 1 \pmod 2$ and $\delta = 0$ otherwise.*

Semiregular relative difference sets with parameters $(n, n, n, 1)$ are of special interest since they can be used to construct quasiregular projective planes of order n, see [39, Chapter 5]. It is conjectured that an $(n, n, n, 1)$ relative difference set exists if and only if n is a prime power and the underlying group is an elementary abelian p-group. The following result from [33] can be interpreted as a result on collineation groups of projective planes.

Theorem 3.16 (Ma, Pott) *If a relative $(p^2, p^2, p^2, 1)$ difference set exists in an abelian group G for p an odd prime, then G must be elementary abelian.*

4. Field descent exponent bounds

Aside from the Theorem 3.12, all exponent bounds we have seen depend on a self-conjugacy argument. This is very restrictive: It can be seen that self-conjugacy "almost never" holds if the group order has many prime divisors, for instance, see [46]. Thus more general exponent bounds are desirable. Such bounds have been obtained by the author of the present paper by the new method of *field descent*. We describe the most important results concerning the field descent here. First we need a definition.

Definition 4.1 Let m, n be positive integers, and let $m = \prod_{i=1}^{t} {p_i}^{c_i}$ be the prime power decomposition of m. For each prime divisor q of n let

$$m_q := \begin{cases} \prod_{p_i \neq q} p_i & \text{if } m \text{ is odd or } q = 2, \\ 4\prod_{p_i \neq 2,q} p_i & \text{otherwise.} \end{cases}$$

Let $\mathcal{D}(n)$ be the set of prime divisors of n. We define $F(m,n) = \prod_{i=1}^{t} {p_i}^{b_i}$ to be the minimum multiple of $\prod_{i=1}^{t} p_i$ such that for every pair (i,q), $i \in \{1,...,t\}$, $q \in \mathcal{D}(n)$, at least one of the following conditions is satisfied:

(a) $q = p_i$ and $(p_i, b_i) \neq (2,1)$;
(b) $b_i = c_i$;
(c) $q \neq p_i$ and $q^{\mathrm{ord}_{m_q}(q)} \not\equiv 1 \pmod{p_i^{b_i+1}}$.

It is worth to note the following important property of $F(m,n)$.

Proposition 4.2 *Let P be a finite set of primes, and let Q be the set of all positive integers which are products of powers of primes in P. Then there is a computable constant $C(P)$ such that*

$$F(m,n) \leq C(P)$$

for all $m, n \in Q$.

Now we are ready to state the field descent result from [46].

Theorem 4.3 (Schmidt) *Assume $X\overline{X} = n$ for $X \in \mathbb{Z}_m$ where n and m are positive integers. Then*

$$X\xi_m^j \in \mathbb{Z}[\xi_{F(m,n)}]$$

for some j.

The field descent leads to the following general exponent bounds proved in [47].

Theorem 4.4 (Schmidt) *Assume the existence of a (v, k, λ, n)-difference set D in a group G. If U is a normal subgroup of G such that G/U is cyclic of order e then*

$$e \leq \frac{vF(e,n)}{2\sqrt{n\varphi(F(e,n))}}$$

where φ denotes the Euler totient function and F is defined as in 4.1.

It is worth to state the abelian case separately.

Theorem 4.5 *Assume the existence of a (v, k, λ, n)-difference set in an abelian group G. Then*

$$\exp G \leq \frac{vF(v,n)}{2\sqrt{n\varphi(F(v,n))}}.$$

In particular, if G is cyclic, then

$$n \leq \frac{F(v,n)^2}{4\varphi(F(v,n))}.$$

For relative difference sets, we get the following exponent bound [47].

Theorem 4.6 (Schmidt) *Assume the existence of an (m, n, k, λ)-difference set in a group G relative to N. Let U be any subgroup of G not containing N such that G/U is cyclic of order e. Then*

$$|U \cap N| \leq \frac{|U|F(e,k)}{2\sqrt{k\varphi(F(e,k))}}.$$

The field descent also can be applied to group invariant weighing matrices and other problems which can be studied by character methods. Here we only mention the most striking applications to difference sets, circulant Hadamard matrices and Barker sequences which can be found in [46, 47].

Theorem 4.7 (Schmidt) *For any finite set P of primes there is a computable constant $C(P)$ such that*

$$\exp G \leq C(P)|G|^{1/2}$$

for any abelian group G containing a Hadamard difference set whose order u^2 is a product of powers of primes in P.

Ryser's conjecture asserts that there is no (v, k, λ, n) difference set with $\gcd(v, n) > 1$ in any cyclic group. It is interesting to check Ryser's conjecture for the parameters of known difference sets with $\gcd(v,n) > 1$. These are the Hadamard, McFarland, Spence, Chen, and Davis–Jedwab parameters; see [7].

Theorem 4.8 (Schmidt) (a) *If there is a Hadamard difference set in a cyclic group of order* $v = 4u^2$ *then* $F(v,u)^2/\varphi(F(v,u)) \geq v$.

(b) *If there is a difference set with McFarland parameters in a cyclic group of order* $q^{d+1}[\frac{q^{d+1}-1}{q-1}+1]$, $q = p^f$, *then* $p > 2$, $d = f = 1$ *and*

$$\frac{p+2}{\varphi(p+2)} \geq 4 - \frac{12}{p+2} \tag{4.1}$$

In particular, $p+2$ *has at least* 20 *distinct prime divisors and* $p > 2 \cdot 10^{28}$.

c) There are no difference sets with Spence or Chen/Davis/Jedwab parameters in any cyclic groups.

By computer search, one can check that Theorem 4.8 implies the following.

Theorem 4.9 (Schmidt) *For* $k \leq 5 \cdot 10^{10}$, *Ryser's conjecture is true for all parameters* (v,k,λ,n) *of known difference sets with the possible exception of* $(v,k,\lambda) = (4u^2, 2u^2-u, u^2-u)$ *with* $u \in \{165, 11715, 82005\}$.

McFarland difference sets with $p = 3$, $f = 2$ and $d = 1$ were studied thoroughly by Arasu and Ma [4]. This case is very difficult since the self-conjugacy condition does not hold. Their result is the following.

Theorem 4.10 (Arasu, Ma) *A McFarland difference set in an abelian group* G *of order* 891 *exists if and only if the Sylow* 3*-subgroup of* G *is elementary abelian.*

In [47], a necessary and sufficient condition for the existence of McFarland difference sets with $f = d = 1$ was obtained. This is the first example of a necessary and sufficient condition for a (presumably) infinite family of difference sets known in the literature which does not rely on the self-conjugacy argument.

Theorem 4.11 (Schmidt) *Let* p *be an odd prime such that* $p+2$ *is square-free and*

$$\frac{p+2}{\varphi(p+2)} < 4 - \frac{12}{p+2}. \tag{4.2}$$

Then a $(p^2(p+2), p(p+1), p+1)$*-difference set in an abelian group* G *exists if and only if*

$$G \cong (\mathbb{Z}/p\mathbb{Z})^2 \times (\mathbb{Z}/(p+2)\mathbb{Z}).$$

Ryser's conjecture implies two further longstanding conjectures, namely, the Barker and the circulant Hadamard matrix conjecture. A *circulant Hadamard matrix of order* v is a matrix of the form

$$H = \begin{pmatrix} a_1 & a_2 & \cdots & a_v \\ a_v & a_1 & \cdots & a_{v-1} \\ \cdots & \cdots & \cdots & \cdots \\ a_2 & a_3 & \cdots & a_1 \end{pmatrix}$$

with $a_i = \pm 1$ and $HH^t = vI$ where I is the identity matrix. It is conjectured that no circulant Hadamard matrix of order $v > 4$ exists. A sequence $(a_i)_{i=1}^v$, $a_i = \pm 1$, is called a *Barker sequence of length* v if $|\sum_{i=1}^{v-j} a_i a_{i+j}| \leq 1$ for $j = 1, ..., v-1$. The *Barker conjecture* asserts that there are no Barker sequences of length $v > 13$. Storer and Turyn [48] proved the Barker conjecture for all odd v. It is known [7, VI, Section 14] that the existence of a Barker sequence of even length v implies the existence of a circulant Hadamard matrix of order v which in turn is equivalent to the existence of a Hadamard difference set in a cyclic group of order v. Together with other known results, Theorem 4.8 (a) implies the following.

Theorem 4.12 (Schmidt) (a) *There is no circulant Hadamard matrix of order* v, $4 < v \leq 10^{11}$, *with the possible exceptions* $v = 4u^2$, $u \in \{165, 11715, 82005\}$.

(b) *There is no Barker sequence of length* v *with* $13 < v < 4 \cdot 10^{12}$.

References

[1] K.T. Arasu and Y.Q. Chen, A difference set in $(\mathbb{Z}/4\mathbb{Z})^3 \times \mathbb{Z}/5\mathbb{Z}$, *Des. Codes Cryptogr.*, to appear.

[2] K.T. Arasu, J.A. Davis and J. Jedwab, A nonexistence result for abelian Menon difference sets using perfect binary arrays, *Combinatorica* **15** (1995), 311–317.

[3] K.T. Arasu, J.A. Davis, J. Jedwab, S.L. Ma and R.L. McFarland, Exponent bounds for a family of abelian difference sets, *Groups, Difference Sets, and the Monster* (eds. K.T. Arasu et al.), De Gruyter, Berlin, 1996, 129–143.

[4] K.T. Arasu and S.L. Ma, Abelian difference sets without self-conjugacy, *Des. Codes Cryptogr.* **15** (1998), 223–230.

[5] K.T. Arasu and S.K. Sehgal, Difference sets in abelian groups of p-rank two, *Des. Codes Cryptogr.* **5** (1995), 5–12.

[6] L.D. Baumert, *Cyclic Difference Sets*, Lecture Notes in Math. **182**, Springer, Berlin, 1971.

[7] T. Beth, D. Jungnickel and H. Lenz, *Design Theory* (2nd edition), Cambridge University Press, Cambridge, 1999.

[8] Z.I. Borevich and I.R. Shafarevich, *Number Theory*, Academic Press, New York, 1966.

[9] W.K. Chan, Necessary conditions for Menon difference sets, *Des. Codes Cryptogr.* **3** (1993), 147–154.

[10] W.K. Chan, S.L. Ma and M.K. Siu, Non-existence of certain perfect arrays, *Discrete Math.* **125** (1994), 107–113.

[11] Y.Q. Chen, On the existence of abelian Hadamard difference sets and a new family of difference sets, *Finite Fields Appl.* **3** (1997), 234–256.

[12] C.W. Curtis and I. Reiner, *Representation Theory of Finite Groups and Associative Algebras*, Wiley, New York, 1962.

[13] J.A. Davis, Difference sets in abelian 2-groups, *J. Combin. Theory Ser. A* **57** (1991), 262–286.

[14] J.A. Davis, Constructions of relative difference sets in p-groups, *Discrete Math.* **103** (1992), 7–15.

[15] J.A. Davis and J. Iiams, Hadamard difference sets in non-abelian 2-groups with high exponent, *J. Algebra* **199** (1998), 62–87.

[16] J.A. Davis and J. Jedwab, A unifying construction of difference sets, *J. Combin. Theory Ser. A* **80** (1997), 13–78.

[17] J.A. Davis and J. Jedwab, Nested Hadamard difference sets, *J. Statist. Plann. Inference* **62** (1997), 13–20.

[18] J.F. Dillon, Variations on a scheme of McFarland for noncyclic difference sets, *J. Combin. Theory Ser. A* **40** (1985), 9–21.

[19] S. Eliahou and M. Kervaire, Barker sequences and difference sets, *L'Enseignement Math.* **38** (1992), 345–382.

[20] S. Eliahou, M. Kervaire and B. Saffari, A new restriction on the length of Golay complementary sequences, *J. Combin. Theory Ser. A* **55** (1990), 49–59.

[21] M. Hall, A survey of difference sets, *Proc. Amer. Math. Soc.* **7** (1956), 975–986.

[22] J.E. Iiams, On difference sets in groups of order $4p^2$, *J. Combin. Theory Ser. A* **72** (1995), 256–276.

[23] J.E. Iiams, A note on certain 2-groups with Hadamard difference sets, submitted.

[24] K. Ireland and M. Rosen, *A Classical Introduction to Modern Number Theory*, Graduate Texts in Math. **84**, Springer, New York, 1990.

[25] D. Jungnickel, Difference Sets, *Contemporary Design Theory: A Collection of Surveys* (eds. J.H. Dinitz and D.R. Stinson), Wiley, New York 1992, 241–324.

[26] D. Jungnickel and B. Schmidt, Difference sets: an update, *Geometry, Combinatorial Designs and Related Structures*, (eds. J.W.P. Hirschfeld et al.), Cambridge University Press, Cambridge, 1997, 89–112.

[27] D. Jungnickel and B. Schmidt, Difference sets: a second update, *Rend. Circ. Mat. Palermo Suppl.* **53** (1998), 89–118.

[28] R.G. Kraemer, Proof of a conjecture on Hadamard 2-groups, *J. Combin. Theory Ser. A* **63** (1993), 1–10.

[29] E.S. Lander, *Symmetric Designs: An Algebraic Approach*, London Math. Soc. Lecture Notes Series **75**, Cambridge University Press, Cambridge, 1983.

[30] R.A. Liebler and K.W. Smith, On difference sets in certain 2-groups, *Coding Theory, Design Theory, Group Theory: Proceedings of the Marshall Hall Conference*, Wiley, New York, 1993, 195–212.

[31] S.L. Ma, Polynomial addition sets, Ph.D. thesis, University of Hong Kong, 1985.

[32] S.L. Ma, Planar functions, relative difference sets and character theory, *J. Algebra* **185** (1996), 342–356.

[33] S.L. Ma and A. Pott, Relative difference sets, planar functions and generalized Hadamard matrices, *J. Algebra* **175** (1995), 505–525.

[34] S.L. Ma and B. Schmidt, On (p^a, p, p^a, p^{a-1})-relative difference sets, *Des. Codes Cryptogr.* **6** (1995), 57–72.

[35] S.L. Ma and B. Schmidt, A sharp exponent bound for McFarland difference sets with $p = 2$, *J. Combin. Theory Ser. A* **80** (1997), 347–352.

[36] S.L. Ma and B. Schmidt, Relative (p^a, p^b, p^a, p^{b-a})-difference sets: a unified exponent bound and a local ring construction, *Finite Fields Appl.*, to appear.

[37] R.L. McFarland, A family of difference sets in non-cyclic groups, *J. Combin. Theory Ser. A* **15** (1973), 1–10.

[38] R.L. McFarland, Difference sets in abelian groups of order $4p^2$, *Mitt. Math. Sem. Giessen* **192** (1989), 1–70.

[39] A. Pott, *Finite Geometry and Character Theory*, Lecture Notes in Math. **1601**, Springer, Berlin, 1995.

[40] A. Pott, A survey on relative difference sets, *Groups, Difference Sets and the Monster*, Eds. K.T. Arasu et al., De Gruyter, Berlin, 1996, 195–233.

[41] D.K. Ray-Chaudhuri and Q. Xiang, New necessary conditions for abelian Hadamard difference sets, *J. Statist. Plann. Inference* **62** (1997), 69–79.

[42] H.J. Ryser, *Combinatorial Mathematics*, Wiley, New York, 1963.

[43] B. Schmidt, Differenzmengen und relative Differenzmengen, Dissertation, Verlag Dr. Wißner, Augsburg, 1995.

[44] B. Schmidt, On (p^a, p^b, p^a, p^{a-b})-relative difference sets. *J. Algebraic Combin.* **6** (1997), 279–297.

[45] B. Schmidt, Nonexistence Results on Chen and Davis–Jedwab Difference Sets. *J. Algebra* **202** (1998), 404–413.

[46] B. Schmidt, Cyclotomic integers and finite geometry, *J. Amer. Math. Soc.* **12** (1999), 929–952.

[47] B. Schmidt, Towards Ryser's conjecture. Proc. Third European Congress in Mathematics, (2000).

[48] J. Storer and R. Turyn, On binary sequences, *Proc. Amer. Math. Soc.* **12** (1961), 394–399.

[49] R.J. Turyn, Character sums and difference sets, *Pacific J. Math.* **15** (1965), 319–346.

[50] R.J. Turyn, Sequences with small correlation, *Error Correcting Codes* (ed. H.B. Mann), Wiley, New York, 1969, 195–228.

[51] K. Yamamoto, Decomposition fields of difference sets, *Pacific J. Math.* **13** (1963), 337–383.

ON SYMMETRIES AND TRANSLATION GENERALIZED QUADRANGLES

Koen Thas

Department of Pure Mathematics and Computer Algebra
Ghent University
Galglaan 2, B-9000 Ghent
Belgium
kthas@cage.rug.ac.be

Abstract

Let $\mathcal{S}$ be a generalized quadrangle (GQ) of order (s,t), $s,t \neq 1$, and suppose p is a point of $\mathcal{S}$. A *whorl about* p is a collineation of $\mathcal{S}$ which fixes p linewise, and a line L is an *axis of symmetry* if there is a full group of collineations of $\mathcal{S}$ of size s of which each element is a whorl about every point on L. Suppose $\mathcal{S}$ is a generalized quadrangle of order (s,t), with $s,t \neq 1$, and suppose that L is a line of $\mathcal{S}$. Then we prove that L is an axis of symmetry if and only if L is regular and if there is a point p on L, a group H of whorls about p, and a line M with $M \sim L \mathrm{I} p \not\mathrm{I} M$, such that H acts transitively on the points of $M \setminus \{M \cap L\}$. As a corollary, we prove that a line of a generalized quadrangle $\mathcal{S}$ through an elation point p is an axis of symmetry if and only if it is a regular line, and that the group of symmetries about such a line is always completely contained in any elation group corresponding to an elation point on this line. This proves the converse of a theorem of [8]. Using these results, we prove new characterizations of translation generalized quadrangles, and we considerably improve some characterizations of translation generalized quadrangles of Chen and Frohardt [2], and Hachenberger [5]; we will prove that an elation generalized quadrangle $\mathcal{S}$ with elation point p and elation group G is a translation generalized quadrangle with translation point p and translation group G if and only if p is incident with at least two regular lines, and in the case where there is an odd number of points on a line, one regular line is sufficient. Also, as an application, we prove a new characterization theorem of the classical generalized quadrangles arising from a quadric.

A. Blokhuis et al. (eds.), Finite Geometries, 333–345.

1. Introduction

A (finite) *generalized quadrangle* (GQ) *with parameters* (s,t) *(or of order* (s,t)*)*, $s,t \geq 1$, is an incidence structure $\mathcal{S} = (P,B,I)$ in which P and B are disjoint (non-empty) sets of objects called *points* and *lines* respectively, and for which I is a symmetric point-line *incidence relation* satisfying the following axioms:

(GQ1) each point is incident with $t+1$ lines ($t \geq 1$) and two distinct points are incident with at most one line;

(GQ2) each line is incident with $s+1$ points ($s \geq 1$) and two distinct lines are incident with at most one point;

(GQ3) if p is a point and L is a line not incident with p, then there is a unique point-line pair (q,M) such that $pIMIqIL$.

If $s=t$, then we also say that $\mathcal{S}$ is of *order* s.

Generalized quadrangles were introduced by Tits [17] in his celebrated work on triality. The main results, up to 1983, on finite generalized quadrangles are contained in the monograph *Finite Generalized Quadrangles* (FGQ) by S.E. Payne and J.A. Thas [8]. A survey of some 'new' developments on this subject in the period 1984-1992, can be found in the article *Recent developments in the theory of finite generalized quadrangles* [11]. We also refer to [12] for a survey on the subject.

Let $\mathcal{S} = (P,B,I)$ be a (finite) generalized quadrangle of order (s,t), $s \neq 1 \neq t$. Then $|P| = (s+1)(st+1)$ and $|B| = (t+1)(st+1)$. Also, $s \leq t^2$ and, dually, $t \leq s^2$, and $s+t$ divides $st(s+1)(t+1)$.

There is a point-line duality for GQ's of order (s,t) for which in any definition or theorem the words "point" and "line" are interchanged and also the parameters. Normally, we assume without further notice that the dual of a given theorem or definition has also been given. Sometimes, a line will be identified with the set of points incident with it.

Let p and q be (not necessarily distinct) points of the GQ $\mathcal{S}$; we write $p \sim q$ and say that p and q are *collinear*, provided that there is some line L such that $pILIq$ (so $p \not\sim q$ means that p and q are not collinear). Dually, for $L, M \in B$, we write $L \sim M$ or $L \not\sim M$ according as L and M are *concurrent* or *non-concurrent*. If $p \neq q \sim p$, the line incident with both is denoted by pq, and if $L \sim M \neq L$, the point which is incident with both is sometimes denoted by $L \cap M$. If a point p and a line L are not incident, we will denote this by $p\not{I}L$.

For $p \in P$, put $p^\perp = \{q \in P \parallel q \sim p\}$ (note that $p \in p^\perp$). For a pair of distinct points $\{p,q\}$, the *trace* of $\{p,q\}$ is defined as $p^\perp \cap q^\perp$, and we denote this set by $\{p,q\}^\perp$. Then $|\{p,q\}^\perp| = s+1$ or $t+1$, according as $p \sim q$ or $p \not\sim q$. For $p \neq q$, the *span* of the pair $\{p,q\}$ is $sp(p,q) = \{p,q\}^{\perp\perp} = \{r \in P \parallel r \in s^\perp \text{ for all } s \in \{p,q\}^\perp\}$. When $p \not\sim q$, then $\{p,q\}^{\perp\perp}$ is also called *the hyperbolic line* defined by p and q, and $|\{p,q\}^{\perp\perp}| = s+1$ or $|\{p,q\}^{\perp\perp}| \leq t+1$ according as $p \sim q$ or $p \not\sim q$. If $p \sim q$, $p \neq q$, or if $p \not\sim q$ and $|\{p,q\}^{\perp\perp}| = t+1$, we say that the pair $\{p,q\}$ is *regular.* The point p is *regular* provided $\{p,q\}$ is regular for every $q \in P \setminus \{p\}$. Regularity for lines is defined dually. A point p is *coregular* provided each line incident with p is regular. One easily proves that $t \leq s$ if $\mathcal{S}$ has a regular pair of non-collinear points, with $s \neq 1$, see FGQ.

If $\mathcal{S}$ is a GQ, then by $\mathrm{Aut}(\mathcal{S})$ we denote the group of all *automorphisms* of $\mathcal{S}$ (see FGQ). If p is a point of $\mathcal{S}$ and L a line not through p, then the unique line as defined by (GQ3) is sometimes denoted by $[p, L]$.

2. Preliminary results

2.1. The GQ's $\mathcal{Q}(d,q)$

The GQ $\mathcal{Q}(d,q)$, $d \in \{3,4,5\}$, is the generalized quadrangle which arises by taking the points and lines of a nonsingular quadric with Witt index 2 (that is, with projective index 1) in the d-dimensional projective space $\mathsf{PG}(d,q)$ over $GF(q)$. Respectively, the orders are $(q,1)$, (q,q) and (q,q^2). The GQ's $\mathcal{Q}(d,q)$ are so called *classical generalized quadrangles* as defined by Tits; see, for example, FGQ.

2.2. EGQ's and TGQ's

A *whorl about the point* p of $\mathcal{S}$ is a collineation of $\mathcal{S}$ which fixes each line through p. An *elation about the point* p is a whorl about p that fixes no point of $P \setminus p^\perp$. By definition, the identical permutation is also an elation (about every point). If p is a point of the GQ $\mathcal{S}$, for which there exists a group of elations G about p which acts regularly on the points of $P \setminus p^\perp$, then $\mathcal{S}$ is said to be an *elation generalized quadrangle* (EGQ) with *elation point* p and *elation group* G, and we sometimes write $(\mathcal{S}^{(p)}, G)$ for $\mathcal{S}$. A *symmetry about a point* p of the GQ $\mathcal{S}$ (of order (s,t) with $s,t \neq 1$) is an elation about p which fixes every point of $p^\perp$. A group of symmetries about a point can have at most t elements, and if this bound is reached, the point is called a *centre of symmetry.* Dually, we define the notion *axis of symmetry.* Any centre or axis of symmetry is regular; see FGQ. If a GQ $(\mathcal{S}^{(p)}, G)$ is an EGQ with elation point p, and if every line incident with p is an axis of symmetry, then we say that

$\mathcal{S}$ is a *translation generalized quadrangle* (TGQ) with *translation point* p and *translation group* G. In such a case, G is uniquely defined, G is generated by all symmetries about every line incident with p, and G is the set of all elations about p; see FGQ.

TGQ's were introduced by J.A. Thas in [9] for the case $s = t$.

Theorem 2.1 (FGQ, 8.3.1) *Let $\mathcal{S} = (P, B, I)$ be a GQ of order (s,t), $s,t \geq 1$. Suppose each line through some point p is an axis of symmetry, and let G be the group generated by the symmetries about the lines through p. Then G is elementary abelian and $(\mathcal{S}^{(p)}, G)$ is a TGQ.*

The following theorem is an improvement of this result for the case $s \neq t$.

Theorem 2.2 (K. Thas [13]) *Let $\mathcal{S} = (P, B, I)$ be a GQ of order (s,t), $s \neq t$ and $s,t \neq 1$. Suppose that there are more than $t - s + 2$ axes of symmetry through a point p, and let G be the group generated by the symmetries about these lines. Then G is elementary abelian and $(\mathcal{S}^{(p)}, G)$ is a TGQ.*

For the case $s = t$, we have the following result of FGQ, see also [13, 15] for several alternative short proofs.

Theorem 2.3 (FGQ, 11.3.5) *Let $\mathcal{S} = (P, B, I)$ be a GQ of order s, with $s \neq 1$. Suppose that there are at least three axes of symmetry through a point p, and let G be the group generated by the symmetries about these lines. Then G is elementary abelian and $(\mathcal{S}^{(p)}, G)$ is a TGQ.*

Theorem 2.4 (FGQ, 8.2.3 and 8.5.2) *Suppose $(\mathcal{S}^{(q)}, G)$ is an EGQ of order (s,t), $s \neq 1 \neq t$. Then $(\mathcal{S}^{(q)}, G)$ is a TGQ if and only if G is an (elementary) abelian group. Also in such a case there is a prime p and there are natural numbers n and k, where k is odd, such that either $s = t = p^n$ or $s = p^{nk}$ and $t = p^{n(k+1)}$. It follows that G is a p-group.*

Suppose $(\mathcal{S}^{(p)}, G)$ is an EGQ of order (s,t), $s,t \neq 1$, with elation point p and elation group G, and let q be a point of $P \setminus p^{\perp}$. Let $L_0, L_1, \cdots, L_t$ be the lines incident with p, and define r_i and M_i by $L_i I r_i I M_i I q$, $0 \leq i \leq t$. Put $G_i = \{\theta \in G \parallel M_i^{\theta} = M_i\}$, $G_i^* = \{\theta \in G \parallel r_i^{\theta} = r_i\}$, and $\mathcal{J} = \{G_i \parallel 0 \leq i \leq t\}$. Then $|G| = s^2 t$ and $\mathcal{J}$ is a set of $t+1$ subgroups of G, each of order s. Also, for each i, G_i^* is a subgroup of G of order st containing G_i as a subgroup. Moreover, the following two conditions are satisfied:

(K1) $G_i G_j \cap G_k = \mathbf{1}$ for distinct i, j and k;

(K2) $G_i^* \cap G_j = \mathbf{1}$ for distinct i and j.

If we denote the set $\{G_0^*, G_1^*, \cdots, G_t^*\}$ by $\mathcal{J}^*$, then $(\mathcal{J}, \mathcal{J}^*)$ is said to be a *4-gonal family of type* (s, t) *in* G.

It was first shown by W. M. Kantor in [7] that, conversely, a group of order s^2t, where $s, t \neq 1$, admitting a 4-gonal family is an elation group of a suitable elation generalized quadrangle.

We now have the following interesting properties.

Theorem 2.5 (FGQ, 8.2.2) *G_i is a group of symmetries about the line L_i if and only if $G_i \lhd G$ (and hence $\mathcal{S}$ is a TGQ if and only if $G_i \lhd G$ for each i), only if L_i is a regular line if and only if $G_iG_j = G_jG_i$ for all $G_j \in \mathcal{J}$.*

Theorem 2.6 (X. Chen and D. Frohardt [2]) *Let G be a group of order s^2t admitting a 4-gonal family $(\mathcal{J}, \mathcal{J}^*)$ of type (s, t). If there exist two distinct members in $\mathcal{J}$ which are normal subgroups of G, then s and t are powers of a prime p and G is an elementary abelian p-group.*

Theorem 2.7 (D. Hachenberger [5]) *Let G be a group of order s^2t admitting a 4-gonal family $(\mathcal{J}, \mathcal{J}^*)$ of type (s, t). If G is a group of even order, and if there exists a member of $\mathcal{J}$ which is a normal subgroup of G, then s and t are powers of 2 and G is an elementary abelian 2-group.*

In geometrical terms, Theorem 2.6 reads as follows: "Let $(\mathcal{S}^{(x)}, G)$ be an EGQ of order (s, t), $s, t \neq 1$, and suppose that there are at least two axes of symmetry L and M through the elation point x, such that the full groups of symmetries about L and M are completely contained in G. Then s and t are powers of the same prime number p and G is an elementary abelian p-group." Hence, by Theorem 2.4, $\mathcal{S}^{(x)}$ is a translation generalized quadrangle. In geometrical terms, we have the following for Theorem 2.7: "Let $(\mathcal{S}^{(x)}, G)$ be an EGQ of order (s, t), with $s, t \neq 1$ and s or t even, and suppose that there is at least one axis of symmetry L through the elation point x, such that the full group of symmetries about L is completely contained in G. Then s and t are powers of 2 and G is an elementary abelian 2-group." Thus, $\mathcal{S}^{(x)}$ is a translation generalized quadrangle.

3. Some lemmas

A *panel* of a generalized quadrangle $\mathcal{S} = (P, B, I)$ is an element (p, L, q) of $P \times B \times P$ for which $pILIq$ and $p \neq q$. Dually, one defines *dual panels*. Let (p, L, p') be a panel of the GQ $\mathcal{S}$ of order (s, t), $s \neq 1 \neq t$. A (p, L, p')*-collineation* is a whorl about p, L and p'.

Lemma 3.1 (FGQ, 9.2.1) *Suppose (p, L, p') is a panel of the GQ $\mathcal{S}$ of order (s,t), $s \neq 1 \neq t$, and let θ be a (p, L, p')-collineation. Then θ is a symmetry about L if L is regular.*

Corollary 3.2 *Suppose $\mathcal{S}$ is a GQ of order (s,t), $s,t \neq 1$, and suppose that L is a line which contains a panel (p, L, p') for which there is a full group of (p, L, p')-collineations of size s. Then L is an axis of symmetry if and only if L is regular.*

Lemma 3.3 *Suppose $\mathcal{S}$ is a GQ of order (s,t), $s,t \neq 1$, with a whorl θ about a point p, and that L and M are lines such that $M \sim LIp\not{I}M$, and for which $M^\theta = M$. Moreover, suppose that L is a regular line, and let q be a point on L, different from p and not on M. Then we have one of the following possibilities:*

(1) *q is not fixed by θ;*

(2) *q is fixed by θ, and then also every line through q.*

Proof Suppose that q is fixed by θ, and consider a line QIq, $Q \neq L$. Also, consider an arbitrary point rIM and not on L, and the line $[r, Q]$. Let NIp be such that $N \sim [r, Q]$, and let the intersection point be r'. Now consider the point r^θ. Since L is a regular line, there holds that the line $[r^\theta, Q]$ also intersects N, say in r''. It is clear that $r'^\theta = r''$, and that $[r, Q]^\theta = [r^\theta, Q]$. Since the point q is fixed, it follows easily that the line Q also is fixed by θ, and hence the lemma follows. □

Corollary 3.4 *Suppose $\mathcal{S}$ is a GQ of order (s,t), $s,t \neq 1$, and let p be a point, and L and M lines, such that $M \sim LIp\not{I}M$. Also, suppose that L is a regular line. Assume that θ is a whorl about p fixing M, and fixing a point q on L, different from p and not on M. Then* every line *through $r = L \cap M$ is fixed by θ.*

Proof By Lemma 3.3, every line through q is fixed by θ. Since M is fixed, also r is fixed, and by the same lemma the proof follows. □

Lemma 3.5 *Let $\mathcal{S}$ be a GQ of order (s,t), $s,t \neq 1$, and suppose $\theta \neq$* **1** *is a whorl about distinct collinear points p and q. If moreover, the line pq is regular, then θ is a symmetry about L.*

Proof Suppose M is an arbitrary line of $(pq)^\perp$ not through p or q. Suppose U and U' are distinct lines through p and different from pq; define the lines V and V' as the unique lines of $\{U, M\}^{\perp\perp}$ and $\{U', M\}^{\perp\perp}$ respectively, which are incident with q. Then $\{U, V\}^{\perp\perp} \cap \{U', V'\}^{\perp\perp} =$

$\{M\}$, $[\{U,V\}^{\perp\perp}]^\theta = \{U,V\}^{\perp\perp}$, and $[\{U',V'\}^{\perp\perp}]^\theta = \{U',V'\}^{\perp\perp}$. Hence $M^\theta = M$, and thus every line of $(pq)^\perp$ is fixed by θ. Hence θ is a symmetry about pq. □

Lemma 3.6 *Suppose $\mathcal{S}$ is a GQ of order (s,t), with $s \neq 1 \neq t$. Let p be a point of the GQ, and suppose that L and M are lines such that $M \sim LIp \not\sim M$, with L a regular line. Suppose that θ is a whorl about p fixing M, and suppose q is a point on L, different from p and not on M, which is also fixed by θ. Then θ is a symmetry about L.*

Proof By Lemma 3.3, all the lines through q (and also $L \cap M$) are fixed by θ. From Lemma 3.5, the result follows. □

Now suppose that $\mathcal{S}$ is a GQ with order (s,t), $s,t \neq 1$. Also, assume that p is a point and LIp a regular line, and again that $M \sim L$ is a line not through p. Suppose θ is a whorl about p which fixes M, and such that $\langle\theta\rangle$ acts semi-regularly on the points of M not on L. Hence $|\langle\theta\rangle|$ is a divisor of s. Consider an arbitrary nontrivial element ϕ of $\langle\theta\rangle$ of prime-power order, and consider the action of $\langle\phi\rangle$ on the points of $X = L \setminus [\{p\} \cup \{L \cap M\}]$. Since $|X| = s-1$ and because of the fact that $s-1$ and s are coprime, there follows immediately that there is a point $x \in X$ for which $x^{\langle\phi\rangle} = \{x\}$ (so every element of $\langle\phi\rangle$ fixes x). By Lemma 3.6, this implies that $\langle\phi\rangle$ is a group of symmetries about L. Since a finite group is generated by its elements of prime-power order and since ϕ was arbitrary, it follows that also $\langle\theta\rangle$ is a group of symmetries about L; the product of two symmetries about the same line is clearly again a symmetry about this line.

Theorem 3.7 *Suppose $\mathcal{S}$ is a GQ of order (s,t), $s,t \neq 1$, and suppose that L is a regular line through the point p. Let M be a line for which $L \sim M \not\sim p$. If H is a group of whorls about p which fixes M and which acts transitively on the points of M which are not L, then H contains a full group of symmetries about L of order s; that is, L is an axis of symmetry.*

Proof If $|H| = s$, then the statement follows directly from the preceding observation, hence suppose that $|H| > s$. Put $X = M \setminus \{L \cap M\}$. If θ is a nontrivial element of H which fixes at least two points of X, then by 8.1.1 of FGQ it follows that $t < s$, a contradiction since L is a regular line. Hence the permutation group (X, H) satisfies the following conditions: (a) H acts transitively on X, (b) for every $x \in X$, we have that $|H_x| \neq 1$, and (c) the only element of H which fixes at least two elements of X is the trivial element. Thus (X, H) is a *Frobenius group*;

see, for example, [6]. Hence by the theorem of Frobenius, H contains a normal subgroup N which acts regularly on X, and by the preceding observation, this group is a full group of symmetries about L. □

There is a very nice corollary of Theorem 3.7.

Theorem 3.8 (1) *Suppose $\mathcal{S}$ is a GQ of order (s, t), $s, t \neq 1$, and suppose that L is a regular line through the point p. If G is a group of whorls about p which acts transitively on the points of $P \setminus p^{\perp}$, then L is an axis of symmetry. In particular, suppose that $(\mathcal{S}^{(p)}, G)$ is an EGQ of order (s, t), $s, t \neq 1$, with elation point p and elation group G, and suppose $L\mathrm{I}p$ is a regular line. Then L is an axis of symmetry.*

(2) *Suppose $\mathcal{S}$ is an EGQ of order (s, t), $s, t \neq 1$, with elation point p. Then $L\mathrm{I}p$ is a regular line* if and only if *L is an axis of symmetry.*

In Section 8.1 of FGQ, the following property is noted:

- *the set of all symmetries about some line L of a GQ $\mathcal{S}$ of order (s, t), $s, t \neq 1$, is always a group, and every symmetry about L is an elation about L and about every point on L, hence the group has at most size s; thus, if the group of symmetries about L has size s, then this group is unique.*

We have the following theorem.

Theorem 3.9 *Suppose $\mathcal{S}$ is a GQ of order (s, t), $s, t \neq 1$, and suppose G is a group of whorls about the point p which acts transitively on $P \setminus p^{\perp}$.*

(1) *If L is a regular line through p, then L is an axis of symmetry, and the full group of symmetries of size s about L is completely contained in G.*

(2) *If $L\mathrm{I}p$ is an axis of symmetry, with G_L the full group of symmetries about L, then G_L is completely contained in G.*

In particular, the same statement holds for elation generalized quadrangles.

Proof (1) From the proof of Theorem 3.7, it follows that G contains a group G_L of size s of symmetries about L and, because of the previous remark, this group is unique.

(2) Directly from (1) and the fact that an axis of symmetry is a regular line. □

We are ready to state the converse of Theorem 2.5.

Theorem 3.10 *Let $\mathcal{S} = (P, B, I)$ be an EGQ of order (s, t) with elation point p and where $s, t \neq 1$, and suppose q is a point of $P \setminus p^{\perp}$. Suppose $L_0, \cdots, L_t$ are the lines through p, and suppose M_i are lines such that $L_i \sim M_i I q$. Let G_i be the subgroup of the elation group G which fixes M_i, for all i, and put $\mathcal{J} = \{G_0, \cdots, G_t\}$. Then G_i is a group of symmetries about the line L_i if and only if $G_i \lhd G$ (and hence $\mathcal{S}$ is a TGQ if and only if $G_i \lhd G$ for each i)* if and only if *L_i is a regular line if and only if $G_iG_j = G_jG_i$ for all $G_j \in \mathcal{J}$.*

We can now give an alternative, more geometrical, definition of translation generalized quadrangles without the use of symmetries, abelian groups or Galois geometries (see FGQ). Suppose $\mathcal{S}$ is a generalized quadrangle of order (s, t), $s \neq 1 \neq t$, with a point p for which there is a group of whorls which acts transitively on $P \setminus p^{\perp}$. Then $\mathcal{S}$ is a TGQ with translation point p if and only if every line through p is a regular line.

Remark 3.11 This result was proved by J.A. Thas for $s = t$ in [9], see also [8, 8.3.3].

Using Theorems 2.2 and 2.3, we immediately have the following theorem.

Theorem 3.12 *Let $\mathcal{S}$ be a GQ of order (s, t) with $s \neq 1 \neq t$, and suppose p is a point of $\mathcal{S}$.*

(1) *If $s = t$, then $\mathcal{S}$ is a TGQ with translation point p if and only if p is incident with three regular lines L_1, L_2, L_3 for which there are lines M_1, M_2, M_3 such that $L_i \sim M_i \not{I} p$ and such that there are groups G_i of whorls about p which act transitively on the points of $M_i \setminus \{M_i \cap L_i\}$.*

(2) *If $s \neq t$, then $\mathcal{S}$ is a TGQ with translation point p if and only if p is incident with $t - s + 3$ regular lines $L_1, \cdots, L_{t-s+3}$ for which there are lines $M_1, \cdots, M_{t-s+3}$ such that $L_i \sim M_i \not{I} p$ and such that there are groups G_i of whorls about p which act transitively on the points of $M_i \setminus \{M_i \cap L_i\}$.*

4. A characterization of the classical generalized quadrangles arising from a quadric

A panel (p, L, q) of a GQ of order (s, t), $s, t \neq 1$, is called *Moufang* if there is a group of (p, L, q)-collineations of size s. A line L is *Moufang* if every panel of the form (p, L, q) is Moufang. A GQ is *half Moufang* if every panel is Moufang or every dual panel is Moufang, and a GQ is a *Moufang GQ* if every panel and every dual panel is Moufang.

If p is a point of a GQ $\mathcal{S} = (P, B, I)$ such that there is a group of whorls about p which acts transitively on $P \setminus p^\perp$, then we call p a *centre of transitivity.* If p is a point of a GQ $\mathcal{S} = (P, B, I)$ for which there is a group of automorphisms of $\mathcal{S}$ which fixes p and which acts transitively on the points of $P \setminus p^\perp$, then p is a *point of transitivity.*

Theorem 4.1 *Suppose $\mathcal{S}$ is a GQ of order (s, t), $s, t \neq 1$. Then $\mathcal{S}$ is isomorphic to $\mathcal{Q}(4, s)$ or $\mathcal{Q}(5, s)$ if and only if $\mathcal{S}$ contains a centre of transitivity p, a collineation θ of $\mathcal{S}$ for which $p^\theta \not\sim p$, and a regular line.*

Proof If $\mathcal{S}$ is isomorphic to one of $\mathcal{Q}(4, s)$, $\mathcal{Q}(5, s)$, then any point of $\mathcal{S}$ is a translation point, and hence every point is a centre of transitivity. Also, every line is regular.

Conversely, suppose that $\mathcal{S}$ contains a centre of transitivity p with a group G_p of whorls about p acting transitively on $P \setminus p^\perp$, a collineation θ of $\mathcal{S}$ for which $p^\theta \not\sim p$ and a regular line. Then it is very easy to see that every point of $\mathcal{S}$ is a centre of transitivity, and that $\mathrm{Aut}(\mathcal{S})$ acts transitively on the lines of $\mathcal{S}$. Hence, any line of $\mathcal{S}$ is a regular line, and by Theorems 3.2.1 and 5.2.1 of FGQ, we have one of the following possibilities:

(i) $\mathcal{S}$ is isomorphic to $\mathcal{Q}(4, s)$;

(ii) $t > s$.

Now consider the last case. Since every line is regular and every point is a centre of transitivity, we have by Theorem 3.8 that every line of $\mathcal{S}$ is an axis of symmetry. It is clear that an axis of symmetry L is Moufang for any panel on L, and hence $\mathcal{S}$ is half Moufang. Now, by the main theorem of J.A. Thas, S.E. Payne and H. Van Maldeghem [10], it follows that $\mathcal{S}$ is a Moufang generalized quadrangle, and hence, by Fong and Seitz [3, 4], $\mathcal{S}$ must be classical or dual classical. Since the only two classical or dual classical quadrangles with regular lines are $\mathcal{Q}(4, s)$ and $\mathcal{Q}(5, s)$ [8], the proof is complete. □

Remark 4.2 We have classified the GQ's satisfying the first and the last condition of Theorem 4.1, and for which there is a collineation θ such that $p \neq p^\theta \sim p$, and where the regular line must be an element of $(pp^\theta)^\perp \setminus \{pp^\theta\}$, see [14].

4.1. Strong elation generalized quadrangles

We now introduce the notion of *strong elation generalized quadrangles.* A *strong elation generalized quadrangle* (SEGQ) is a generalized quadrangle $\mathcal{S}$ for which each point is an elation point.

Theorem 4.3 *If an SEGQ $\mathcal{S}$ of order (s,t), $s,t \neq 1$, has a regular line, then we have one of the following cases:*

(1) $\mathcal{S} \cong \mathcal{Q}(4,s)$;

(2) $\mathcal{S} \cong \mathcal{Q}(5,s)$.

Proof This follows immediately from Theorem 4.1. □

It is easy to prove that the automorphism group of a strong elation generalized quadrangle acts transitively on the pairs of non-concurrent lines, see [16]. Therefore it is by Theorem 4.3 sufficient to demand that an SEGQ has a regular pair of lines to conclude that it arises from a quadric. A similar remark can be made about Theorem 4.1, and hence we can restate Theorem 4.1 as follows.

Theorem 4.4 *Suppose $\mathcal{S}$ is a GQ of order (s,t), $s,t \neq 1$. Then $\mathcal{S}$ is isomorphic to $\mathcal{Q}(4,s)$ or $\mathcal{Q}(5,s)$ if and only if $\mathcal{S}$ contains a centre of transitivity p, a collineation θ of $\mathcal{S}$ for which $p^\theta \not\sim p$, and a regular pair of lines .*

Conjecture 4.5 Every strong elation generalized quadrangle is classical or dual classical.

Using the classification of finite simple groups, Buekenhout and Van Maldeghem [1] have showed that any finite *distance transitive generalized polygon* is Moufang, and hence classical by Fong and Seitz [3, 4]. As a corollary for finite generalized quadrangles, it follows that every SEGQ indeed is classical or dual classical. We are however interested in a proof which does not use the classification.

Note The problem of classifying the SEGQ's was already suggested by H. Van Maldeghem in [18].

5. Improvements of results of Chen and Frohardt, and Hachenberger

In Theorems 2.6 and 2.7, the group theoretical descriptions have one major restriction: they demand that the groups of symmetries about the lines must be completely contained in the elation group. By our preceding theorems and observations, we can now omit these "contained in" conditions completely; moreover, we do not even ask that the lines are axes of symmetry, but only that they are regular!

Theorem 5.1 *Let $(\mathcal{S}^{(x)}, G)$ be an EGQ of order (s,t), $s,t \neq 1$. If there are two distinct regular lines through the point x, then s and t are powers* *of the same prime number p, G is an elementary abelian p-group and hence $\mathcal{S}^{(x)}$ is a TGQ with translation group G.*

Proof Suppose L and M are regular lines through the point x. Then by Theorem 3.9, we know that L and M are axes of symmetry of which the full groups of symmetries of size s are completely contained in G. Hence, by Theorem 2.6, the proof follows immediately. □

Theorem 5.2 *Let $(\mathcal{S}^{(x)}, G)$ be an EGQ of order (s,t), $s,t \neq 1$. If there is a regular line through the point x and G is a group of even order, then s and t are powers of 2, and G is an elementary abelian 2-group; so $\mathcal{S}^{(x)}$ is a TGQ with translation group G.*

Proof Suppose L is a regular line through the point x. Then by Theorem 3.9, L is an axis of symmetry of which the full group of symmetries of size s is contained in G. Hence, by Theorem 2.7, the proof is complete. □

Remark 5.3 By results of Chapter 8 of FGQ (see also Section 2.2 of the present paper), it follows that if a GQ $(\mathcal{S}^{(x)}, G)$ satisfies the conditions of Theorem 5.1 or Theorem 5.2, that G is the *complete set of elations about x*.

Acknowledgements

The author is a research fellow supported by the Flemish Institute for the promotion of Scientific and Technological Research in Industry (IWT), grant no. IWT/SB/991254/ Thas. The author also wishes to thank J.A. Thas and H. Van Maldeghem for some useful remarks on the subject.

References

[1] F. Buekenhout and H. Van Maldeghem, Finite distance transitive generalized polygons, *Geom. Dedicata* **52** (1994), 41–51.

[2] X. Chen and D. Frohardt, Normality in a Kantor family, *J. Combin. Theory Ser. A* **64** (1993), 130–136.

[3] P. Fong and G.M. Seitz, Groups with a (B,N)-pair of rank 2, I, *Invent. Math.* **21** (1973), 1–57.

[4] P. Fong and G.M. Seitz, Groups with a (B,N)-pair of rank 2, II, *Invent. Math.* **21** (1974), 191–239.

[5] D. Hachenberger, On finite elation generalized quadrangles with symmetries, *J. London Math. Soc.* **53** (1996), 397–406.

[6] N. Jacobson, *Basic Algebra II*, Freeman, San Francisco, 1980.

[7] W.M. Kantor, Generalized quadrangles associated with $G_2(q)$, *J. Combin. Theory Ser. A* **29** (1980), 212–219.

[8] S.E. Payne and J.A. Thas, *Finite Generalized Quadrangles*, Research Notes in Mathematics **110**, Pitman, London, 1984.

[9] J.A. Thas, Translation 4-gonal configurations, *Atti Accad. Naz. Lincei Rend.* **56** (1974), 303–314.

[10] J.A. Thas, S.E. Payne and H. Van Maldeghem, Half Moufang implies Moufang for finite generalized quadrangles, *Invent. Math.* **105** (1991), 153–156.

[11] J.A. Thas, Recent developments in the theory of finite generalized quadrangles, *Acad. Analecta, Med. Kon. Acad. Wet. Lett. Sch. K. Belg., Kl. Wet.* (1994).

[12] J. A. Thas, *Generalized Polygons*, Chapter 9 of *Handbook of Incidence Geometry* (ed. F. Buekenhout), Amsterdam, North-Holland (1995), 383–431.

[13] K. Thas, *Symmetrieën in Eindige Veralgemeende Vierhoeken*, Master thesis, Ghent University (1999), 1–186.

[14] K. Thas, The classification of generalized quadrangles with two translation points, submitted.

[15] K. Thas, On generalized quadrangles with some concurrent axes of symmetry, submitted.

[16] K. Thas, Half pseudo Moufang generalized quadrangles, manuscript in preparation.

[17] J. Tits, Sur la trialité et certains groupes qui s'en déduisent, *Inst. Hautes Etudes Sci. Publ. Math.* **2** (1959), 14–60.

[18] H. Van Maldeghem, Some consequences of a result of Brouwer, *Ars Combin.* **48** (1998), 185–190.

COMPLETE CAPS IN PROJECTIVE SPACE WHICH ARE DISJOINT FROM A SUBSPACE OF CODIMENSION TWO

David L. Wehlau
Department of Mathematics and Computer Science
Royal Military College
Kingston, Ontario
Canada K7K 7B4
wehlau@rmc.ca

Abstract Working over the field of order 2 we consider those complete caps which are disjoint from some codimension 2 subspace of projective space. We derive restrictive conditions which such a cap must satisfy in order to be complete. Using these conditions we obtain explicit descriptions of complete caps which do not meet every hyperplane in at least 5 points. In particular, we determine the set of cardinalities of all such complete caps in all dimensions.

1. Introduction

A subset of the projective space $\Sigma = \mathrm{PG}(n,2)$ is a *cap* if no three points of S are collinear. A cap in Σ is said to be *complete* if it is not properly contained in any other cap of Σ. A cap S in Σ is called *large* if $|S| \geq 2^{n-1}+1$; otherwise S is said to be *small.* Much is known about the structure of large complete caps. In particular, it was shown in [5] that if $S \subset \Sigma$ is a large complete cap then $|S| = 2^{n-1} + 2^i$ for some $i \in \{0,1,2,\ldots,n-3,n-1\}$. In [5] it is also shown that if S is a cap in Σ satisfying $|S| \geq 2^{n-1}+2$ then there exists a codimension 2 subspace $H_\infty \cong \mathrm{PG}(n-2,2)$ of Σ such that $S \cap H_\infty = \varnothing$. In [3] the same result was shown to hold for all large caps. Conversely in [1] it was shown that given a fixed $r \geq 1$ then for all sufficiently large n there exists a cap S in $\mathrm{PG}(n,2)$ such that S meets every codimension r subspace $H \cong \mathrm{PG}(n-r,2)$ of $\mathrm{PG}(n,2)$.

Here we will consider complete caps, $S \subset \Sigma = \mathrm{PG}(n,2)$ with the property that some codimension 2 subspace $H_\infty \cong \mathrm{PG}(n-2,2)$ of Σ

A. Blokhuis et al. (eds.), *Finite Geometries,* 347–361.

satisfies the condition $S \cap H_\infty = \varnothing$. Fix such an H_∞ and denote the three hyperplanes of Σ which contain H_∞ by K_A, K_B and K_C. Furthermore we write $H_A = K_A \setminus H_\infty$, $H_B = K_B \setminus H_\infty$ and $H_C = K_C \setminus H_\infty$. Write $A = H_A \cap S = K_A \cap S$, $B = H_B \cap S = K_B \cap S$ and $C = H_C \cap S = K_C \cap S$. Finally define $A' = H_A \setminus A$, $B' = H_B \setminus B$, and $C' = H_C \setminus C$.

If $C = \varnothing$ then S is complete if and only if $S = \Sigma \setminus K_C$. Henceforth we will assume that none of the three sets A, B, and C is empty.

Consider the subspace $\widetilde{F} \cong \mathrm{PG}(r,2)$ of K_C generated by C. Put $F = \widetilde{F} \cap H_\infty$ and $\widehat{C} = \widetilde{F} \setminus F$. Then $F \cong \mathrm{PG}(r-1,2)$ and $\widehat{C} \cong \mathrm{AG}(r,2)$. Consider the decomposition of H_A and of H_B into cosets of $\widetilde{C}$. Each such coset F' consists of 2^r points such that $F' \sqcup F \cong \mathrm{PG}(r,2)$. There are 2^{n-r-1} such cosets in each of H_A and H_B. Each of these cosets is isomorphic to $\widetilde{F} \setminus F \cong \mathrm{AG}(r,2)$. We will denote the cosets in H_A by $H_A(1), H_A(2), \ldots, H_A(2^{n-r-1})$. Similarly we denote the cosets in H_B by $H_B(1), H_B(2), \ldots, H_B(2^{n-r-1})$ ordered so that $H_B(i) = \widehat{C} + H_A(i)$ for all i. Write $A(i) = H_A(i) \cap A$, $A'(i) = H_A(i) \cap A'$, $B(i) = H_B(i) \cap B$ and $B'(i) = H_B(i) \cap B'$.

The cap S is complete if and only if

$$A + B = C', \quad A + C = B', \quad B + C = A'; \tag{1.1}$$

$$(A \oplus A) \sqcup (B \oplus B) \sqcup (C \oplus C) = H_\infty. \tag{1.2}$$

In particular, if S is complete we must have

$$A(i) + C = B'(i) \text{ and } B(i) + C = A'(i) \tag{1.3}$$

for all $i = 1, 2, \ldots, 2^{n-r-1}$.

One way to satisfy Equation (1.3) is to take $A(i) = H_A(i)$ and $B(i) = \varnothing$. Similarly we may take $A(i) = \varnothing$ and $B(i) = H_B(i)$. We will call these two solutions the *trivial* solutions. A second way to satisfy this equation is to take $A(i) = \{a_1\}$ and $B(i) = H_B(i) \setminus (a_1 + C)$. Symmetrically we may take $B(i) = \{b_1\}$ and $A(i) = H_A(i) \setminus (b_1 + C)$. If either $|A| = 1$ or $|B| = 1$ then we call the solution a singleton solution.

Once we have solutions to Equation (1.3) we still will need the extra conditions of $A + B = C'$ together with Equation 1.2. In Section 3 we explicitly describe all complete caps with $|C| = 1$. Using these results we are able in Section 4 to give detailed sufficient conditions on a cap S which satisfies Equation 1.3 for all i, for S to be complete.

In Sections 6, 7 and 8 we consider complete caps which meet at least one hyperplane in less than 5 points. For such caps we are able to find all solutions to Equation (1.3) and thus to explicitly describe such complete caps.

In particular, we are able to construct many small complete caps. Also note that the Plotkin doubling construction (described in Section 2), as well as its generalization the Black/White lift (described in [4]) can be used to build a great many more small complete caps from the small complete caps constructed here.

2. Preliminaries

Let $\mathbb{F}_2^{n+1}$ denote the $(n+1)$-dimensional vector space over $\mathbb{F}_2$, the field of order 2. The elements of $\mathrm{PG}(n,2)$ are the one dimensional subspaces of $\mathbb{F}_2^{n+1}$. Each such subspace may be uniquely represented by the non-zero vector it contains. Fix a basis $\{e_0, e_1, \ldots, e_n\}$ of $\mathbb{F}_2^{n+1}$ and consider a point $x \in \mathrm{PG}(n,2)$. For ease of notation, we write $x = s_1 s_2 \cdots s_r$ if $e_{s_1} + e_{s_2} + \ldots + e_{s_t}$ is the non-zero vector in x where $s_1, s_2, \ldots, s_r$ are distinct elements of $\{0, 1, \ldots, n\}$. For example, 013 denotes the element of $\mathrm{PG}(n,2)$ comprised by the one-dimensional subspace containing $e_0 + e_1 + e_3$. From this point of view, the three points of $\mathrm{PG}(n,2)$ represented by x, y and z are collinear if and only if x, y and z lie in a plane in $\mathbb{F}_2^{n+1}$ which occurs if and only if $x + y = z$ in $\mathbb{F}_2^{n+1}$. The usual inner product on $\mathbb{F}_2^{n+1}$ induces an inner product on $\mathrm{PG}(n,2)$. We will write $(x)^\perp$ to denote the hyperplane of points in $\mathrm{PG}(n,2)$ which are orthogonal to x with respect to this inner product. On occasion we will identify $\mathbb{F}_2^n$ with the affine space $\mathrm{AG}(n,2)$. In that setting we will use $\varnothing$ to denote the zero vector in $\mathrm{AG}(n,2) = \mathbb{F}_2^n$.

Let X and Y be two subsets of $\Sigma = \mathrm{PG}(n,2)$ and let $z \in \mathrm{PG}(n,2)$. We write $X \oplus Y$ to denote the set $\{x + y \mid x \in X, y \in Y, x \neq y\} \subseteq \Sigma$. If $X \cap Y = \varnothing$ we also write $X + Y = X \oplus Y$. If $z \notin X$ we write $z + X = \{z + x \mid x \in X\} = \{z\} + X$.

A subset X of $\mathrm{PG}(n,2)$ is said to be *periodic* if there exists a point $v \in \mathrm{PG}(n,2)$ such that $v + X = X$. If such a point v exists it is called a *vertex* of X. The (possibly empty) set of all vertices of X is denoted $V(X)$. Note that if X is periodic then necessarily $|X|$ is even.

Let X be a subset of $\Sigma = \mathrm{PG}(n,2)$. Embed Σ in $\widetilde{\Sigma} \cong \mathrm{PG}(n+1,2)$ and let $v \in \widetilde{\Sigma} \setminus \Sigma$. We define the *Plotkin double* of X (from the vertex v) by

$$\phi(X) = X \sqcup \{v + x \mid x \in X\}.$$

Then $\phi(X)$ is a periodic subset of $\widetilde{\Sigma}$ with $V(\phi(X)) = \phi(V(X)) \sqcup \{v\}$. It is straightforward to verify that $\phi(X)$ is a cap in $\widetilde{\Sigma}$ if and only if X is a cap in Σ; also, $\phi(X)$ is complete if and only if X is.

A slightly more general version of the following lemma may be found in [3, Lemma 3.9]. We include a proof here for the reader's convenience.

Lemma 2.1 *Let $S \subset \Sigma$ be complete. Suppose there exists a hyperplane L of Σ such that $S \cap L$ is periodic. Then S is periodic. Furthermore, if v is a vertex for $S \cap L$ then v is also a vertex for S.*

Proof Let v be a vertex for $S \cap L$ and assume, by way of contradiction, that v is not a vertex for S. Then there exists a point x_1 of S such that $x_2 = x_1 + v \notin S$. Clearly $x_1 \notin L$ since $v \in L$ and thus $x_2 \notin L$. Since S is complete, x_2 lies on a secant of S: say $x_2 = y_1 + z_1$ where $y_1, z_1 \in S$. Without loss of generality $y_1 \in L$ since every line of Σ meets L. Therefore $y_2 = v + y_1 \in L \cap S$. But $y_2 = v + y_1 = v + x_2 + z_1 = x_1 + z_1$ and thus the line $\{y_2, x_1, z_1\}$ is fully contained in S, a contradiction. □

Lemma 2.2 *Let X be a subset of $\Sigma = \mathrm{PG}(n, 2)$ and let $V(X)$ denote the (possibly empty) set of vertices for X. Then $V(X)$ is a projective subspace of Σ.*

Proof We must prove that $V(X)$ is closed under addition, that is, if u, v are two distinct points of $V(X)$ then $u+v \in V(X)$. Accordingly let u, v be two distinct points of $V(X)$. We must show that $w = v + u$ is also a vertex of X. Let $p \in X$. Since u is a vertex, $u+p \in X$. Similarly, since $v \in V(X)$ we have $w + p = v + (u + p) \in X$. Hence $w + X \subseteq X$ and thus w is a vertex of X. □

3. Caps having a tangent hyperplane

In this section we characterize those complete caps S which have a tangent hyperplane. Many of the results of this section are related to results found in [3].

We suppose that K_C is a hyperplane which is tangent to S, that is, that C consists of a single point, $C = \{c_o\}$. Choose a codimension 2 subspace $H_\infty \cong \mathrm{PG}(n-2, 2)$ contained in K_C and disjoint from S. Clearly if S is complete we must have $B' = c_0 + A$ and $A' = c_0 + B$. Thus S is determined by A (or by B). We wish to investigate conditions on the set A which guarantee that S is complete.

Define $E = \Sigma \setminus (S \sqcup (S \oplus S))$ and $T = S \sqcup E$.

Lemma 3.1 *T is a complete cap in $\Sigma = \mathrm{PG}(n, 2)$.*

Proof Assume, by way of contradiction that T contains a line, ℓ Clearly ℓ must contain at least two points of E, say e, e_1. Clearly $E \subset K_C$ and thus $|\ell \cap K_C| \geq 2$ from which it follows that $\ell \subset K_C$. Thus ℓ contains at least one point of H_∞, e say. Therefore, $e + A = A'$ and $e + B = B'$. Now if $e_1 \in H_\infty$ then $e_1 + A = A'$ and $e_1 + B = B'$. Conversely if $e_1 \in H_C$ then $e_1 + A = B'$ and $e_1 + B = A'$. In both of these cases we have $e + e_1 + (A \sqcup B) = (A \sqcup B)$. This shows that $e + e_1 \in (A \sqcup B) \oplus (A \sqcup B) \subset S \oplus S$. Therefore $e + e_1 \notin T$. This contradiction proves the lemma. □

Proposition 3.2 *Let A and B be as defined above. Then $V(A) = V(B)$. In particular, A is periodic if and only if B is periodic.*

Proof Suppose A is periodic, that is, suppose there exists $v \in H_\infty$ such that $v + A = A$. Since $v + H_A = H_A$ we have $v + A' = A'$. Now $c_0 + B = A'$ and $c_0 + A = B'$. Therefore $v + B = v + c_o + A' = c_0 + v + A' = c_0 + A' = B$. □

Next we give a sufficient condition on A to guarantee that the set $S = \{c_0\} \sqcup A \sqcup B$ is a *complete* cap. The following theorem is similar to [3, Theorem 4.1].

Theorem 3.3 *Let S be a cap in $\Sigma = \mathrm{PG}(n, 2)$ with $|C| = 1$ and $B' = c_0 + A$. Suppose that $|A| \neq 2^{n-2}$. If A (or B) is not periodic then S is a complete cap.*

Proof Clearly every point of $\Sigma \setminus K_C$ lies on a secant to S through c_0. Let $\alpha = |A|$ and $\beta = |B|$. Then $\alpha + \beta = 2^{n-1}$. Since $\alpha \neq 2^{n-1}$ one of α or β, say without loss of generality α, exceeds 2^{n-2}. Therefore $A \oplus A = H_\infty$, that is, every point of H_∞ lies on a secant to A.

Finally we prove that if S is not complete (and $|A| \neq 2^{n-2}$) then A is periodic. Thus we suppose that S is not complete. By the foregoing this implies that there exists $x \in H_C$ with $x \neq c_0$ such that $x \notin A + B$. Then $x + A = B'$ and $c_0 + A = B'$. Therefore $v + A = x + c_0 + A = x + B' = A$. Therefore A is periodic with vertex v. □

Remark 3.4 Note that the hypothesis that $|A| \neq 2^{n-2}$ is required. See Example 3.12.

Suppose that S is a cap with $|C| = 1$ and $B' = c_0 + A$. Then $|S| = 2^{n-1} + 1$. Since $|S|$ is odd, S cannot be periodic. Thus by Lemma 2.1, we have that both A and B are not periodic. Hence we have proved the following result.

Theorem 3.5 *In $\Sigma = \mathrm{PG}(n, 2)$, let S be a cap which meets a hyperplane K_C of Σ in a single point, c_0. Choose a codimension 2 subspace $H_\infty \cong \mathrm{PG}(n-2, 2)$ contained in K_C and disjoint from S. Suppose $B' = c_0 + A$ and that $|A| \neq 2^{n-2}$. Then S is complete if and only if A is not periodic if and only if B is not periodic.*

Next we consider in detail the case where $|A| = |B| = 2^{n-2}$.

Suppose first that $A \oplus A \neq H_\infty$, that is, suppose there exists $p \in H_\infty \setminus (A \oplus A)$. Then $p + A = A'$. Therefore $A \oplus A = (p + A') \oplus (p + A') = A' \oplus A' = (c_0 + A') \oplus (c_0 + A') = B \oplus B$. In particular, $p \notin B \oplus B$. Note that we have shown that whenever $|A| = |B| = 2^{n-2}$, we have $A \oplus A = B \oplus B = A' \oplus A' = B' \oplus B'$.

Lemma 3.6 *If* $|E \cap H_\infty| \geq 2$ *then* A, B, A' *and* B' *are each periodic.*

Proof Let $e_1, e_2 \in E \cap H_\infty$. Then $e_1 + e_2 + A = e_1 + A' = A$ and thus A is periodic. Similarly $e_1 + e_2$ is a vertex for B, A' and B'. □

Suppose A is not periodic. By the preceding lemma, $|E| \leq 1$. Hence either S is complete or $T = S \sqcup \{e\}$. Thus we have proved the following theorem.

Theorem 3.7 *In* $\Sigma = \mathrm{PG}(n, 2)$, *let* S *be a cap which meets a hyperplane* K_C *of* Σ *in a single point,* c_0. *Choose a codimension 2 subspace* $H_\infty \cong \mathrm{PG}(n-2, 2)$ *contained in* K_C *and disjoint from* S. *Suppose* $B' = c_0 + A$. *If* S *is complete then* A *and* B *are not periodic. If* A *(or* B*) is not periodic then either*

(i) *S is complete or*

(ii) $|A| = |B| = 2^{n-2}$ *and* $\Sigma = S \sqcup (S \oplus S) \sqcup \{e\}$ *where* $e \in H_\infty$ *and the complete cap* $T = S \sqcup \{e\}$ *is periodic with* $V(T) = \{e + c_0\}$.

Remark 3.8 If $n \leq 4$ and $|A| = 2^{n-2}$ then $A \oplus A \neq H_\infty$ since then $|A \oplus A| \leq \binom{|A|}{2} < |H_\infty|$.

The following four examples show that $A \oplus A$ may or may not equal H_∞ and A may or may not be periodic in all possible combinations when $|A| = 2^{n-2}$.

Example 3.9 In $\mathrm{PG}(9, 2)$ take

$$\begin{aligned}
A_0 &= \{8\}, \\
A_1 &= \{08, 18, 28, 38, 48, 58, 68, 78\}, \\
A_5 &= \{ijklm8 \mid 0 \leq i < j < k < l < m \leq 7\}, \\
A_8 &= \{012345678\}, \\
A^\sharp &= A_0 \sqcup A_1 \sqcup A_5 \sqcup A_8.
\end{aligned}$$

With $K_A = (9)^\perp$ and $K_B = (8)^\perp$ we have $A^\sharp \oplus A^\sharp = H_\infty$ and $|A^\sharp| = 66$. Now the point $v = 01 \in H_\infty$ lies on 16 secants to $A^\sharp$ and thus $|A^\sharp \sqcup (v + A^\sharp)| = 116$. Hence we may easily extend $A^\sharp$ to a set $A \subset (K_A \setminus H_\infty)$

with $|A| = 128$ and $v + A = A$ by adding the 12 points of 6 new secant lines of v to $A^\sharp$. Then $A \oplus A = H_\infty$ and A is periodic.

Example 3.10 In PG(5, 2) take

$$A = \{4, 04, 14, 24, 34, 01234, 0134, 0234\},$$

where $K_A = (5)^\perp$ and $K_B = (4)^\perp$. Then A is not periodic and $A \oplus A = H_\infty$. Thus $S = A \sqcup (45 + A) \sqcup \{45\}$ is complete.

Example 3.11 In PG(5, 2) take $A = \{4, 04, 14, 014, 24, 024, 34, 034\}$ where $K_A = (5)^\perp$ and $K_B = (4)^\perp$. Then A is periodic (with vertex 0) and $A \oplus A \neq H_\infty$. Here if $c_0 = 45$ then $E = \{045, 123, 0123\}$.

Example 3.12 In PG(4, 2) take $A = \{3, 03, 13, 23\}$ where $K_A = (4)^\perp$ and $K_B = (3)^\perp$. Then A is not periodic and $A \oplus A \neq H_\infty$. Thus $H_\infty \setminus (A \oplus A) = \{e = 012\}$ and $S = A \sqcup (c_0 + A) \sqcup \{c_0, e\}$ is a complete cap.

4. General case

Next we suppose that the sets A and B are such that Equation 1.3 is satisfied for all i. Under this hypothesis we want to investigate conditions on A and B which are sufficient to guarantee that S is a *complete* cap.

We suppose that $\{1, 2, \ldots, t\} = \{i \mid H_A(i) \subset S\}$ for some t with $0 \leq t \leq 2^{n-r-1}$. Thus $\{1, 2, \ldots, t\} = \{i \mid H_B(i) \cap S = \varnothing\}$. Further suppose that $\{t+1, t+2, \ldots, t+u\} = \{i \mid H_A(i) \cap S = \varnothing\}$ for some u with $0 \leq u \leq 2^{n-r-1} - t$. Then $\{t+1, t+2, \ldots, t+u\} = \{i \mid H_B(i) \subset S\}$. Thus the pair of cosets corresponding to $i = 1, 2, \ldots, t+u$ are precisely those having a trivial solution to Equation 1.3.

Let $\overline{\Sigma} \cong \mathrm{PG}(n-r, 2)$ denote the quotient geometry of Σ by F. This is just the projective geometry associated to the quotient of the vector space associated to $\Sigma = \mathrm{PG}(n, 2)$ by the vector space associated to F.

Let $\overline{a_i}$ and $\overline{b_i}$ denote the points of $\overline{\Sigma}$ corresponding to the cosets $H_A(i)$ and $H_B(i)$ respectively. Write $\overline{C}$ for the point of $\overline{\Sigma}$ corresponding to $\widetilde{F}$. Similarly we define $\overline{H_A}$, $\overline{H_B}$, etc. for the subsets of $\overline{\Sigma}$ corresponding to H_A, H_B, etc. respectively.

Corresponding to the full, empty and non-empty cosets of A and B we define the following subsets of $\overline{\Sigma}$. Put

$$\begin{aligned}\overline{A_f} &= \{\overline{a_1}, \overline{a_2}, \ldots, \overline{a_t}\},\\ \overline{B_f} &= \{\overline{b_{t+1}}, \overline{b_{t+2}}, \ldots, \overline{b_{t+u}}\},\\ \overline{A_e} &= \{\overline{a_{t+1}}, \overline{a_{t+2}}, \ldots, \overline{a_{t+u}}\},\\ \overline{B_e} &= \{\overline{b_1}, \overline{b_2}, \ldots, \overline{b_t}\},\\ \overline{A_{ne}} &= \overline{H_A} \setminus \overline{A_e},\\ \overline{B_{ne}} &= \overline{H_B} \setminus \overline{B_e}.\end{aligned}$$

Recall that $\widehat{C} = \widetilde{F} \setminus F$.

Theorem 4.1 *Suppose that A and B are such that Equation 1.3 is satisfied for all $i = 1, 2 \ldots, 2^{n-r-1}$. Suppose $0 < t + u < 2^{n-r-1}$. Suppose that either $\overline{A_f}$ is non-periodic and $t \neq 2^{n-r-2}$ or that $\overline{B_f}$ is non-periodic and $u \neq 2^{n-r-2}$. Further suppose that $\widehat{C} \setminus C \subseteq \cup_{i=1}^{2^{n-r-1}} (A(i) + B(i))$. Then S is a complete cap in Σ.*

Proof By construction, S is a cap in Σ. Also Equation 1.3 guarantees that every point of A' and every point of B' lies on at least one secant of S through a point of C. Thus we need to show that every point of H_∞ and every point of $H_C \setminus C$ lies on a secant of S.

Without loss of generality suppose that $\overline{A_f}$ is non-periodic and $t \neq 2^{n-r-2}$. Consider the cap $\overline{S_1}$ in $\overline{\Sigma}$ given by $\overline{S_1} = \overline{A_f} \sqcup \overline{B_{ne}} \sqcup \{\overline{C}\}$. Applying Theorem 3.5 we see that $\overline{S_1}$ is a complete cap. Thus for every point $\overline{c_k}$ of $\overline{H_C}$ different from $\overline{C}$ there exists $i \leq t$ and $j \geq t+1$ such that $\overline{a_i} + \overline{b_j} = \overline{c_k}$. Since the coset $H_A(i)$ is entirely contained in S, and $H_B(j) \cap S \neq \varnothing$, this means that every point of the coset $H_C(k) = H_A(i) + H_B(j)$ lies on a secant to S.

By assumption, every point of $\widehat{C} \setminus C$ is contained in $A + B$.

Hence it only remains to prove that every point of H_∞ lies on at least one secant to S. Note that every point of F lies in $H_A(1) \oplus H_A(1)$ (or in $H_B(1) \oplus H_B(1)$ if $t = 0$). Thus we consider a point $z \in H_\infty \setminus F$ and write $\overline{z}$ for the point of $\overline{\Sigma}$ corresponding to the coset of F generated by z. Clearly if $\overline{z} \in \overline{A_f} + \overline{B_{ne}}$ or $\overline{z} \in \overline{A_{ne}} + \overline{B_f}$ then z lies on a secant line to S. Thus we suppose that $\overline{z} + \overline{A_f} \subset \overline{B_e}$ and $\overline{z} + \overline{B_f} \subset \overline{A_e}$. Since $\overline{C} + \overline{A_f} = \overline{B_e}$ this implies $\overline{z} + \overline{A_f} \subset \overline{C} + \overline{A_f}$. Hence $\overline{z} + \overline{C} + \overline{A_f} \subset \overline{A_f}$ and thus $(\overline{z} + \overline{C}) + \overline{A_f} = \overline{A_f}$. Therefore $\overline{A_f}$ is periodic contradicting our hypothesis. □

Remark 4.2 Note that if there exists at least one value of i for which a singleton solution occurs then $\widehat{C} \setminus C \subseteq \cup_{i=1}^{2^{n-r-1}} (A(i) + B(i))$. Conversely if $t+u = 2^{n-r-1}$ then $A(i)+B(i) = \varnothing$ for all i and thus $\widehat{C}\setminus C \not\subseteq \cup_{i=1}^{2^{n-r-1}}(A(i)+B(i))$. Hence if $t+u = 2^{n-r-1}$ then S is not complete. Furthermore, in order that $1 \leq t+u \leq 2^{n-r-1}-1$ we must have $r \leq n-2$.

Remark 4.3 Note that, if t or u is odd, then $\overline{A_f}$ or $\overline{B_f}$ respectively is non-periodic. Thus for a fixed value of $t+u$ with $1 \leq t+u \leq 2^{n-r-1}-1$ we may take $t=1$, for example, to arrange that $\overline{A_f}$ is non-periodic.

Remark 4.4 There do exist complete caps for which $t+u=0$ (and thus $\overline{A_f} = \overline{B_f} = \varnothing$). See Example 5.4.

5. A family of examples

In this section we consider the case where $|C| = 2^r - 1$, that is, where $\widehat{C}\setminus C$ is a single point $\{c_0\}$. If $r=0$ then C is empty contrary to our earlier assumption. If $r=1$ then $|C|=1$ and we are in the case considered in Section 3. Thus we will assume that $r \geq 2$. Accordingly, we must have $n \geq 3$.

Using the cap property of S we see that if $|S \cap H_A(i)| \geq 2$ then $S \cap H_B(i) = \varnothing$. Furthermore if $S \cap H_A(i)$ is a single point, $\{\alpha_i\}$, then using the completeness of S we must have $S \cap H_B(i) = \{\beta_i = \alpha_i + c_0\}$. Finally, if $S \cap H_A(i) = \varnothing$, then the completeness of S implies that $H_B(i) \subset S$.

Thus the only solutions to Equation 1.3 are the trivial solutions and the singleton solutions. Let s denote the number of these pairs of cosets which have the singleton solution.

Then $|S| = (2^r-1) + 2s + 2^r(2^{n-r-1}-s) = 2^{n-1} + 2^r - 1 - (2^r - 2)s$. By Remark 4.2, s cannot equal 0. Furthermore by Theorem 4.1 (and Remark 4.3) there do exist complete caps of this form for all $s = 1, 2, \ldots, 2^{n-r-1}-1$. Thus we find complete caps of this form of all cardinalities: $2^{n-r} + k(2^r-2)+1$ for $k = 2, 3, \ldots, 2^{n-r-1}$ where $2 \leq r \leq n-2$.

The only remaining possibility is that $s = 2^{n-r-1}$, that is, $t=u=0$. Thus we now consider caps with $\widehat{C}\setminus C = \{c_0\}$ and with $|H_A(i)| = |H_B(i)| = 1$ for all $i = 1, 2, \ldots, 2^{n-r-1}$.

Lemma 5.1 *Let $2 \leq r \leq n-1$ and suppose that $\widehat{C}\setminus C$ is a single point $\{c_0\}$ and that $|A(i)| = |B(i)| = 1$ for all $i = 1, 2, \ldots, 2^{n-r-1}$. Then S is complete cap if and only if $A \oplus A = H_\infty \setminus F$.*

Proof Note that $A \oplus A = B \oplus B$ since $\alpha_i + \alpha_j = (\alpha_i + c_0) + (\alpha_j + c_0) = \beta_i + \beta_j$. Furthermore $A \oplus A \subseteq H_\infty \setminus F$ and $C \oplus C = F$. Thus $H_\infty \subset S \oplus S$ if and only if $A \oplus A = H_\infty \setminus F$.

Suppose now that $A \oplus A = H_\infty \setminus F$. By construction $A+C = B'$ and $B+C = A'$ thus S is complete if and only if $A+B = C'$. Now $A+B = A + (c_0 + A) = \{c_0\} \sqcup (c_0 + (A \oplus A)) = \{c_0\} \sqcup (c_0 + (H_\infty \setminus F)) = C'$. Hence we conclude that S is complete if and only if $A \oplus A = H_\infty \setminus F$. □

We will now determine those sets $A = \{\alpha_1, \alpha_2, \ldots, \alpha_{2^{n-r-1}}\}$ for which $A \oplus A = H_\infty \setminus F$. The remainder of this section will be devoted to answering this question. We will reduce this question to an equivalent geometric question in $\overline{K_A}$ where, as above, $\overline{K_A}$ denotes a hyperplane in $\overline{\Sigma}$, the quotient geometry of Σ with respect to F.

To do this we will fix some subspace $L \cong \mathrm{PG}(n-r-1, 2) \subset K_A$ such that $L \cap F = \varnothing$. Choose an identification of one of the cosets $H_A(i)$, say $H_A(1)$, with $\mathrm{AG}(r,2)$. Choose this identification so that the unique point $a_0 \in H_A(1) \cap L$ is the point identified with the zero vector in $\mathrm{AG}(r,2)$. Note that with this identification if $u, v \in \mathrm{AG}(r,2)$ correspond to $x, y \in H_A(1)$ respectively then $u + v \in \mathrm{AG}(r,2)$ corresponds to $x + y + a_0 \in H_A(1)$.

Now for each point a of $K_A \setminus F$ there exists a unique point x of $\mathrm{AG}(r,2)$ such that $x + a \in L$. For each $x \in \mathrm{AG}(r,2)$ define $(K_A)_x = \{a \in (K_A \setminus F) \mid x + a \in L\}$. This partitioning of $K_A \setminus F$ induces a partitioning of A: $A = \sqcup_{x \in \mathrm{AG}(r,2)} A_x$ where $A_x = A \cap (K_A)_x$. Similarly we have the partitioning of H_∞: $H_\infty = \sqcup_{x\in \mathrm{AG}(r,2)} (H_\infty)_x$ where $(H_\infty)_x = H_\infty \cap (K_A)_x \cong \mathrm{PG}(n-r-2)$ for all $x \in \mathrm{AG}(r,2)$.

Example 5.2 Take $n = 7$ and $r = 2$. Put $K_A = (7)^\perp$, $K_B = (6)^\perp$ and $K_C = (67)^\perp$. Then $H_\infty \cong \mathrm{PG}(5,2)$ is the projective subspace generated by $\{0,1,2,3,4,5\}$. Take $C = \{067, 167, 0167\}$ so that $c_0 = 67$. Then $\widehat{C} = \{067, 167, 0167, 67\}$, $\widetilde{F} \cong \mathrm{PG}(2,2)$ is generated by $\{0,1,67\}$ and F is the line $\{0,1,01\}$. Choose $L \cong \mathrm{PG}(4,2)$ to be the subspace of K_A generated by $\{2,3,4,5,6\}$. Then $L \cap F = \varnothing$ as required. We take $H_A(1)$ to be the coset containing 6 say. Thus $H_A(1) = \{6, 06, 16, 016\}$ and $a_0 = 6 \in H_A(1) \cap L$. Then, for example, if we choose our identification so that $u = 1 \in \mathrm{AG}(2,2)$ corresponds to $16 \in H_A(1)$, we have

$$(H_\infty)_u = (H_\infty)_{16} = \{1, 12, 13, 14, 15, 123, 124, 125, 134, 135, 145, 1234, 1235, 1245, 1345, 12345\}.$$

Now the useful property of this partition is that for $u, w \in \mathrm{AG}(r,2)$ we have $A_u \oplus A_v \subseteq (H_\infty)_{u+v}$. Thus $A \oplus A = H_\infty \setminus F$ if and only if for every $w \in \mathrm{AG}(r,2)$ we have $\sqcup_{u \in \mathrm{AG}(r,2)} (A_u \oplus A_{u+w}) = (H_\infty)_w \setminus F$. Note that the partition of A yields a partition of $\overline{A} = \overline{H_A}$:

$$\overline{H_A} = \sqcup_{x \in \mathrm{AG}(r,2)} \overline{A_x}.$$

Furthermore $A \oplus A = H_\infty \setminus F$ if and only if for every $w \in \mathrm{AG}(r,2)$ we have

$$\cup_{u\in \mathrm{AG}(r,2)} \left(\overline{A_u} \oplus \overline{A_{u+w}}\right) = \overline{(H_\infty)}. \tag{5.1}$$

Thus we have proved the following proposition.

Proposition 5.3 *Let* $C \cong \mathrm{AG}(r,2)\setminus\{c_0\}$ *and suppose* $|A(i)| = |B(i)| = 1$ *for all* $i = 1, 2, \ldots, 2^{n-r-1}$. *Then* S *is a complete cap if and only if the partition of* $\mathrm{AG}(n-r-1,2) = \overline{H_A}$ *induced by* A *satisfies Equation 5.1 for all* $w \in \mathrm{AG}(r,2)$.

Example 5.4 We continue with the notation of the previous example. Thus $\overline{H_A} \cong \mathrm{AG}(4,2) \subset \overline{K_A} \cong \mathrm{PG}(4,2)$. Hence to obtain a complete cap S we require a partition of $\mathrm{AG}(4,2)$ into 4 subsets, $\overline{A_0}, \overline{A_1}, \overline{A_{01}}, \overline{A_\varnothing}$ satisfying Equation 5.1 for all $w \in \mathrm{AG}(2,2) = \{0, 1, 01, \varnothing\}$. It is not too difficult to find such partitions. For example,

$$\begin{aligned}
\overline{A_\varnothing} &= \varnothing, \\
\overline{A_0} &= \{6, 246, 346, 256, 2346, 2456\}, \\
\overline{A_1} &= \{236, 356, 456, 2356, 3456\} \\
\overline{A_{01}} &= \{26, 36, 46, 56, 23456\}.
\end{aligned}$$

This partition corresponds to the complete cap $S = A \sqcup B \sqcup C$ where

$$\begin{aligned}
A &= \{1236, 1356, 1456, 12356, 13456, 06, 0246, 0346, 0256, 02346, \\
&\qquad 02456, 0126, 0136, 0146, 0156, 0123456\}, \\
B &= c_0 + A = 01 + A.
\end{aligned}$$

Lemma 5.5 *Let* $\mathrm{AG}(k,2) \subset \Lambda = \mathrm{PG}(k,2)$ *denote the complement of a hyperplane* H *of* Λ. *Suppose there exists a partition of* $\mathrm{AG}(k,2)$ *indexed by* $\mathrm{AG}(r,2) : \mathrm{AG}(k,2) = \sqcup_{w\in \mathrm{AG}(r,2)} X_w$ *such that*

$$\cup_{u\in \mathrm{AG}(r,2)} \left(\overline{X_u} \oplus \overline{X_{u+w}}\right) = H \tag{5.1}$$

for all $w \in \mathrm{AG}(r,2)$. *Then* $r \leq k-1$.

Proof The partition of $\mathrm{AG}(k,2)$ induces a partition of the secant lines of $\mathrm{AG}(k,2)$ as follows. If $a_1 \in X_u$ and $a_2 \in X_v$ then we say the secant line through a_1 and a_2 is of type $u+v \in \mathrm{AG}(r,2)$. Equation 5.1 is the condition that every point of H lies on at least one secant line of type w for all $w \in \mathrm{AG}(r,2)$. Since $|H| = 2^k - 1$ and since there are $\binom{2^k}{2} = 2^{k-1}(2^k-1)$ secant lines we must have $2^{k-1}(2^k-1) \geq 2^r(2^k-1)$. Thus $k-1 \geq r$. □

Remark 5.6 Clearly, if a partition of $\mathrm{AG}(k,2)$ exists which satisfies Equation 5.1 for some value r_0 of r then such a partition exists for all values of r less than r_0.

6. $|C| = 2$

Lemma 6.1 *Let S be a complete cap in $\Sigma = \mathrm{PG}(n, 2)$ satisfying $|C| = 2$. Then $S = \phi(S')$ where $S' = S \cap \Sigma'$ is a complete cap in a hyperplane $\Sigma' \cong \mathrm{PG}(n-1, 2)$. In particular, $|S| = 2^{n-1} + 2$.*

Proof Write $C = S \cap K_C = \{c_1, c_2\}$. Then C is periodic with vertex $v = c_1 + c_2 \in H_\infty$. By Lemma 2.1, S is also periodic with vertex v. Choose any hyperplane Σ' of Σ not containing v and put $S' = S \cap \Sigma'$. Then $S = \phi(S')$ (with respect to the vertex v). Since S is complete in Σ, it follows that S' is complete in Σ'. Also the hyperplane $K_C \cap \Sigma'$ of Σ' is a tangent hyperplane for S' and thus $|S'| = 2^{n-2} + 1$. Hence $|S| = 2|S'| = 2(2^{n-2} + 1) = 2^{n-1} + 2$. □

7. $|C| = 3$

Now suppose that S is a complete cap meeting a hyperplane K_C in 3 points. Choose a codimension 2 subspace $H_\infty \cong \mathrm{PG}(n-2, 2)$ contained in K_C and disjoint from S. Write $C = \{c_1, c_2, c_3\}$. Then $F = \{c_1 + c_2, c_1 + c_3, c_2 + c_3\}$ is a line in H_∞. Put $c_0 = c_1 + c_2 + c_3$. Then $\widetilde{F} = F \cup C \cup \{c_0\} \cong \mathrm{PG}(2, 2)$ and $\widehat{C} = C \cup \{c_0\} \cong \mathrm{AG}(2, 2)$. Thus $|C| = 3$ implies that we are in the case described in Section 5 with $r = 2$.

Thus by Theorem 4.1 (and Remark 4.3) we have maximal caps with $|C| = 3$ of all cardinalities $3 + 2m + 4(2^{n-3} - m) = 2^{n-1} - 2m + 3$ where $1 \leq m \leq 2^{n-3} - 3$ or $m = 2^{n-3} - 1$, that is, of cardinalities $2^{n-2} + 5, 2^{n-2} + 9, 2^{n-2} + 11, 2^{n-2} + 13, \ldots, 2^{n-1} + 1$ for $n \geq 4$.

Now we show that there for all $n \geq 7$ there exist complete caps with $|C| = 3$ and with $|A(i)| = |B(i)| = 1$ for all $i = 1, 2, \ldots, 2^{n-3}$. By Proposition 5.3 we must find a partition of $\mathrm{AG}(n-3, 2)$ into 4 subsets indexed by $\mathrm{AG}(2, 2)$ satisfying Equation 5.1. There are very many ways to do this. We give one construction here. We proceed by induction on $k = n - 3$. By Lemma 5.5 we have $k \geq r + 1$ and thus the first case is $k = 3$.It is easy to show that there is no partition of $AG(3, 2)$ into subsets indexed by $AG(2, 2)$ satisfying Equation (5.1). Thus we begin with $k = 4$. We exhibited such a partition for $k = 4$ in Example 5.2.

Now we will inductively construct from this example a partition for all values of $k \geq 5$. Suppose we have a partition of $\mathrm{AG}(k, 2)$ into four sets $\overline{A_\varnothing}$, $\overline{A_0}$, $\overline{A_1}$ and $\overline{A_{01}}$ where we further suppose that $\overline{A_\varnothing} = \varnothing$. Write $\mathrm{PG}(k+1, 2) = \phi(\mathrm{PG}(k, 2))$ with respect to some vertex $z \in PG(k+1, 2) \setminus \mathrm{PG}(k, 2)$. Then $\mathrm{AG}(k+1, 2) = \phi(\mathrm{AG}(k, 2))$. Let $\widetilde{H} = \mathrm{PG}(k+1, 2) \setminus \mathrm{AG}(k+1, 2) = \phi(H)$. We partition $\mathrm{AG}(k+1, 2)$ into 4 sets: $\mathrm{AG}(k+1, 2) = \widetilde{A_\varnothing} \sqcup \widetilde{A_0} \sqcup \widetilde{A_1} \sqcup \widetilde{A_{01}}$ by defining $\widetilde{A_w} = \phi(\overline{A_w}) = \overline{A_w} \sqcup (z + \overline{A_w})$. This partition almost satisfies Equation 5.1. However the point $z \in \widetilde{H}$ is not contained in any of the three sets $\widetilde{A_0} + \widetilde{A_1}$, $\widetilde{A_0} + \widetilde{A_{01}}$ and $\widetilde{A_1} + \widetilde{A_{01}}$. To overcome this defect we modify the partition slightly. Note that every point of $\widetilde{H}$ different from z lies on at least two secants of each type $u \in \mathrm{AG}(2,2)$. Consider the three points of $\mathrm{AG}(k,2)$: $a_0 = 6$, $a_1 = 236$ and $a_{01} = 26$ and the line in H: $z_{01} = a_0 + a_1$, $z_1 = a_0 + a_{01}$ and $z_0 = a_1 + a_{01}$. Each of the three points of this line lies on secants of all types which do not utilize a_0, a_1 nor a_{01}. Specifically we have $z_{01} = 23 = 256 + 356 = 2346 + 46 = 456 + 23456$, $z_1 = 2 = 2456 + 456 = 246 + 46 = 356 + 56$ and $z_0 = 3 = 256 + 2356 = 346 + 46 = 356 + 56$. We shift $z + a_0$ into $\widetilde{A_1}$, $z + a_1$ into $\widetilde{A_{01}}$ and $z + a_{01}$ into $\widetilde{A_0}$. This modified partition now satisfies Equation 5.1. Also note that we have not shifted any of the points a_0, a_1, a_{01} nor indeed any of the points of $\mathrm{AG}(k,2)$ and so our induction proceeds using the same points a_0, a_1, a_{01} at each stage.

In conclusion we see that there are complete caps, S in $\mathrm{PG}(n,2)$ with $|C| = 3$ of cardinality $2^{n-2} + 3$ for all $n \geq 7$. Thus we have the following.

Proposition 7.1 *Let S be a complete cap in* $\mathrm{PG}(n,2)$ *with $n \geq 4$. Further, suppose that S meets a hyperplane K_C of* $\mathrm{PG}(n,2)$ *in exactly* 3 *points. Then $|S|$ is one of the numbers: $2^{n-2} + 5, 2^{n-2} + 9, 2^{n-2} + 11, 2^{n-2} + 13, \ldots, 2^{n-1} + 1$ or, if $n \geq 7$, $|S|$ may also be $2^{n-2} + 3$, and complete caps of all these sizes exist for all corresponding values of n.*

8. $|C| = 4$

In this section, we assume that S is a complete cap meeting a hyperplane K_C in 4 points. As usual we choose a codimension 2 subspace $H_\infty \cong \mathrm{PG}(n-2,2)$ contained in K_C and disjoint from S. Note that when $|K_C \cap S| = 4$ there are two possibilities; either these four points lie in a Fano plane or they do not. Now if they do lie in a Fano plane they form a periodic set and it follows as in Section 6 that if S complete then $|S| = 2^{n-1} + 4$ and S is large.

It remains to consider the case where $S \cap K_C$ consists of four points spanning a three dimensional projective subspace of K_C. In order to determine all solutions to Equation 1.3 we first assume that this three dimensional subspace is K_C and thus that $n = 4$. Without loss of generality, $C = \{04, 14, 24, 34\}$, $K_C = (01234)^\perp$, $K_B = (4)^\perp$ and $K_A = (0123)^\perp$. Thus $H_\infty = \{01, 02, 03, 12, 13, 23, 0123\}$ and $H_A = H_A(1) = \{4, 014, 024, 034, 124, 134, 234, 01234\}$.

Let G denote the group of projectivities of $\Sigma = \mathrm{PG}(4,2)$ which stabilize, C, K_A, K_B and K_C. Then G contains the subgroup L of projectivities which permute the points $0, 1, 2$ and 3 and which also fix the point 4. The group G also contains the projectivities f and h defined by: $f(0) = 0$, $f(1) = 0$, $f(2) = 0124$, $f(3) = 0134$ and $f(4) = 014$ and $h(0) = 123$, $h(1) = 023$, $h(2) = 013$, $h(3) = 012$, and $h(4) = 01234$.

Suppose that A and B satisfy Equation 1.3 and further that A and B do not correspond to either trivial or singleton solutions of that equation. Thus $|A| \geq 2$ and $|B| \geq 2$. Since G acts transitively on H_A we may assume without loss of generality that $01234 \in A$. Now the subgroup of G fixing 01234 contains the group L. The action of L decomposes the remaining points of H_A into two orbits: $\{4\}$ and $\{014, 024, 034, 124, 134, 234\}$. Thus without loss of generality we have two cases to consider: (i) $A \supset \{01234, 4\}$ and (ii) $A \supset \{01234, 014\}$. These two cases are distinguished by the fact that in the former case the point $z_0 = c_1 + c_2 + c_3 + c_3 = 04 + 14 + 24 + 34 = 0123$ lies on the secant through the two given points of A while in the latter case it does not.

In the first case, we see that $B' = A + C \supset \{01234, 4\} + C = H_B$ and thus we have the trivial solution $A = H_A$ and $B = \emptyset$. In the second case $B' = A + C \supset \{01234, 014\} + C = H_B \setminus \{2, 3\}$. Since $|B| \geq 2$, we must have $B = \{2, 3\}$ and $A = \{01234, 014\}$.

Thus we find only one new solution to Equation 1.3. This solution is determined by $A(i) = \{a_1, a_2\}$, $a_1 + a_2 \neq z_0$, and $B(i) = \{b_1, b_2\} = H_B(i) \setminus (A(i) + C)$ where $z_0 = c_1 + c_2 + c_3 + c_4$. Note that $b_1 + b_2 = a_1 + a_2$. Thus we have five solutions in total: the two trivial solutions, the two singleton solutions and this new solution. Notice that for the new solution and for both of the singleton solutions we have $z_0 \notin A(i) \oplus A(i)$ and $z_0 \notin B(i) \oplus B(i)$. Since $z_0 \notin C \oplus C$, if S is complete there must be at least one i for which the trivial solution occurs. In particular, we may apply Theorem 4.1 to obtain complete caps S with $|C| = 4$ and C not periodic.

Suppose that S is a complete cap with $|C| = 4$. If C is periodic then $|S| = 2^{n-1} + 4$. Suppose then that C is not periodic and that m of the pairs of cosets $H_A(i), H_B(i)$ utilize the new solution, s of them utilize singleton solutions and $t + u = 2^{n-4} - s - m$ of them utilize the trivial solution. Note that if $s = 0$ then we must have $m \geq 2$ in order that $\widehat{C} \setminus C \subseteq \cup_{i=1}^{2n-r-1} (A(i) + B(i))$. The cardinality of S is then given by $|S| = 4 + 4m + 5s + 8(2^{n-4} - s - m) = 2^{n-1} - 3(s + m) - m + 4$ where $0 \leq m, s \leq 2^{n-4} - 1$ and $1 \leq m + s \leq 2^{n-4} - 1$ (and $(m, s) \neq (1, 0)$). Thus if $|C| = 4$ and C is not periodic then $n \geq 5$ and we find complete caps S of all cardinalities satisfying $2^{n-2} + 8 \leq |S| \leq 2^{n-1} - 2$. In summary we have the following.

Proposition 8.1 *Let S be a complete cap in* PG$(n, 2)$ *with $n \geq 5$. Further suppose that S meets a hyperplane K_C of* PG$(n, 2)$ *in exactly 4 points. Then $2^{n-2} + 8 \leq |S| \leq 2^{n-1} - 2$ or else $|S| = 2^{n-1} + 1$. Moreover complete caps of all these sizes exist for all $n \geq 5$.*

Acknowledgements

I thank Aiden Bruen for many useful conversations. This research was supported by grants from ARP and NSERC.

References

[1] A.E. Brouwer, A.A. Bruen and D.L. Wehlau, There exist caps which block all subspaces of fixed codimension in PG$(n, 2)$, *J. Combin. Theory Ser. A* **73** (1996) 168–169.

[2] A.A. Bruen, L. Haddad and D.L. Wehlau, Binary codes and caps, *J. Combin. Des.* **6** (1998) 275–284.

[3] A.A. Bruen and D.L. Wehlau, Long binary linear codes and large caps in projective space, *Des. Codes Cryptogr.* **17** (1999) 37–60.

[4] A.A. Bruen and D,L. Wehlau, New codes from old; a new geometric construction, *J. Combin. Theory Ser. A*, to appear.

[5] A.A. Davydov and L.M. Tombak, Quasiperfect linear binary codes with distance 4 and complete caps in projective geometry, *Problems Inform. Transmission* **25** (1990), 265–275.

Participants

Australia
M.R. Brown
N. Hamilton
M. Law
T. Penttila
C.E. Praeger

Austria
A. Blunck

Belgium
F. Buekenhout
P. Cara
K. Coolsaet
B. De Bruyn
F. De Clerck
A. Delandtsheer
A. Devillers
J. Doyen
E. Govaert
C. Huybrechts
D. Luyckx
L. Storme
J.A. Thas
K. Thas
H. Van Maldeghem

Bulgaria
I.N. Landjev

Canada
J.M.N. Brown
D.L. Wehlau

Germany
J. Eisfeld
A. Enge
D. Jungnickel
B. Schmidt

Hungary
A. Gács
T. Szőnyi

Israel
R. Meshulam

Italy
L. Bader
A. Cossidente
M.J. de Resmini
D. Ghinelli
M. Giulietti
G. Korchmáros
A. Pasini
O. Polverino

Netherlands
R.J. Blok
A. Blokhuis
A.E. Brouwer
R. Gramlich
W.H. Haemers
M. Lavrauw
E.R. van Dam

United Kingdom
S. Ball
P.J. Cameron
A.R. Camina
L. Giuzzi
R. Hill
J.W.P. Hirschfeld
J. Jedwab
O.H. King
S.-L. Ng
A.R. Prince
R. Shaw
B.S. Webb

United States
A.R. Calderbank
G.L. Ebert
R.A. Liebler
S.E. Payne
M. Ronan
Q. Xiang

Talks

L. Bader, Flat flocks

S. Ball, Maximal multi-arcs in projective planes

R.J. Blok, Point-line geometries whose generating rank depends on the underlying field

A. Blunck, Derivation of affine spaces

M.R. Brown, On the full embedding of semipartial geometries in AG$(4, q)$

B. De Bruyn, Near polygons with three points on a line

F. Buekenhout, Generalized elliptic cubic curves

P.J. Cameron, Fixed points and cycles

A. R. Calderbank, Orthogonal designs and space-time codes for wireless communication

P. Cara, Independent generating sets and RWPRI geometries

K. Coolsaet, Towards an algebraic construction of the Ree octagons

A. Cossidente, Curves of projective space, tangent developables and partial spreads

A. Devillers, A classification of finite homogeneous semilinear spaces

G.L. Ebert, Regular hyperbolic fibrations

J. Eisfeld, Minimal covers of the Klein quadric

A. Enge, A fast subexponential algorithm for computing hyperelliptic logarithms

L. Giuzzi, Groups stabilizing classical unital intersections

E. Govaert, Hermitian spreads in the generalized hexagon $H(q)$

R. Gramlich, Local recognition of graphs related to Tits buildings

W.H. Haemers, The pseudo-geometric graphs for generalized quadrangles of order $(3, t)$

N. Hamilton, Perp-systems of polar spaces and partial geometries

C. Huybrechts, Dual polar spaces and fully gated graphs

J. Jedwab, Applying combinatorial design theory in an industrial research lab

G. Korchmáros, Maximal curves of degree $q+1$ in the intersection of a Hermitian surface and a quadric in $\mathrm{PG}(3, \mathbf{F}_{q^2})$

I.N. Landjev, The geometric approach to linear codes

M. Lavrauw, Eggs or translation generalised quadrangles

R.A. Liebler, The arithmetic invariants of the incidence relations between the points and k-flats in $\mathrm{PG}_n(q)$

D. Luyckx, Flocks and locally hermitian 1-systems of $Q(6, q)$

R. Meshulam, A bound on the size of caps

A. Pasini, On flag-transitive finite $L.L^*$-geometries where planes are projective spaces and residues of points are dual projective spaces

S.E. Payne, The Law–Penttila q-clan geometries

T. Penttila, A new infinite family of flocks in characteristic 2

C.E. Praeger, Implications of line-transitivity for designs

B. Schmidt, All two-weight irreducible cyclic codes?

R. Shaw, Partial spreads in $\mathrm{PG}(4, 2)$

L. Storme, Partial t-spreads in $\mathrm{PG}(N, q)$

T. Szőnyi, Multiple blocking sets

K. Thas, Generalized quadrangles with axes of symmetry

E.R. van Dam, A spectral characterization of generalized polygons

H. Van Maldeghem, Construction and reconstruction of buildings

D. Wehlau, New constructions of small complete caps in $\mathrm{PG}(n, 2)$

Q. Xiang, A trace conjecture and flag-transitive affine planes

Developments in Mathematics

1. Alladi et al. (eds.): *Analytic and Elementary Number Theory*. 1998 ISBN 0-7923-8273-0
2. S. Kanemitsu and K. Győry (eds.): *Number Theory and Its Applications*. 1999 ISBN 0-7923-5952-6
3. A. Blokhuis, J.W.P. Hirschfeld, D. Jungnickel and J.A. Thas (eds.): *Finite Geometries*. Proceedings of the Fourth Isle of Thorns Conference. 2001 ISBN 0-7923-6994-7

KLUWER ACADEMIC PUBLISHERS – DORDRECHT / BOSTON / LONDON